Edited by

John A. Manthey
David E. Crowley
Douglas G. Luster

Biochemistry of Metal Micronutrients in the Rhizosphere

LEWIS PUBLISHERS

Boca Raton   Ann Arbor   London   Tokyo

Library of Congress Cataloging-in-Publication Data

Biochemistry of metal micronutrients in the rhizosphere / edited by
  John A. Manthey, David E. Crowley, Douglas G. Luster.
     p. cm.
  Includes bibliographical references and index.
  ISBN 0-87371-942-5
  1. Soils--Trace element content. 2. Trace elements in plant
nutrition. 3. Plant-soil relationships.   4. Soil biochemistry.
I. Manthey, John A.   II. Crowley, David E.   III. Luster, Douglas G.
S592.6.T7B56   1994
631.4′ 22--dc20   93-47294
                                                                                          CIP

CONTRIBUTORS

Jean E.L. Arceneaux, Department of Microbiology, University of Mississippi Medical Center, 2500 North State Street, Jackson, Mississippi 39216-4505

Dallas B. Aronson, Department of Chemistry, College of Environmental Science and Forestry, State University of New York, Syracuse, New York 13210

Larry L. Barton, Department of Biology, University of New Mexico, Albuquerque, New Mexico 87131

Gregory L. Boyer, Department of Chemistry, College of Environmental Science and Forestry, State University of New York, Syracuse, New York 13210

John C. Brown, Agronomy and Horticulture Department, Brigham Young University, Provo, Utah 84602

Jeffrey S. Buyer, USDA/ARS, Soil Microbial Systems Laboratory, Building 318 BARC-East, Beltsville, Maryland 20705

B. Rowe Byers, Department of Microbiology, University of Mississippi Medical Center, 2500 North State Street, Jackson, Mississippi 39216-4505

Rufus L. Chaney, USDA/ARS, Environmental Chemistry Laboratory, Beltsville Area Research Center, Beltsville, Maryland 20705-2350

David E. Crowley, Department of Soil and Environmental Sciences, 2208 Geology Building, University of California, Riverside, California 92521

Eckhard George, Institute of Plant Nutrition (330), Hohenheim University, 70593 Stuttgart, Germany

Paul R. Gill, Jr., Department of Food Microbiology, University College, Cork, Ireland

Robin D. Graham, Department of Plant Science, Waite Agricultural Research Institute, University of Adelaide, Glen Osmond, S.A. 5064, Australia

Dirk Gries, Department of Soil and Environmental Sciences, University of California, Riverside, California 92521

Claude Grignon, Laboratoire de Biochimie et Physiologie Végétales, CNRS (URA 573), ENSA.M/INRA, Place Viala, F34060 Montpellier cedex 1 France

Mary Lou Guerinot, Department of Biological Sciences, Dartmouth College, Hanover, New Hampshire 03755

S. Hasegawa, Institute of Whole Body Metabolism, 340-2 Nauchi, Shiroi, Inba, Chiba, Japan

Monica Höfte, Laboratory of Phytopathology and Phytovirology, Faculty of Agricultural and Applied Biological Sciences, University of Gent, Coupure Links 653, B-9000 Gent, Belgium

Marcia J. Holden, USDA/ARS, Climate Stress Laboratory, Beltsville Area Research Center, Beltsville, Maryland 20705-2350

B.N. Johri, Department of Microbiology, G.B. Pant University of Agriculture and Technology, Pantnagar 263145, U.P., India

Von D. Jolley, Agronomy and Horticulture Department, Brigham Young University, Provo, Utah 84602

Marian G. Kratzke, USDA/ARS, Soil Microbial Systems Laboratory, Building 318 BARC-East, Beltsville, Maryland 20705

Stuart H. Laurie, Chemistry Department, De Montfort University, Leicester LE1 9BH, United Kingdom

Richard H. Loeppert, Soil and Crop Sciences Department, Texas A&M University, College Station, Texas 77843

Douglas G. Luster, USDA/ARS, Foreign Disease-Weed Science Research Unit, Fort Detrick, Building 1301, Frederick, Maryland 21702

John A. Manthey, USDA/ARS, Fruit and Vegetable Chemistry Laboratory, 263 South Chester Avenue, Pasadena, California 91106

Horst Marschner, Institute of Plant Nutrition (330), Hohenheim University, 70593 Stuttgart, Germany

G.W. Miller, Biology Department, Utah State University, Logan, Utah 84322-5305

Satoshi Mori, Laboratory of Plant Nutrition and Fertilizer, Faculty of Agriculture, University of Tokyo, Yayoi, Bunkyo-ku, Tokyo-113, Japan

J.B. Neilands, Division of Biochemistry and Molecular Biology, University of California, Berkeley, California 94720

William R. Ocumpaugh, TAES, Texas A&M University, Beeville, Texas 78102

Judith F. Pedler, Department of Plant Science, Waite Agricultural Research Institute, University of Adelaide, Glen Osmond, S.A. 5064, Australia

Robert E. Redmann, Department of Crop Science and Plant Ecology, University of Saskatchewan, Saskatoon, SK S7N 0W0, Canada

H.M. Reisenauer, Department of Land, Air and Water Resources, University of California, Davis, California 95616

Zdenko Rengel, Department of Plant Science, Waite Agricultural Research Institute, University of Adelaide, Glen Osmond, S.A. 5064, Australia

Volker Römheld, Institute of Plant Nutrition (330), Hohenheim University, 70593 Stuttgart, Germany

Jennifer A. Saleeba, Department of Biological Sciences, Dartmouth College, Hanover, New Hampshire 03755

Hervé Sentenac, Laboratoire de Biochimie et Physiologie Végétales, CNRS (URA 573), ENSA.M/INRA, Place Viala, F34060 Montpellier cedex 1 France

A.K. Sharma, Department of Biological Sciences, Grambling State University, Grambling, Louisiana 71245

A. Shigematsu, Institute of Whole Body Metabolism, 340-2 Nauchi, Shiroi, Inba, Chiba, Japan

Lawrence J. Sikora, USDA/ARS, Soil Microbial Systems Laboratory, Building 318 BARC-East, Beltsville, Maryland 20705

P.C. Srivastava, Department of Soil Science, G.B. Pant University of Agriculture and Technology, Pantnagar 263145, U.P., India

Charles G. Suhayda, Department of Crop Science and Plant Ecology, University of Saskatchewan, Saskatoon, SK S7N 0W0, Canada

Jean-Baptiste Thibaud, Laboratoire de Biochimie et Physiologie Végétales, CNRS (URA 573), ENSA.M/INRA, Place Viala, F34060 Montpellier cedex 1 France

Willy Verstraete, Laboratory of Microbial Ecology, Faculty of Agricultural and Applied Biological Sciences, University of Gent, Coupure Links 653, B-9000 Gent, Belgium

Craig R. Vester, Department of Biology, University of New Mexico, Albuquerque, New Mexico 87131

Xiaoyan Wang, Department of Crop Science and Plant Ecology, University of Saskatchewan, Saskatoon, SK S7N 0W0, Canada

Michael J. Webb, Department of Plant Science, Waite Agricultural Research Institute, University of Adelaide, Glen Osmond, S.A. 5064, Australia

Liang-Chou Wei, Soil and Crop Sciences Department, Texas A&M University, College Station, Texas 77843

G.W. Welkie, Biology Department, Utah State University, Logan, Utah 84322-5305

Marleen Vande Woestyne, Laboratory of Microbial Ecology, Faculty of Agricultural and Applied Biological Sciences, University of Gent, Coupure Links 653, B-9000 Gent, Belgium

Ying Yi, Department of Biological Sciences, Dartmouth College, Hanover, New Hampshire 03755

TABLE OF CONTENTS

CHAPTER 1

Biochemistry of Metal Micronutrients in the Rhizosphere: An Introduction

John A. Manthey, Douglas G. Luster, and David E. Crowley

INTRODUCTION

The critical importance of trace mineral nutrition of plants in agriculture cannot be overstated. Not only are nutritional aspects of the crop at stake, but also cultivation efficiency and crop productivity (Figure 1).[1] The availability to plants of metal ion micronutrients is strongly influenced by a number of soil factors — particularly pH and carbonate content.[2-5] Alkaline high carbonate soils, common in key agricultural areas, are primary causes of micronutrient deficiencies in many economically important crops.[6,7] While other elements are also involved, iron, manganese, and zinc deficiencies have the greatest impact worldwide. Recent advances in our understanding of the biochemistry of iron, zinc, and manganese suggest that there are unifying concepts that integrate biological and chemical events influencing the availability and uptake of micronutrients by plants and microorganisms. It is the objective of this book to integrate what we know about the many aspects of micronutrient chemistry and absorption by plants and microorganisms to obtain a better understanding of the vastly complicated biological and chemical events that occur in the plant rhizosphere.

The biological activities of microorganisms in the rhizosphere to a large degree mediate the solubility, and hence the availability, of metal ions at root surfaces. The complexities of the interactions between soil, microorganisms, and plants are obvious and new experimental techniques are needed to study these highly sophisticated biological systems. Future research will need to more closely account for soil chemical dynamics as affected by microbial and plant growth. To date, much of what has been learned about these biological interactions has involved artificial systems, perhaps representing only the first level of experimental design. With regard to the use of nutrient media, however,

Figure 1. Desert agriculture, similar to that occurring in the Coachella Valley, provides a study of many of the rhizosphere-related topics covered in this book. Many agronomic problems are ultimately linked to soil characteristics and to the biological interactions between roots and rhizosphere microorganisms. The high carbonate soils of the Valley severely limit the solubilities of a number of essential metal ion micronutrients. Iron deficiencies are particularly acute, and often dictate rootstock and cultivar selections in citrus and other fruit and vegetable crops. The need to replant citrus with more iron-deficiency-tolerant cultivars is illustrated in this figure. Soil characteristics and the need for extensive irrigation in the Coachella Valley often lead to salinization, which dramatically limits crop productivity in affected areas. Impaired root growth, causing diminished root exudation, directly affects microbial activities in the rhizosphere and hence influences micronutrient uptake and disease resistance.

significant advances have been made in controlling micronutrient cation activities through the application of chelate buffering.[8-16] Use of these systems has allowed for the critical evaluation of the essential levels for plant growth of a number of micronutrients.

A number of models describing the possible ecological significance of microorganisms in affecting micronutrient availability in the rhizosphere are presented in this book. Furthermore, as reviewed by Laurie and Manthey in Chapter 12, no study can neglect the basic coordination chemistry that controls the solubility and redox poise of these metal ions in the rhizosphere. Chelation, which so remarkably mediates the chemical properties of metal ions, is crucial in controlling metal ion availability to living systems. This book provides an overview of the biochemical schemes by which microorganisms and plants mediate metal ion solubilities — either by affecting the localized pH at the root surface, by changing the redox state of the metal ions, or by chelation. These advances in the understanding of the biochemistry of rhizosphere activities suggest that, in most cases, metal ion supply in the rhizosphere is linked to the activities of both the plant and the associated microflora.

IRON AND RHIZOSPHERE MICROFLORA

There have been a number of reports on the effects of redox activities of soilborne microorganisms on the solubilities of both manganese and iron at root surfaces.[18-20] While these activities play significant roles in certain cases, the most significant metal ion solubilization activity of microorganisms is the nearly universal production of Fe^{3+}-specific siderophore molecules.[21-23] This essential role of microbially produced siderophores in controlling the availability of iron in the rhizosphere is a topic that is dealt with extensively in this book. Significant progress has been made in linking the regulation of iron availability in the rhizosphere to siderophore production. However, in spite of a great deal of progress made in the characterization of siderophores in soil,[23-26] much remains to be learned about the ecology of rhizosphere microflora[27-32] and the levels and composition of siderophores in the root zone.

The Fe uptake by both plants and microorganisms is tightly regulated, and while progress is being made in understanding Fe regulation of siderophores in certain microbial systems, we still know very little about the regulation of Fe uptake in plants. In bacterial systems, as outlined in Chapter 2 by Neilands and Chapter 3 by Byers and Arceneaux, and reviewed elsewhere,[22,35,36] Fe uptake is regulated by the *fur* gene; with Fe^{2+} binding to the gene product. As in so many other instances, this work with microbial systems will very likely lead the work ultimately done with plants. Similarly, the discussion in Chapter 3 by Byers and Arceneaux of the role of Fe in the virulence of pathogens suggests important parallels that can apply to microbe/microbe interactions in controlling pathogenicity and rhizosphere ecology. Crowley and Gries in Chapter 14 and Buyer et al. in Chapter 6 propose roles for siderophores in which these compounds regulate and contribute to microbial adaptation and competition in the rhizosphere. Based on these models, siderophore regulation by Fe determines microflora ecology and micronutrient supply to both the microorganisms and the plant roots. Directly related to this Fe regulation of rhizosphere ecology are studies of the plant growth-promoting rhizobacteria described in Chapter 7 by Höfte et al. Clearly, the implications for this level of biocontrol of both plant pathogens and mineral nutrition in modern agriculture are immense, and the study of the manipulation of rhizosphere microorganisms will be an important direction in future agricultural research.

Iron availability in the rhizosphere is also a key element in the growth of nitrogen-fixing symbiotic bacteria including the genera *Azorhizobium*, *Bradyrhizobium*, and *Rhizobium*. Several of the key enzymes involved with N_2 fixation have Fe-containing active sites, thus these microorganisms have an absolute need for Fe, both as the symbiont and as free-living bacteria.[40] Studies have shown, however, that not all nodule-producing bacteria produce siderophores.[41-43] This raises questions concerning the possible physiological roles of both Fe and siderophores in the various events involved in nodulation and the symbiotic relationship between the plant host and the bacteria.[44,45] Earlier work by Tang et al.[46] has shown that the stage of nodulation most sensitive to iron deficiency is nodule initiation. It has also been learned that nodule-producing bacteria generally have receptors for a number of widely different Fe siderophores.[47-49] This ability to use Fe siderophores from other soil microorganisms potentially provides a considerable ecological advantage to the root nodule bacteria. Additionally, ferric citrate appears to be a major source of Fe for these bacteria. Because of ferric citrate transport by plants,[50] and the elevated production of citrate in Fe-deficient roots of certain Fe-efficient dicots,[51,52] the use of ferric citrate as a primary Fe source by the nodule-producing bacteria has important implications in the symbiotic relationship between these bacteria and their plant hosts.[43] This also raises important questions concerning the physiological roles of the various other siderophores that are produced by certain bacteria of the *Rhizobia*. Do these siderophores help the organism selectively compete against other microflora in the rhizosphere? Related to this is the discussion in Chapter 5 by Barton et al. of the production of the chemically distinct siderophore, rhizobactin 1021, produced by iron-deficient *R. meliloti*. Is the associated Fe complex of rhizobactin directly used by the bacteria during symbiosis, or are there other less well-understood but ecologically important roles for this compound in the rhizosphere?

Directly related to this discussion of the siderophore production by root nodule bacteria is Chapter 4 by Boyer and Aronson, which reports on siderophore production by the symbiotic filamentous bacterium *Frankia* from the Actinomycetales. This symbiont also reduces N_2, and produces the novel siderophore frankobactin in a manner similar to the production of the unique siderophore rhizobactin by *Rhizobium*. These studies represent important advances in our understanding of the possible roles of these unique siderophores in the function and growth of the vastly important diazotrophic bacteria inhabiting the rhizosphere.

ZINC, MANGANESE, AND RHIZOSPHERE MICROFLORA

This study of the biochemistry of trace metal micronutrients includes an examination by Webb in Chapter 13 of zinc and manganese uptake by plants, and a detailed evaluation by Reisenauer in Chapter 11 of the interactions between iron and manganese in the rhizosphere. Influences of metal chelation on micronutrient absorption are discussed in Chapter 12 by Laurie and Manthey. An important aspect of manganese and zinc availability to plants is the biological activity of soil microorganisms in the rhizosphere. Chapter 10 by Rengel et al. provides a very interesting study of the control of manganese solubility and supply to plants by Mn^{2+}-oxidizing activities of soilborne microorganisms. This study provides an example of microorganisms using a metal micronutrient other than iron to mediate virulence, and thus points to a potentially important control of plant disease resistance.

Related to the effects of rhizoflora on the supply of metal micronutrients to plants are the possible roles that mycorrhiza play in supplying micronutrients to plants. The increased uptake of phosphorus by mycorrhiza-inoculated plants is well documented.[56,57] However, there is a renewed interest in the influence of mycorrhizal fungi on the uptake of Fe, Cu, Mn, and Zn by plants. Inoculation by mycorrhiza has been shown to increase the uptake of zinc,[58,59] copper,[60] and in ericoid plants, iron.[61,62] In contrast, mycorrhizae function to decrease manganese uptake[20,63] and alleviate manganese toxicity.[64] The biochemistry of these processes, especially with regard to the interactions between the metal ions, phosphate compounds, and the extramatrical hyphae are poorly understood.[60] However, as discussed in Chapter 8 by George et al., many of these influences may be results of the changes that occur in the rhizosphere ecology due to the interactions between the host and mycorrhiza. Lindermann[66] earlier described the effects of extramatrical hyphae of mycorrhizal fungi on soil microorganisms. This effect, termed the "mycorrhizosphere effect", enhances the growth of certain populations of microorganisms near the fungal structures and root surfaces. Furthermore, mycorrhizae are known to strongly affect root exudation, which plays a significant role in the growth of microorganisms in the rhizosphere. Kothari et al.[20] observed profound differences in the properties of microbial populations between mycorrhizal and nonmycorrhizal plants. These differences were reflected in the changes in the relative levels of manganese-reducing and -oxidizing microorganisms which, in turn, controlled the solubility of manganese at the root surface. Chapter 10 by Rengel et al. provides another example of this control of the manganese supply to plants involving manganese oxidation by microorganisms in the rhizosphere. What similar roles these reducing and oxidizing microorganisms might possibly have in controlling the redox poise of iron, and hence iron solubility, also warrants consideration. Hence, the complexity as well as the underlying unity of the biochemistry of metal micronutrients at root surfaces emerges from the consideration of the many factors that control microbial populations in the rhizosphere.

MECHANISMS OF PLANT MICRONUTRIENT SUPPLY

While research has shown that microorganisms have a profound influence on the availability of trace metal micronutrients in the rhizosphere, plants also exert considerable influence on metal ion solubility at the root surfaces. Perhaps the biggest distinction between plants in this regard exists between the graminaceous monocots and certain low-iron stress-tolerant dicotyledonous plants.[67,68] Graminaceous monocots produce a series of nicotianamine-derived compounds, termed phytosiderophores.[69,70] In Chapter 15, Mori provides a detailed review of the recent advances made in understanding the biosynthesis and regulation of the release of phytosiderophores into the rhizosphere by these plants during Fe deficiency. These compounds, released in high quantities in spatially discrete portions of actively growing roots represent potential carbon sources for microorganisms in the rhizosphere,[37,71] and it is possible that as carbon sources these phytosiderophores dramatically contribute to an increased growth of microorganisms in the rhizosphere. This increased microbial growth has important implications with regard to the nutritional competition for trace metal micronutrients, especially iron.[71,72] The resulting microbial siderophore production in the rhizosphere influences — via iron-regulation — not only microbe/microbe interactions as discussed earlier, but now also microbe/plant interactions. Hence, a better understanding of mineral nutrition and rhizosphere ecology will require the examination of the roles of root exudation in controlling microbial populations in the rhizosphere. Finally, consideration needs to be given to the possible involvement of microbial siderophores as either a direct source of soluble iron for monocot uptake,[25] or in ligand exchange coupled with phytosiderophores.

While the graminaceous monocots successfully respond to iron deficiency by the production of phytosiderophores, most monocots and the dicotyledonous plants do not. Rather, iron-deficiency tolerance in dicots is a function of the expression of a number of induced responses in the roots.[8,67,73,74] These physiological responses primarily include:

1. Increased rates of electron transfer reactions at root surfaces, capable of driving Fe^{3+} reduction reactions
2. Increased rates of rhizosphere acidification
3. Increased release of phenolic compounds, i.e., caffeic acid and chlorogenic acid
4. The accumulation of citric acid in the root tissue

The most extensively studied of the dicot responses is the induction of Fe^{3+} reduction at the root surfaces. While this physiological response has been extensively characterized in a number of Fe-efficient dicots, little is known about this response on a molecular level. However, significant advances in the characterization of the plasma membrane enzyme(s) responsible for the Fe^{3+} reduction catalysis are reported in Chapter 18 by Holden et al. while a molecular genetic analysis of the enzyme is discussed in Chapter 19 by Yi et al. The chapter by Holden et al. describes the purification of the induced plasma membrane reductase from Fe-deficient tomato (*Lycopersicon esculentum* Mill.). This represents the first purification of an enzyme responsible for the Fe^{3+} reduction catalysis induced by iron deficiency. This accomplishment represents an important early step in implementing genetic engineering approaches toward addressing iron-deficiency susceptibilities in certain key agricultural crops. A second molecular genetic approach toward the characterization of the biological events involved during Fe deficiency in dicots is the study of the induced responses in *Arabidopsis thaliana* (L.), a model plant system for molecular genetic studies. This chapter by Yi et al. involves two approaches: (1) molecular genetic analysis through the use of hybridization probes from analogous yeast genes, and (2) genetic analysis of rapidly generated mutants of *Arabidopsis*. These characterizations of the plasma membrane-associated enzymes will directly contribute to our understanding of the molecular events involved in one of the primary Fe deficiency responses.

While the graminaceous monocots and the dicots have two general Fe-deficiency responses, it is an interesting fact that very little is actually known about how these systems use, and are ultimately affected by, the microbial siderophores.[24,25,77-80] Based on the above discussion, and reported elsewhere,[24,25] siderophores are likely to strongly influence the speciation of a large percentage of the iron available to plants. A number of studies have demonstrated that the microbial Fe siderophores are extremely poor substrates for root-associated reductive processes.[78] As reported in Chapter 17 by Miller et al. Fe-deficient tomato plants acquire iron from the iron-rhodotorulate complex (Fe_2RA_3) by reductively releasing Fe^{2+} from the ferrisiderophore complex. Bar-Ness et al.[82] have shown reductive iron absorption by cotton (*Gossypium* spp.) from ferrioxamine and a fluorescently labeled ferrioxamine derivative. This iron uptake by cotton, as well as by maize (*Zea mays* L.), was significantly affected by the presence of microorganisms. Considering the importance of siderophores in the plant rhizosphere, and the occurrence of the monocot and dicot responses, our understanding of iron uptake mechanisms will need to focus more closely on the chemical and biological stabilities of the ferrisiderophores in the rhizospheres. This work needs to explore what other roles microorganisms and other root cell surface components contribute to metal micronutrient solubilization at the cell surface and finally to the transmembrane transport.[82]

Finally, much of what has been learned about these induced responses has come from studies of plants grown in hydroponic nutrient solutions. Yet these biological activities associated with roots in nature operate in concert with the vastly complex and varying properties of the rhizosphere. Included among these factors are the plant responses to deficiencies and excesses of other soil minerals and salts. Chapter 22 by Loeppert et al. reviews many of these soil factors, including a careful evaluation of the importance of different active carbonate fractions. In soils that are high in carbonate and pH, several mineral deficiencies are likely to occur. Little is known about how plants integrate their responses to more than one trace metal deficiency. In addition, rapidly increasing amounts of agricultural land are being adversely affected by salinity. These problems will certainly affect molecular events at the root surfaces. While these effects are poorly understood, in Chapter 21, Suhayda et al. report on salinity effects on trace mineral uptake in two barley species *Hordeum vulgare* and *H. jubatum*. Included in this chapter is an important consideration of the roles of the cell wall in cation uptake mechanisms. Cell walls, being in close contact with the soil solution, have a key role, through metal ion binding and electron transfer reactions, in the overall micronutrient uptake by roots. Related to this is Chapter 20 by Thibaud et al., which provides new insights into the effects of pH differentials at the plasma membrane on the activities of the associated Fe^{3+} reductase systems. Proton fluxes will certainly affect metal ion solubilities and other kinetic and thermodynamic equilibria controlling the metal micronutrient availability in plant rhizospheres. As proposed by Thibaud et al., pH may directly influence the catalytic properties of the plasma membrane-bound reductases. Careful consideration must be given to the joint effects of rhizosphere acidification and electron transfer to iron substrates at the root cell surface.

REFERENCES

1. Mortvedt, J.J., Cox, F.R., Shuman, L.M., and Welch, R.M., Eds., *Micronutrients in Agriculture,* 2nd ed., Soil Science Society of America, Madison, WI, 1991.
2. Loeppert R.H., Wei, L.C., and Ocumpaugh, W.R., Soil factors influencing the mobilization of iron in calcareous soils, in *The Biochemistry of Metal Micronutrients in the Rhizosphere,* Manthey, J.A., Crowley, D.E., and Luster, D.G., Eds., Lewis Publishers, Chelsea, MI, 1993 (Chapter 22, this volume).
3. Schwertmann, U., Solubility and dissolution of iron oxides, *Plant Soil,* 130, 1, 1991.
4. Lindsay, W.L., Inorganic equilibria affecting micronutrients in soil, in *Micronutrients in Agriculture,* 2nd ed., Mortvedt, J.J., Cox, F.R., Shuman, L.M., and Welch, R.M., Eds., Soil Science Society of America, Madison, WI, 1991, 89.
5. Uren, N.C., Forms, reactions and availability of iron in soils, *J. Plant Nutr.,* 7, 165, 1984.
6. Vose, P.B., Iron nutrition in plants: a world overview, *J. Plant Nutr.,* 5, 233, 1982.
7. Clark, R.B., Iron deficiency in plants grown in the great plains of the U.S., *J. Plant Nutr.,* 5, 251, 1982.
8. Chaney, R.L., Chen, Y., Green, C.E., Holden, M.J., Bell, P.F., Luster, D.G., and Angle, J.S., Root hairs on chlorotic tomatoes are an effect of chlorosis rather than part of the adaptive Fe-stress response, *J. Plant Nutr.,* 15, 1857, 1992.
9. Norvell, W.A. and Welch, R.M., Growth and nutrient uptake by barley (*Hordeum vulgare* L. cv Herta): studies using an N-(2-hydroxyethyl)ethylenedinitrolotriacetic acid-buffered nutrient solution technique. I. Zinc ion requirements, *Plant Physiol.,* 101, 619, 1993.
10. Norvell, W.A. and Welch, R.M., Growth and nutrient uptake by barley (*Hordeum vulgare* L.cv Herta): studies using an N-(2-hydroxyethyl)ethylenedinitrolotriacetic acid-buffered nutrient solution technique. II. Role of zinc in the uptake and root leakage of mineral nutrients, *Plant Physiol.,* 101, 627, 1993.

11. Parker, D.R., Chaney, R.L., and Norvell, W.A., Chemical equilibrium models: applications to plant nutrition research, in *Chemical Equilibrium and Reaction Models,* Loeppert, R.H., Ed., Soil Science Society of America, Madison, WI, (in press).

12. Bell, P.F., Chaney, R.L., and Angle, J.S., Free metal activity and total metal concentrations as indices of micronutrient availability to barley (Hordeum vulgare L. 'Klages'), in *Iron Nutrition and Interactions in Plants,* Chen, Y. and Hadar, Y., Eds., Kluwer Academic, Boston, 1991, 69.

13. Chaney, R.L, Bell, P.F., and Coloumbe, B.A., Screening strategies for improved nutrient uptake and use by plants, *Hortic. Sci.,* 24, 565, 1989.

14. Laurie, S.H., Tancock, N.P., McGrath, S.P., and Sanders, J.R., Influences of complexation on the uptake by plants of iron, manganese, copper, and zinc. I. Effects of EDTA in a multi-metal and computer simulation study, *J. Exp. Bot.,* 42, 509, 1991.

15. Laurie, S.H., Tancock, N.P., McGrath, S.P., and Sanders, J.R., Influences of complexation on the uptake by plants of iron, manganese, copper, and zinc. II. Effects of DTPA in a multi-metal and computer simulation study, *J. Exp. Bot.,* 42, 515, 1991.

16. Webb, M.J., Norvell, W.A., Welch, R.M., and Graham, R.D., Using a chelate-buffered nutrient solution to establish the critical solution activity of Mn^{2+} required by barley (*Hordeum vulgare* L.), *Plant Soil,* 153, 195, 1993.

17. Laurie, S.L. and Manthey, J.A., The chemistry and role of metal ion chelation in plant uptake processes, in *The Biochemistry of Metal Micronutrients in the Rhizosphere,* Manthey, J.A., Crowley, D.E., and Luster, D.G., Eds., Lewis Publishers, Chelsea, MI, 1993 (Chapter 12 in this volume).

18. Rengel, Z., Pedler, J.F., and Graham, R.D., Control of Mn status in plants and rhizosphere: genetic aspects of host and pathogen effects in the wheat take-all interaction, in *The Biochemistry of Metal Micronutrients in the Rhizosphere,* Manthey, J.A., Crowley, D.E., and Luster, D.G., Eds., Lewis Publishers, Chelsea, MI, 1993 (Chapter 10 in this volume).

19. Marschner, P., Ascher, J.S., and Graham, R.D., Effect of manganese-reducing rhizosphere bacteria on the growth of *Gaeumannonyces graminis* var. tritici and on manganese uptake by wheat (*Triticum aestivum* L.), *Biol. Fertil. Soils,* 12, 33, 1991.

20. Kothari, S.K., Marschner, H., and Römheld, V., Effect of a vesicular-arbuscular mycorrhizal fungus and rhizosphere micro-organisms on manganese reduction in the rhizosphere and manganese concentration in maize (*Zea mays* L.) *New Phytol.,* 117, 645, 1991.

21. Winkelmann, G., Specificity of iron transport in bacteria and fungi, in *CRC Handbook of Microbial Iron Chelates,* Winkelmann, G., Ed., CRC Press, Boca Raton, FL, 1991, 65.

22. Neilands, J.B., Konopka, K., Schwyn, B., Coy, M., Francis, R.T., Paw, B.H., and Bagg, A., Comparative biochemistry of microbial iron assimilation in iron transport, in *Iron Transport in Microbes, Plants and Animals,* Winkelmann, G., van der Helm, D., and Neilands, J.B., Eds., VCH Publishers, Weinheim, Germany, 1987, 3.

23. Höfte, M., Classes of microbial siderophores, in *Iron Chelation in Plants and Soil Microorganisms,* Barton, L.L. and Hemming, B.C., Eds., Academic Press, New York, 1993, 3.

24. Crowley, D.E., Wang, Y.C., Reid, C.P.P., and Szaniszlo, P.J., Mechanisms of iron acquisition from siderophores by microorganisms and plants, *Plant Soil,* 130, 179, 1991.

25. Crowley, D.E., Reid, C.P.P., and Szaniszlo P.J., Microbial siderophores as iron sources for plants, in *Iron Transport in Plant, Microbes, and Animals*, Winkelmann, G., van der Helm, D., and Neilands, J.B., Eds., VCH Publishers, Weinheim, Germany, 1987, 375.

26. Reid, R.K., Reid, C.P.P., Powell, P.E., and Szaniszlo, P.J., Comparison of siderophore concentrations in aqueous extracts of rhizosphere and adjacent soils, *Pediobiologia*, 26, 263, 1984.

27. Hemming, B.C., Microbial-iron interactions in the plant rhizosphere. An overview, *J. Plant Nutr.*, 9, 505, 1986.

28. Bossier, P. and Verstraete, W., Ecology of *Arthrobacter* JG-9-detectable hydroxamate siderophores in soils, *Soil Biol. Biochem.*, 18, 487, 1986.

29. Bossier, P., Höfte, M., and Verstraete, W., Ecological significance of siderophores in soil, *Adv. Microb. Ecol.*, 10, 385, 1988.

30. Loper, J.E. and Buyer, J.S., Siderophores in microbial interactions on plant surfaces, *Mol. Plant-Microbe Interfact*, 4, 5, 1991.

31. Handelsman, J. and Parke, J.L., Mechanisms in biocontrol of soilborne plant pathogens, in *Plant Microbe Interactions*, 3, 27, 1989.

32. Höfte, M., Boelens, J., and Verstraete, W., Survival and root colonization of mutants of plant growth-promoting *Pseudomonas* affected in siderophore biosynthesis or regulation of siderophore production, *J. Plant Nutr.*, 15, 2253, 1992.

33. Neilands, J.B., Iron in biology, in *The Biochemistry of Metal Micronutrients in the Rhizosphere*, Manthey, J.A., Crowley, D.E., and Luster, D.G., Eds., Lewis Publishers, Chelsea, MI, 1993 (Chapter 2 in this volume).

34. Byers, B.R. and Arceneaux, J.E.L., Iron acquisition and virulence of the bacterial genus *Aeromonas*, in *The Biochemistry of Metal Micronutrients in the Rhizosphere*, Manthey, J.A., Crowley, D.E., and Luster, D.G., Eds., Lewis Publishers, Chelsea, MI, 1993 (Chapter 3 in this volume).

35. Braun, V., Hantke, K., Eick-Helmerich, K., Koster, W., Presler, U., Sauer, M., Schaffer, S., Schoffler, H., Staudenmaier, H, and Zimmermann, L., Iron transport systems in *Escherichia coli*, in *Iron Transport in Microbes, Plants, and Animals*, Winkelmann, G., van der Helm, D., and Neilands, J.B., Eds., VCH Publishers, Weinheim, Germany, 1987, 53.

36. Crosa, J.H., Genetics and molecular biology of siderophore-mediated iron transport in bacteria, *Microbiol. Rev.*, 53, 517, 1989.

37. Crowley, D.E. and Gries, D., Modeling of iron availability in the plant rhizosphere, in *The Biochemistry of Metal Micronutrients in the Rhizosphere*, Manthey, J.A., Crowley, D.E., and Luster, D.G., Eds., Lewis Publishers, Chelsea, MI, 1993 (Chapter 14 in this volume).

38. Buyer, J.S., Kratzke, M.G., and Sikora, L.J., Microbial siderophores and rhizosphere ecology, in *The Biochemistry of Metal Micronutrients in the Rhizosphere*, Manthey, J.A., Crowley, D.E., and Luster, D.G., Eds., Lewis Publishers, Chelsea, MI, 1993 (Chapter 6 in this volume).

39. Höfte, M., Woestyne, M.V., and Verstraete, W., Role of siderophores in plant growth promotion and plant protection by fluorescent *Pseudomonads*, in *The Biochemistry of Metal Micronutrients in the Rhizosphere*, Manthey, J.A., Crowley, D.E., and Luster, D.G., Eds., Lewis Publishers, Chelsea, MI, 1993 (Chapter 7 in this volume).

40. Guerinot, M.L., Iron and the nodule, in *Iron Chelation in Plants and Soil Microorganisms,* Barton, L.L. and Hemming, B.C., Eds., Academic Press, New York, 1993, 197.

41. Bosch, I., Meidl, E.J., Hoult, M., Plessner, O., and Guerinot, M.L., Iron uptake and metabolism in the Bradyrhizobium/soybean symbiosis, in *Nitrogen Fixation: Hundred Years After,* Bothe, H., de Bruijn, F.J., and Newton, W.E., Eds., Gustav Fischer, New York, 1988, 652.

42. Carillo-Castaneda, G. and Peralta, J.R.V., Siderophore-like activities in *Rhizobium phaseoli, J. Plant Nutr.,* 11, 935, 1988.

43. Guerinot, M.L., Iron uptake and metabolism in the rhizobia/legume symbiosis, in *Iron Nutrition and Interactions in Plants,* Chen, Y. and Hadar, Y., Eds., Kluwer Academic, Boston, 1991, 239.

44. Barton, L.L., Vester, C.R., Gill, P.R., Jr., and Neilands, J.B., The role of siderophores in symbiotic nitrogen-fixation by legume root nodule bacteria, in *The Biochemistry of Metal Micronutrients in the Rhizosphere,* Manthey, J.A., Crowley, D.E., and Luster, D.G., Eds., Lewis Publishers, Chelsea, MI, 1993 (Chapter 5 in this volume).

45. Neilands, J.B., Overview of bacterial iron transport and siderophore systems in *Rhizobia,* in *Iron Chelation in Plants and Soil Microorganisms,* Barton, L.L. and Hemming, B.C., Eds., Academic Press, New York, 1993, 179.

46. Tang, C., Robson, A.D., and Dilworth, M.J., The role of iron in the (Brady)rhizobium legume symbiosis, *J. Plant Nutr.,* 15, 2235, 1992.

47. Smith, M.J., Shoolery, J.N., Schwyn, B., Holden, I., and Neilands, J.B., Rhizobactin, a structurally novel siderophore from *Rhizobium meliloti, J. Am. Chem. Soc.,* 107, 1739, 1985.

48. Modi, M., Shah, K.S., and Modi, V.V., Isolation and characterization of catechol-like siderophores from cowpea *Rhizobium* RA-I, *Arch. Microbiol.,* 141, 156, 1985.

49. Skorupska, A., Derylo, M., and Lordiewicz, Z., Siderophore containing 2,3-dihydroxybenzoic acid and threonine formed by *Rhizobium trifolii, Acta Biochim. Pol.,* 35, 119, 1988.

50. Tiffin, L.O., Translocation of iron citrate and phosphorus in xylem exudate of soybean, *Plant Physiol.,* 45, 280, 1970.

51. Landsberg, E.C., Function of rhizodermal transfer cells in Fe stress response mechanism of *Capsicum annuum* L., *Plant Physiol.,* 82, 511, 1986.

52. Sijmons, P.C. and Bienfait, H.F., Development of Fe^{3+} reduction activity and H^+ extrusion during growth of iron-deficient soybean plants in a rhizostat, *Biochem. Physiol. Pflanz.,* 181, 283, 1986.

53. Boyer, G.L. and Aronson, D.B., Iron uptake and siderophore formation in the actinorhizal symbiont *Frankia,* in *The Biochemistry of Metal Micronutrients in the Rhizosphere,* Manthey, J.A., Crowley, D.E., and Luster, D.G., Eds., Lewis Publishers, Chelsea, MI, 1993 (Chapter 4 in this volume).

54. Webb, M.J., Mn and Zn absorption and translocation, in *The Biochemistry of Metal Micronutrients in the Rhizosphere,* Manthey, J.A., Crowley, D.E., and Luster, D.G., Eds., Lewis Publishers, Chelsea, MI, 1993, (Chapter 13 in this volume).

55. Reisenauer, H.M., The interactions of manganese and iron, in *The Biochemistry of Metal Micronutrients in the Rhizosphere,* Manthey, J.A., Crowley, D.E., and Luster, D.G., Eds., Lewis Publishers, Chelsea, MI, 1993, (Chapter 11 in this volume).

56. Rovira, A.D., Bowen, G.D., and Foster, R.C., The significance of rhizosphere microflora and mycorrhizas in plant nutrition, in *Inorganic Plant Nutrition,* Lauchli, A. and Bieleski, R.L., Eds., Springer-Verlag, New York, 1983, 61.

57. Smith, S.E., Robson, A.D., and Abbott, L.K., The involvement of mycorrhizas in assessment of genetically dependent efficiency of nutrient uptake and use, *Plant Soil,* 146, 169, 1992.

58. Sharma, A.K., Srivastiva, P.C., and Johri, B.N., Contributions of VA-mycorrhiza to zinc uptake in plants, in *The Biochemistry of Metal Micronutrients in the Rhizosphere,* Manthey, J.A., Crowley, D.E., and Luster, D.G., Eds., Lewis Publishers, Chelsea, MI, 1993, (Chapter 9 in this volume).

59. Sharma, A.K., Srivastava, P.C., Rathmore, V.S., and Johri, B.N., Kinetics of zinc uptake by mycorrhizal (VAM) and nonmycorrhizal corn (*Zea mays* L.) roots, *Biol. Fertil. Soils,* 13, 206, 1992.

60. Li, X.I., Marschner, H., and George, E., Acquisition of phosphorus and copper by VA-mycorrhizal hyphae and root-to-shoot transport in white clover, *Plant Soil,* 136, 49, 1991.

61. Leake, J.R., Shaw, G., and Read, D.J., The biology of mycorrhiza in the Ericaceae. XVI. Mycorriza and iron uptake in *Calluna vulgaris* (L.) Hull in the presence of two calcium salts, *New Phytol.,* 114, 651, 1990.

62. Shaw, G., Leake, J.R., Baker, A.J.M., and Read, D.J., The biology of mycorrhiza in the Ericaceae. XVII. The role of mycorrhizal infection in the regulation of iron uptake by ericaceous plants, *New Phytol.,* 115, 251, 1990.

63. Arines, J., Vilarino, A., and Sainz, M., Effect of different inocula of vesicular-arbuscular mycorrhizal fungi on manganese content and concentration in red clover (*Trifolium pratense* L.) plants, *New Phytol.,* 112, 215, 1989.

64. Bethlenfalvay, G.J. and Franson, R.L., Manganese toxicity alleviated by mycorrhizae in soybean, *J. Plant Nutr.,* 12, 953, 1989.

65. George, E., Römheld, V., and Marschner, H., Contribution of mycorrhizal fungi to micronutrient uptake by plants, in *The Biochemistry of Metal Micronutrients in the Rhizosphere,* Manthey, J.A., Crowley, D.E., and Luster, D.G., Eds., Lewis Publishers, Chelsea, MI, 1993, (Chapter 8 in this volume).

66. Lindermann, R.G., Mycorrhizal interactions with the rhizosphere microflora: the mycorrhizosphere effect, *Phytopathology,* 78, 366, 1988.

67. Römheld, V., Existence of two different strategies for the acquisition of iron in higher plants, in *Iron Transport in Microbes, Plants, and Animals,* Winkelmann, G., van der Helm, D., and Neilands, J.B., Eds., VCH Publishers, Weinheim, Germany, 1987, 353.

68. Jolley, V.D. and Brown, J.C., Genetically controlled uptake and use of iron by plants, in *The Biochemistry of Metal Micronutrients in the Rhizosphere,* Manthey, J.A., Crowley, D.E., and Luster, D.G., Eds., Lewis Publishers, Chelsea, MI, 1993, (Chapter 16 in this volume).

69. Nomoto, K., Sugiura, Y., and Takagi, S., Mugineic acid, studies on phytosiderophores, in *Iron Transport in Microbes, Plants and Animals,* Winkelmann, G., van der Helm, D., and Neilands, J.B., Eds., VCH Publishers, Weinheim, Germany, 1987, 401.

70. Takagi, S.I., Production of phytosiderophores, in *Iron Transport in Microbes and Soil Microorganisms,* Barton, L.L. and Hemming, B.C., Eds., Academic Press, New York, 1993, 111.

71. Mori, S., Mechanism of iron acquisition by graminaceous (strategy II) plants, in *The Biochemistry of Metal Micronutrients in the Rhizosphere,* Manthey, J.A., Crowley, D.E., and Luster, D.G., Eds., Lewis Publishers, Chelsea, MI, 1993, (Chapter 15 in this volume).

72. Crowley, D.E., Römheld, V., Marschner, H., and Szaniszlo, P.J., Root-microbial effects on plant iron uptake from siderophores and phytosiderophores, *Plant Soil,* 142, 1, 1992.

73. Welkie, G.W. and Miller, G.W., Plant iron uptake by nonsiderophore systems, in *Iron Chelation in Plants and Soil Microorganisms,* Barton, L.L. and Hemming B.C., Eds., Academic Press, New York, 1993, 345.

74. Korcak, R.F., Iron deficiency chlorosis, Hortic. Rev., 9, 133, 1987.

75. Holden, M.J., Luster, D.G., and Chaney, R.L., Enzymatic iron reduction at the root plasma membrane: partial purification of the NADH:Fe chelate reductase, in *The Biochemistry of Metal Micronutrients in the Rhizosphere,* Manthey, J.A., Crowley, D.E., and Luster, D.G., Eds., Lewis Publishers, Chelsea, MI, 1993, (Chapter 18 in this volume).

76. Yi, Y., Saleeba, J., and Guerinot, M.L., Iron uptake in *Arabidopsis thaliana,* in *The Biochemistry of Metal Micronutrients in the Rhizosphere,* Manthey, J.A., Crowley, D.E., and Luster, D.G., Eds., Lewis Publishers, Chelsea, MI, 1993, (Chapter 19 in this volume).

77. Bienfait, H.F., Mechanisms of Fe-efficiency reactions of higher plants, *J. Plant Nutr.,* 11, 605, 1988.

78. Bar-Ness, E., Chen, Y., Marschner, H., and Römheld, V., Siderophores of *Pseudomonas putida* as an iron source for dicot and monocot plants, in *Iron Nutrition and Interactions in Plants,* Chen, Y. and Hadar, Y., Eds., Kluwer Academic, London, 1991, 271.

79. Shenker, M., Oliver, I., Helmann, M., Hadar, Y., and Chen, Y., Utilization by tomatoes of iron mediated by a siderophore produced by *Rhizopus arrhizus, J. Plant Nutr.,* 15, 2173, 1992.

80. Jurkevitch, E., Hadar, Y., and Chen, Y., Utilization of the siderophores FOB and pseudobacitin by rhizosphere microorganisms of cotton plants, *J. Plant Nutr.,* 15, 2183, 1992.

81. Miller, G.W., Hasegawa, S., Shigematsu, A., and Welkie, G.W., Mechanisms of iron uptake from rhodotorulate-iron by tomato, in *The Biochemistry of Metal Micronutrients in the Rhizosphere,* Manthey, J.A., Crowley, D.E., and Luster, D.G., Eds., Lewis Publishers, Chelsea, MI, 1993, (Chapter 17 in this volume).

82. Bar-Ness, E., Hadar, Y., Chen, Y., Shanzer, A., and Libman, J., Iron uptake by plants from microbial siderophores. A study with 7-nitrobenz-2 oxa-1,3-diazol-deferrioxamine as fluorescent ferrioxamine B analog, *Plant Physiol.,* 99, 1329, 1992.

83. Suhayda, C.G., Redmann, R.E., and Wang, X., Salinity alters root cell·wall properties and trace metal uptake in barley, in *The Biochemistry of Metal Micronutrients in the Rhizosphere,* Manthey, J.A., Crowley, D.E., and Luster, D.G., Eds., Lewis Publishers, Chelsea, MI, 1993, (Chapter 21 in this volume).

84. Thibaud, J.B., Sentenac, H., and Grignon, C., Role of root apoplast acidification in the H+ pump in mineral nutrition of terrestrial plants, in *The Biochemistry of Metal Micronutrients in the Rhizosphere,* Manthey, J.A., Crowley, D.E., and Luster, D.G., Eds., Lewis Publishers, Chelsea, MI, 1993, (Chapter 20 in this volume).

CHAPTER 2

Iron in Biology

J.B. Neilands

INTRODUCTION

This essay is intended to be a survey in which the main focus will be on high affinity, siderophore-mediated iron absorption, a process that appears to be confined to microbes and some plants. However, in order to place this subject in its proper context, some general remarks will be made about iron and its diverse roles in metabolism of microorganisms, plants, and animals. Some salient reviews and books will be referenced, to which the reader is directed for the source of the facts quoted.

Probably the most important book is the one titled *Iron Transport in Microbes, Plants and Animals,* edited by Winkelmann et al.[1] Iron uptake mechanisms in bacteria have been reviewed by Crosa[2] and by Payne,[3] both of whom place special emphasis on pathogenic species. The book edited by Bullen and Griffiths[4] should be consulted for further details of the relationship between iron and infection. A recent issue of *Biology of Metals* (Vol. 4, No. 1, 1991) has been devoted entirely to iron, and represents a synopsis of papers presented at a meeting in Austin, TX in June 1990. Weinberg[5] has written the definitive review of iron in health and disease, while Winkelmann has edited a useful book titled *Handbook of Microbial Iron Chelates.*[6] All of these books, plus the classic by Warburg,[7] are excellent sources of information on the assimilation, metabolism, and function of iron in biology.

PROPERTIES OF IRON

Element number 26 is built up from argon and, in a thermodynamic sense, enjoys the most stable nucleus in the periodic table. Everything in the universe is destined one day to be converted to iron, and spectroscopic probes show it

0-87371-942-5/94/$0.00+$.50

to be the most prevalent atom in the solar system. Here on planet Earth, it is fourth in abundance among all elements on the surface and ranks just after aluminum among the metals. There are two common oxidation states, designated II and III, although higher oxidation states are possible and are believed to participate in the mechanism of action of cytochrome c oxidase and other heme enzymes. The usual forms encountered in biology are either the ferrous (II) or ferric (III) ions. We can say that the important role of metallic iron in commerce and industry is mirrored in the many functions in which this element has been placed in biology by nature.

The most striking difference between iron II and III is the solubilities of their hydroxides at neutral pH. It is possible to make a 0.1 M solution of Fe(II) at pH 7, although to maintain this level of concentration the solution must be protected from atmospheric oxidation. The solubility product constant for the Fe(III) ion hydroxide is 10^{-38} M, or less. This will fix the maximum concentration of Fe(III) possible at pH 7 as 10^{-17} M, but since biological pH is usually taken as 7.4, the level of ferric ion in equilibrium with the solid trihydroxide is not more than 10^{-18} M. Obviously, biology had to devise some means of keeping Fe(III) in solution in order to make effective use of the two outstanding properties of the element, namely, its abundance and the range of redox potentials of its various coordinated species.

The preferred coordination number of either iron II or III is six, and the ligands are often organized in octahedral or distorted octahedral geometry. However, there is a predictable preference regarding the types of atoms to which these two oxidation states will bond with maximum strength. In the simplified terminology of Pearson, Fe(III) ranks as a "hard" acid since it has a relatively high charge and a small radius. Iron (II) has a significantly lower charge and larger radius. Since coordination stability is based on the ratio of charge/radius, this will affect the values of the formation constants for the two ions. Fe(III) prefers to bond to "hard" base ligands, such as oxygen, whereas Fe(II) will exhibit tighter reaction with nitrogen or sulfur. Since equilibrium constants and reduction potentials are related through the Nernst equation, an Fe(III)-oxygen system will have a low potential and an Fe(II)-nitrogen system will have a high potential. Biologically, potentials range from –0.5 V for certain types of iron sulfur proteins to +0.4 V for cytochromes reacting with oxygen. No other element displays this range of redox potential depending on the nature of its coordination. Doubtless, these are the main reasons why this element has been selected from the periodic table to play such important roles in biology.

Metal ions like to fit into chelate (from the Greek, *chelos*, meaning claw) rings where the ion forms the fifth member of the cycle. This seems to be because the "bite size" of the chelate ring is favorable and competition with competing ligands is discouraged. A ring system with N, O, or S atoms deployed in space and with electrons available for donation to the central metal ion will constitute a good coordination system. Exactly how good can be ascertained by consultation with one of the tables of stability constants. With excess ligand the value of interest will be designated as the beta, representing the equilibrium constant for the stepwise association of the maximum number of bidentate ligands, often three. These values have been compiled by different methods, of which the pH titration method is the most common. An effective chelate for Fe(III) is 8-hydroxyquinoline, while that for Fe(II) is bipyridyl. In the case of the former, both the ligand and its Fe(III) complex are soluble in chloroform and can be extracted into that solvent, the complex being gray-green-black in color while the reagent is colorless. The ferrous bipyridyl complex is bright red, water soluble, and endowed with a 2+ charge. These reagents are valuable for determination of the oxidation state of the metal ion, although it should be kept in mind that they may shift the redox ratio by preferential coordination. In the presence of any electrons as a source of reducing power, a bipyridyl-iron solution will eventually turn all ferrous. Physical methods, such as atomic absorption, give only total iron without reference to the distribution among the various oxidation states.

ENVIRONMENTAL IRON

The planet Earth is estimated reliably to be 4.5 to 4.6×10^9 years in age. Life in one form or another is thought to have existed on the planet for about 3×10^9 years and began with a strictly anaerobic species. Vestiges of this type of metabolism have remained today in all cells, regardless of the extent to which they have adapted to life in the presence of oxygen. So we can surmise that the primordial iron was mainly ferrous, which is the form we see coming out from deep springs and from any other source not attacked by the corrosive gas, oxygen. A true, oxygen-evolving photosynthesis was apparently first performed by the cyanobacteria, or blue-green algae. Many of these bacteria can fix both C and N and are thought to be the precursors of the chloroplasts of higher plants, while the nongreen bacteria became mitochondria. At any rate, the appearance of oxygen in the atmosphere oxidized the surface iron over a period of time, which must have been considerable. In retrospect, we can envisage the surface iron as playing some kind of buffering role against the antibiological effects of the poisonous gas, oxygen. The present situation, in which about a quarter of the atmosphere is oxygen, enabled the evolution and maintenance of eukaryotic life and the generation of high levels of ATP through the reduction of oxygen to water. This reaction required iron, and depended on the development of efficient systems for handling the element in a hostile environment.

IRON IN THE CELL

Preoccupied as we are with human and animal nutrition, the near universal requirement for iron is thought to be for hemoglobin synthesis. Indeed, in higher animals this is a major form of iron, the rest of it being in a storage protein, ferritin, or its quasiequilibrium form, hemosiderin. Only traces can be ascribed to enzymes such as cytochromes and ribonucleotide reductase. At the turn of the century iron had already been identified as a constituent of blood and hence was recognized as essential. Shortly thereafter, Warburg[7] began his experiments in cell physiology, which required development of a practical respirometer and application of the photoelectric spectrophotometer to biological systems. He was able to show that an iron enzyme, the "atmungsferment", performs the terminal stage in respiration by reduction of O_2 to water. He further showed that the enzyme is inhibited by cyanide in the Fe(III) form and by CO in the Fe(II) state. This latter demonstration is considered quite elegant since it did not require isolation of the enzyme — a feat never achieved by Warburg in spite of prodigious efforts. The effectiveness of light of different wavelengths in relieving CO inhibition of respiration enabled him to infer that the enzyme was a hemoprotein. It was known at that time that the CO complex of hemoglobin is photodissociable, and so a model existed for the photochemical absorption spectrum of the atmungsferment. Obviously, CN^- could not be a blood poison since the lethal dose is about a millimole and the average adult human contains about 80 mmol of hemoglobin iron.

Today we recognize that iron is needed for the reduction of O_2, CO_2, and N_2. Of these various functions, the last-named requires special mention. The enzyme nitrogenase is known to occur only in bacterial species and, like cytochrome c oxidase, which contains copper, it may have traces of other metals in addition to iron. But no form of nitrogenase is known which does not have iron. Recently,[8] the molybdenum-iron protein of *Azotobacter vinelandii* nitrogenase has been shown by crystallography to have the metals coordinated to cysteine thiols, inorganic sulfur, imidazole nitrogen, and homocitrate. Other essential enzymes contain this metal. In fact, the sole biological species that has apparently avoided dependence on iron are certain strains of lactobacilli. These bacteria are adapted to flourish in environments exceedingly low in iron, such as milk and all dairy products. They have no cytochromes, peroxidase, catalase, or heme compounds of any type. Their ribonucleotide reductase, which they must have in order to reduce the ribose ring of nucleotide precursors of DNA, contains cobalt in the form of vitamin B_{12}. This appears to take the place of iron in the most usual type of reductase, such as is found in *Escherichia coli* and humans. It has not yet been possible to sufficiently rid media of contaminating iron to get the iron content of the cells below one atom, but the numbers have approached this limiting value. Milk contains lactoferrin, a ferric ion-binding protein that is very effective in denying the element to certain microbes. As an educated guess, this suggests that lactobacilli have learned to live without iron, and this makes the species unique among all organisms.

IN PRAISE OF MICROBES

There are certain basic principles of biology that cut across all species, whether microbial, plant, or animal. In each and every case, the initial discoveries were made with microorganisms and only later extended to plants and animals. Take, for instance, the finding that DNA is the genetic blueprint of the cell. This was proven beyond a reasonable doubt by Avery and his colleagues who were able to show that extracts of smooth, capsule-producing pneumococci could "transform" rough strains of the organism (cited by McCarty[9]). The extract was highly purified and shown not to contain either protein or RNA. It did, however, contain DNA. The genetic code specifying the order at which amino acids are laid down in proteins was worked out following the observation that poly-U coded for polyphenylalanine. This was done with the intestinal bacterium *E. coli*. The precise manner in which this bacterium regulates its synthesis of enzyme proteins was then established following its growth on different sugars. This mode of regulation, with some variation, holds true for other microbes and for higher life forms. Finally, the "restriction" phenomenon whereby cells protect themselves against undesired transformations was noted first in bacteriophages infecting *E. coli*. This gave birth to the practice of recombinant DNA and to the entire field of biotechnology. It is thus apparent that a microbe, *E. coli* in particular, has proven to be a most effective subject in which to investigate some of the processes most fundamental to life.

MICROBIAL MECHANISMS OF IRON UPTAKE

With the possible exception of the lactobacilli, all other bacterial species and doubtless all fungal species require iron. The minimum iron concentration required for maximum growth rates of *E. coli* is said to be about 1.0 μ*M*. The concept of both a high- and a low-affinity iron transport system in *E. coli* arose from studies of a closely related enteric bacterium, *Salmonella typhimurium* Lt-2. Mutants of this latter organism were found to be unable to grow on citrate as a carbon source without addition of very large amounts of iron salts. Some of these mutants excreted into the medium 2,3-dihydroxybenzoic acid and its conjugate with serine. Eventually the wild type cell was found to make enterobactin, the cyclic trimer of 2,3-dihydroxybenzoyl serine. This Fe(III)-specific ligand, subsequently designated a siderophore (Greek, meaning *iron bearer*), apparently scavenges iron and makes it available to the bacterial cell. But synthesis of this high-affinity, siderophore-mediated pathway of iron acquisition is only invoked when the cell is severely stressed for iron. Mutants unable to make enterobactin grow at normal rates on complex media and even on minimal media in the absence of a nonutilized chelating agent, such as citrate. The genes for enterobactin synthesis and transport are clustered in a small part of the chromosome and defects in this region could be shown to be correctable with ascorbic acid. Actually, the vitamin was reducing the iron and making it available to the cell without need of a siderophore.

Siderophores are commonly hydroxamic acids or catechols, although there are a few instances in which neither of these very effective functional groups for Fe(III) are present. Chemical tests for both of these groups are available. In the case of catechols the ring is nitrosated and treated with molybdate which, in alkaline media, gives a relatively potent pink coloration known as the Arnow reaction. In the Czaky test for hydroxamic acids, the oxidized N is converted ultimately to nitrite, which is assayed by diazotization. The most sensitive chemical test is based on the removal of iron from the blue Chrome Azurol S complex to yield the orange-colored iron-free form of the dye. This test, known as CAS,[10] does not depend on any particular chemical function, and because of the exceedingly high extinctions of the dye it is capable of detecting very low levels of siderophores.

As may be seen in Table 1, a number of microbial species appear not to make siderophores. We presume that most of these microorganisms do need some iron for growth. How they get the element is not understood. An organism that makes no siderophores may reduce the iron at the surface of the cell and take up the element as Fe^{2+}. It may use heme, which is abundant in mammalian tissues. It may extract the iron from iron transferrin or it may make a receptor for this glycoprotein. Obviously, quite a number of species do not make a siderophore, but many of them maintain a well-developed transport system for siderophores that may be encountered in the environment. Thus ferrichrome, which is made by fungi but not by bacteria, is transported by many bacterial species.

RHIZOBIAL IRON ABSORPTION

We have already mentioned the central role of iron in nitrogenase and the fact that this catalyst is strictly a bacterial enzyme. The various species harboring nitrogenase may be classified as either symbiotic or nonsymbiotic. Among the free-living (nonsymbiotic) forms, an impressive array of morphological and physiological types are capable of reduction of the abundant but unreactive N_2 to the soluble and reactive ammonia. These include strict anaerobes *(Clostridia)*, strict aerobes *(Azotobacter)*, Gram-negative enterics *(Klebsiella)*, Gram-positive spore formers *(Bacillus)*, archaebacteria, and even photosynthetics, some of which can reduce both CO_2 and N_2 (cyanobacteria).[11] However, most N_2 fixed in the biosphere is thought to arise from activities of the *rhizobia*, which are remarkably specific as regard the plant species they are able to nodulate. Following infection of the rootlet, bacteria multiplying or differentiating into bacteroids must acquire iron from the plant tissue. In addition to nitrogenase proper, a number of closely connected enzymes and proteins contain iron. At the level of initial fixation of dinitrogen, the energetic

Table 1.   A List of Microorganisms Apparently Not Making Siderophores

Haemophilus influenzae	Saccharomyces cerevisiae
Trichomonas vaginalis	Saccharomyces pombe
Bordetella pertussis	Serratia sp.[a]
Mycoplasma pneumoniae	Listeria sp.
Neisseria gonorrhoeae	Legionella sp.
Neisseria meningitidis	Bacteroides fragilis

[a]   A siderophore could not be detected in Serratia marcescens W225,[27] although most members of the genera make enterobactin.[28]

electrons required for the reduction are carried by ferredoxin, a low molecular weight protein containing several percent iron. Hydrogenase, another enzyme associated with nitrogenase, contains iron. Leghemoglobin and related hemo-proteins are present in nodules where they appear to play some role in modu-lating the supply of O_2 available to the bacteroids. Even after nitrogen is organically bound as glutamine, nonheme iron is still involved, via the various forms of glutamate synthase, in moving the N atom into different amino acids and other metabolites. Iron is also a constituent of the nitrification and denitri-fication steps necessary to complete the nitrogen cycle of the biosphere. Obviously, we need more basic information on the iron economy of the nitrogenase complex if we are to increase the efficiency of N_2 fixation, hope-fully in a manner that is environmentally benign.

The considerations just enunciated have prompted us to investigate iron absorption in different strains of *Rhizobium meliloti,* a species symbiotic with the agronomically important plant alfalfa *(Medicago sativa).* This approach turned up rhizobactin, an entirely novel siderophore that contains neither hydroxamate nor catechol ligands.[12] The rhizobactin core is comprised of molecules of D-alanine and L-lysine linked via an ethylene bridge through their apha nitrogen atoms.[13] The epsilon amino group of the lysine is additionally substituted by a residue of L-malic acid. The result is a genuine siderophore in which the two ethylenediamine nitrogens, the three carboxyl groups, and the lone hydroxyl alpha to the malate carboxyl are available for chelation, probably forming a mononuclear complex.[14] Thus a metal-binding ethylenediamine moiety, an elaboration of the simple bidentate ligand system upon which Alfred Werner built his coordination theory, was found to exist in nature. As expected from the structure, rhizobactin displays somewhat greater avidity for Fe(II) when contrasted with other siderophores.

While a genetic analysis of rhizobactin synthesis in *R. meliloti* DM4 might have been possible, such a study would be facilitated in strain 1021 since the genome of the latter is relatively well characterized. Strain 1021 was found, by application of the CAS assay, to make a siderophore, and the bizarre solubility properties of this compound led us to believe it to be structurally novel.[10] After

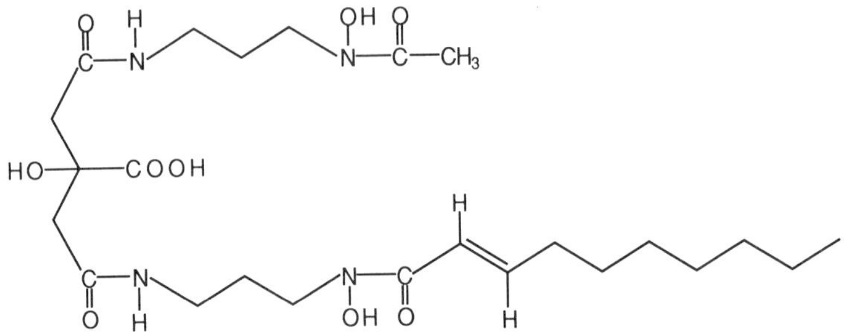

Figure 1.   Structure of rhizobactin 1021 from *Rhizobium meliloti* 1021.[15] This is the fourth known naturally occurring member of the citrate-hydroxamate series of siderophores. In all others characterized to date, namely aerobactin, arthrobactin (Terregens factor), and schizokinen, the acyl moieties of the hydroxamates have been found to be identical rather than, as in the case of rhizobactin 1021, dissimilar.

the constitution had been elucidated through the various modalities of NMR and MS and by chemical degradation, it proved to be a citrate-derived dihydroxamate with the C3 side chains of schizokinen (Figure 1).[15] The novel feature of rhizobactin 1021, which also accounts for its unusual solubility properties, was the presence of dissimilar moieties in the acyl bonds of the hydroxamates, one being the common acetate while the other is the highly unusual C10 alpha-beta *trans* unsaturated fatty acid. The stereochemistry at the quaternary carbon of the citrate, like that of the intermediate in aerobactin synthesis, is unknown at the present time. The short diamine side chains of rhizobactin 1021 require that it complex Fe(III) mainly as a binuclear dimer.[15]

A collection of Tn5-induced mutants of *R. meliloti* 1021 was obtained and classified as either biosynthetic, regulatory, or transport, depending on their response to the CAS assay and their level of growth in the presence of a nonutilized chelator of Fe(III).[16] Preliminary data suggest that an intact system for rhizobactin 1021 is required for N_2 fixation and enhanced plant growth.[17]

Other iron-binding molecular species associated with iron absorption in different *rhizobia* include 2,3-dihydroxybenzoic acid, anthranilic acid, and citric acid.[18] Some *rhizobia* appear to make none of these when grown in synthetic media and stressed for iron. Bacteria have so many diversified modes of acquiring iron that the *rhizobia,* which are quite individualistic, may absorb the element generally via a nonsiderophore pathway.

MOLECULAR GENETICS OF IRON TRANSPORT IN *ESCHERICHIA COLI*

E. coli is known to carry at least five high affinity systems for iron absorption. These include systems for synthesis and transport of enterobactin and aerobactin, as well as systems for uptake of ferrichrome, rhodotorulic acid and coprogen, and ferric citrate. Of these, the aerobactin and enterobactin systems have been particularly well studied. The former occurs commonly in clinical isolates of *E. coli* and is frequently associated with the V (virulence) plasmids, while the latter seems common to nearly all enteric bacteria. Some strains of *Shigella* apparently make the genes for enterobactin, but they are not expressed.

The aerobactin cluster of genes in *E. coli* is the simplest and best understood of the siderophore systems. Cloned versions from pColV-K30 were shown to make only five polypeptides in minicells.[19] On SDS-PAGE these were found to have molecular mass values of 33, 53, 62, 63, and 74 kDa. The last-named was identified as the outer membrane receptor for the ferric complex of the siderophore, leaving the remaining four with a role in the biosynthesis of aerobactin. This siderophore is comprised of citrate substituted on its distal carbons with N^6-hydroxy-N^6-acetyl-L-lysine. This amino acid is made from lysine by oxidation and acetylation, reactions which require, respectively, the 53- and 33-kDa components. The synthetase then completes the assembly of aerobactin in two steps requiring the participation of the 63- and 62-kDa polypeptides. The chiral intermediate, which is highly anionic since it contains two carboxyl groups from citrate and one from the side chain, has an absolute configuration not yet decided. The genes for aerobactin synthesis occur on large plasmids or the chromosome, and they appear to be widely distributed among enteric bacteria. The citrate-hydroxamate type of siderophores also occur in other Gram-negative bacteria and in Gram-positive spore-forming bacilli. Nothing is known about the route of biosynthesis of the siderophores in these other species.

REGULATION OF SIDEROPHORE SYNTHESIS BY IRON

It has been known for many decades that siderophore formation in bacteria and fungi is negatively regulated by iron. Only about 1.0 mg of added iron per liter is enough to shut down extracellular formation of siderophores. The cloning of the aerobactin determinants of *E. coli* (pColV-K30) afforded the opportunity to investigate the molecular mechanism of this regulation. A mutation had been described previously in *S. typhimurium* that led to constitutive expression of all of the high affinity iron uptake pathways of the cell. This mutation, which was obtained by treating the cells with chemical agents, was designated *fur* (*f*erric *u*ptake *r*egulation). The same mutation was generated in *E. coli* and shown to cause the absence of a 17-kDa peptide. The Fur protein was overexpressed in a genetically engineered strain and shown to be a divalent metal ion sensing protein. Fur thus acts as a classical repressor binding an operator sequence to negatively regulate transcription of the siderophore genes. The consensus sequence of the "iron box" operator is 5'-GATAATGATAATCATTATC. Exactly how the Fur protein complexes Fe^{2+}

is a problem that is currently under investigation, as is the mode of attachment of the ferrous Fur to the operator DNA. The Fur protein has only been isolated from *E. coli,* but it may occur generally in bacteria and possibly in fungi.

The *E. coli* Fur repressor contains 147 amino acids of which 12 are histidine and 4 are cysteine. Two of the cysteines, 92 and 95, occur in a pentapeptide represented by .CxyCG., which terminates in glycine. This arrangement is typical of rubredoxins and ferredoxins which, however, are iron metalloproteins rather than iron-binding proteins like Fur. A recent investigation supports the conclusion that the Fe(II) is bound at the C-terminal domain and that this results in a conformational change that affords an enhanced and specific affinity of the N-terminal domain for the operator.[20] As suggested by the palindromic nature of the operator, Fe(II)-Fur very likely binds as a homodimer. It has not yet been possible to crystallize any form of the Fur protein. The metal-activated species may not complex the divalent ion with sufficient rigidity to allow crystallization since the dissociation constant is relatively large, probably about 50 μM. It may be possible to find a divalent metal that binds at the same site as Fe(II) but which has a higher affinity for the protein. Still to be explored is the chance of crystallizing the ternary complex consisting of protein, metal, and operator. This experiment should have high priority since a complex of this type is known to be highly associated.

All known iron-regulated siderophore genes and operons of *E. coli* appear to be governed by the Fur repressor system. Since a null mutation leads to full expression of these genes, it can be concluded that activation plays a minor role in their manifestation. Certainly, the aerobactin promoter is strong in that the −10 and −35 sites are close to the consensus and are separated by the optimum 17 bp. This assures abundant synthesis of the transcript at low iron conditions.

The regulation of the *fur* gene itself is somewhat more baroque in that it is negatively controlled by its own product, Fur, and is positively stimulated by an upstream CAP site.[21]

IRON ABSORPTION IN PLANTS

Plants need iron for development of the photosynthetic apparatus, and without the element they become yellow and suffer from a deficiency known as *chlorosis.* While iron in the soil can be made more available by application of organic sequestering agents, this is expensive and is not regarded as the final solution. Chlorosis occurs in plants growing in every region of the world, but is considered to be an especially important problem in calcareous soils where the pH is relatively high.

It seems that plants have evolved different strategies for assimilation of iron. The thin-leaved varieties, grasses, and monocotyledons, in general, appear to use a siderophore-type compound generically called phytosiderophores. The main example is mugineic acid. These compounds prefer to ligate Fe(III), but show some avidity for Fe(II). Methionine seems to be a major precursor of mugineic acid. So far, no experiments have been reported on the cloning of this type of system. The other major uptake pathway is via some type of reduction mechanism.

GLOBAL IRON REGULATORY CIRCUITS

Just as in microbes, an animal that is replete with respect to iron does not accumulate more of the mineral from the environment. Giving excess iron to animals leads to enhanced synthesis of ferritin and diminished formation of the transferrin receptor. These are important controls since all cellular iron in higher organisms comes via iron transferrin and is stored as ferritin, a protein that both protects the metal against hydrolysis and prevents its reaction with oxygen to yield tissue-destroying radicals. Regulation of ferritin and transferrin receptor occurs mainly at the translational level, and is mediated by similar sequences at the untranslated region of the message. This requires a protein which appears to have some sequence homology with aconitase, an iron enzyme.[22]

CONCLUSION

At this juncture we know relatively quite a lot about one high affinity mode of acquiring iron in one microbial species. Thus the aerobactin system of $E. coli$ requires only four proteins for synthesis of the siderophore. Of these, the gene for the oxygenase converting L-lysine to its N^6-hydroxy form has been sequenced. The acetylase acting on this product as substrate has been isolated, as has the protein catalyzing the addition of this side chain to one of the distal carboxyl groups of citrate. The "unfinished" aerobactin is known to be converted by the enzyme IucC to yield the final form of the siderophore. So out of the four biosynthetic genes one has been sequenced, and among the gene products, two have been isolated in a homogeneous state. Aerobactin is believed to be recycled, and external to the cell it readily associates with Fe(III), either inorganically or bound to transferrin. The transport across the outer membrane requires the product of the fifth gene in the operon, IutA. This protein has been isolated and the sequence of its gene is known. The five-gene operon is regulated by a transcriptional repressor, Fur, coded at 15.7 minutes on the $E. coli$ chromosome. The repressor has been expressed and shown to bind a specific operator sequence following activation by Fe(II) and related first-row divalent transition elements. An NMR study suggests that the repressor is comprised of five "up and down" helical bundles.[23] Yet to be determined are the precise modes of binding ferrous ion and the subsequent tight attachment of the holo-repressor to the operator. Evidence at hand indicates that the metal ion enters the C-terminal domain and the ensuing conformational change then induces the N-terminal part of the repressor to bind the operator.[20] The aerobactin cluster is surrounded by *IS*-1 elements in inverted orientation, and hence it has the general character of a transposon. It is apparent that even in this relatively well-characterized system, a substantial number of details remain to be worked out. The other *E. coli* siderophore systems are less well understood, although the enterobactin complex, which is spread over more than 20 kb of DNA, is starting to be deciphered. The level of the Fur repressor inside the cell is yet another variable, and this is a function of the concentration of cAMP and iron.[21]

An area of which we are nearly totally ignorant is the state and coordination of iron inside *E. coli*. According to present information the interior of the cell must be highly reducing, the O_2 being stopped at the cytoplasmic membrane where it is reduced to water with concomitant generation of ATP. Significantly, the lysine oxygenase catalyzing the first committed step in aerobactin synthesis is located in the membrane. There are very few disulfide proteins in the cytosol — thioredoxin is the only one that comes to mind. Enzymes such as alkaline phosphatase must be dithiol and inactive inside the cell, otherwise essential phosphorus-containing metabolites would be hydrolyzed. We suppose that there is enough Fe(II) in hexa-aquo form or loosely bonded to glutathione and other cellular constituents to afford a pool of the mineral sufficient to activate the Fur repressor.

The manner in which an *E. coli* cell stripped of all siderophore-mediated pathways is able to acquire iron and regulate uptake of the element is still mysterious.

In contrast to the siderophore systems just described, which are mainly regulated by a transcriptional repressor, synthesis of ferritin and the transferrin receptor appears to be controlled at a translational level. An iron-binding protein, which may be identical to cytosolic aconitase, is required.[22] Aconitase has one loosely coordinated iron and so it is possible that this is the means of coupling cellular levels of iron to regulatory efficiency. Nothing is yet known about the molecular genetics of intestinal absorption of iron in the animal.

One thing that has become very apparent in recent years is the fact that iron is both nutritious and noxious. This point has been emphasized by both Emery[24] and Lauffer[25] in current monographs on the topic. Unlike nitrogen, where an excess of the element is discarded as an excretion product, the uptake of iron from the environment is metered with precision. Thus the regulatory processes governing iron assimilation would seem to be a subject deserving of the highest priority for study.

The level of understanding of iron in biology is at this date quite impressive, and is neatly summarized in the recent book by Crichton.[26] Unfortunately, owing to the structure of the modern industrial society, this knowledge not only increases the general enlightenment but makes it possible for the human species to enhance our own perceived welfare at the expense of nature. This arrogant attitude will redound to our detriment. Thus, looking beyond iron, our most important task should be to harmonize basic research with long-term stability of the biosphere.

REFERENCES

1. Winkelmann, G., van der Helm, D., and Neilands, J. B., Eds., *Iron Transport in Microbes, Plants and Animals,* VCH Publications, Weinheim, Germany, 1987.
2. Crosa, J. H., Genetics and molecular biology of siderophore-mediated iron transport in bacteria, *Microbiol. Rev.,* 53, 517, 1989.
3. Payne, S. M., Iron and virulence in the family *Enterobacteriaceae, Crit. Rev. Microbiol.,* 16, 81, 1988.
4. Bullen, J. J. and Griffiths, E., Eds., *Iron and Infection,* John Wiley & Sons, New York, 1987.
5. Weinberg, E. D., Cellular iron metabolism in health and disease, *Drug Metab. Rev.,* 22, 531, 1990.
6. Winkelmann, G., Ed., *Handbook of Microbial Iron Chelates,* CRC Press, Boca Raton, FL, 1991.
7. Warburg, O., *Heavy Metal Prosthetic Groups,* Clarendon Press, Oxford, 1949.
8. Kim, J. and Rees, D. C., Structural models for the metal centers in the nitrogenase molybdenum-iron protein, *Science,* 257, 1677, 1992.
9. McCarty, M., *The Transforming Principle,* Norton, New York, 1985.
10. Schwyn, B. and Neilands, J. B., Universal chemical assay for the detection and determination of siderophores, *Anal. Biochem.,* 160, 47, 1987.
11. Burris, R. H., Nitrogenases, *J. Biol. Chem.,* 266, 9339, 1991.
12. Smith, M. J., Shoolery, J. N., Schwyn, B., Holden, I., and Neilands, J. B., Rhizobactin, a structurally novel siderophore from *Rhizobium meliloti, J. Am. Chem. Soc.,* 107, 1739, 1985.
13. Smith, M. J., Total synthesis and absolute configuration of rhizobactin, a structurally novel siderophore, *Tetrahedron. Lett.,* 30, 313, 1989.
14. Schwyn, B. and Neilands, J. B., Siderophores from agronomically important species of the *Rhizobiacae, Comments Food Agric. Chem.,* 1, 95, 1987.
15. Persmark, M., Pittman, P., Buyer, J. S., Schwyn, B., Gill, P. R., and Neilands, J. B., Isolation and structure of rhizobactin 1021, a siderophore from the alfalfa symbiont *Rhizobium meliloti,* 1021, *J. Am. Chem. Soc.,* 115, 3950, 1993.
16. Gill, P. R. and Neilands, J. B., Cloning a genomic region required for a high affinity iron uptake system in *Rhizobium meliloti,* 1021, *Mol. Microbiol.,* 3, 1183, 1989.
17. Gill, P. R., Barton, L. L., Scoble, M. D., and Neilands, J. B., A high-affinity iron transport system of *Rhizobium meliloti* may be required for efficient nitrogen fixation *in planta, Plant Soil,* 130, 211, 1991.
18. Guerinot, M. L., Iron uptake and metabolism in the rhizobia/legume symbioses, *Iron Nutrition and Interaction in Plants,* Chen, Y. and Hadar, Y., Eds., Kluwer Academic, Dordrecht, 1991, 239.
19. de Lorenzo, V. and Neilands, J. B., Characterization of *iuc*A and *iuc*C genes of the aerobactin system of plasmid ColV-K30 in *Escherichia coli, J. Bacteriol.,* 167, 350, 1986.
20. Coy, M. and Neilands, J. B., Structural dynamics and functional domains of the Fur protein, *Biochemistry,* 30, 8201, 1991.
21. de Lorenzo, V., Herrero, M., Giovannini, F., and Neilands, J. B., Fur (ferric uptake regulation protein) and CAP (catabolite activator protein) modulate transcription of *fur,* gene in *Escherichia coli, Eur. J. Biochem.,* 173, 337, 1988.

22. Rouault, T., Stout, C. D., Kaptain, S., Harford, J. B., and Klausner, R. D., Structural relationship between an iron-regulated RNA-binding protein (IRE-BP) and aconitase: functional implications, *Cell,* 64, 881, 1991.

23. Saito, T., Wormald, M. R., and Williams, R. J. P., Some structural features of the iron-uptake regulation protein, *Eur. J. Biochem.,* 197, 29, 1991.

24. Emery, T. F., *Iron and Your Health: Facts and Fallacies,* CRC Press, Boca Raton, FL, 1991.

25. Lauffer, R., *Iron Balance,* St. Martin's Press, New York, 1991.

26. Crichton, R. R., *Inorganic Biochemistry of Iron Metabolism,* Ellis Horwood, Chichester, 1991.

27. Angerer, A., Klupp, B., and Braun, V., Iron transport systems of *Serratia marcescens, J. Bacteriol.,* 174, 1378, 1991.

28. Reissbrodt, R. and Rabsch, W., Further differentiation of *Enterobacteriaceae,* by means of siderophore pattern analysis, *Zentralbl. Bakt. Hyg.,* A268, 306, 1988.

CHAPTER 3

Iron Acquisition and Virulence of the Bacterial Genus *Aeromonas*

B. Rowe Byers and Jean E.L. Arceneaux

VERTEBRATE IRON MANAGING AND IRON WITHHOLDING

The two stable valences of iron catalyze a range of chemical reactions which (when directed by enzymes) make iron an indispensable cofactor for most biological systems; however, excess or misplaced iron can be fatal.[1] The undisciplined presence of the metal may predispose an animal to infection by promoting microbial growth and may generate toxic oxygen radicals which can cause lipid peroxidation and DNA damage, thus increasing the risk of diseases such as cancer.[1-5] The hazard of iron's reactivity for critical molecules usually is surmounted by special iron or heme binding substances (transferrin, lactoferrin, hemopexin, and haptoglobin in mammals) whose purposes are the safe management and movement of iron- and heme-containing molecules. The term heme refers to iron-protoporphyrin complexes. The distinctive property of the plasma β-globulin transferrin is the reversible binding of iron, which enables it to transport and deliver iron to vertebrate cells.[6] Hemopexin (a plasma β-glycoprotein) and haptoglobin (an α_2-globulin) bind heme and hemoglobin, respectively, recycling iron through the liver and the reticuloendothelial system and serving as protective antioxidants.[7-9] These iron and heme binding substances lower the concentration of free iron in animal sera to 10^{-18} M, which likely assists in protecting the organism from iron-mediated damage and which is much below the level (10^{-6} M) that usually must be available in the environment to support growth of microorganisms.[10]

Availability of iron is one of the pivotal factors controlling the outcome of an invasion of a vertebrate host by pathogenic microorganisms. Many laboratory and clinical studies have revealed the existence in vertebrates of an anti-infective defense system that is called "iron withholding" because it attempts to withhold iron from an invading pathogen.[2] The system includes constitutive

0-87371-942-5/94/$0.00+$.50
© 1994 by CRC Press, Inc.

components like the powerful iron-binding proteins transferrin and lactoferrin.[1,2] Lactoferrin is found in most biological fluids including milk, saliva, tears, and mucous secretions, and it is released from activated neutrophils.[16,18-20] Unlike transferrin, lactoferrin binds iron at low pH, making the protein a good iron scavenger in areas where microbial growth may be producing acids.[16] Part of the antimicrobial activity of lactoferrin may be related to iron withholding,[21] but the mechanism is more complex than simply depriving microbes of iron because lactoferrin also has immediate bactericidal activity.[20] The bactericidal property of lactoferrin resides in an antimicrobial domain consisting of a loop of 18 amino acids that is distinct from the iron-binding site.[20] The peptide retains antibacterial activity when isolated by enzymatic digestion of lactoferrin.[20] The lactoferrin molecule may possess two antimicrobial properties of separate function. Hemopexin and haptoglobin also may deny iron from a few microbes.[11-15]

An infectious episode in vertebrates also induces certain additional iron-withholding responses, such as the protective stance called the "hypoferremic" response, which lowers (by as much as 50%) the amount of iron in circulating transferrin.[16,17] This is triggered in part by interleukins 1 and 6, which provoke macrophages that have digested hemoglobin from decaying erythrocytes to retain iron (rather than recycle the metal to transferrin), and storing it in newly synthesized ferritin.

Although iron-withholding defenses can be productive in combating extracellular pathogens (those which live outside of host cells), such shields may be less effective against intracellular pathogens which seek shelter within host cells. Recent evidence, summarized by Weinberg,[22] suggests that when invaded by microbes, vertebrate cells still attempt to restrict availability of iron. Host cells that have been invaded by microorganisms undergo a net loss of nonheme iron. Moreover, such cells may downregulate expression of surface receptors for transferrin, thereby lowering the level of incoming iron. Nitric oxide may be the signal for such an "iron-depletion" defense strategy.

MICROBIAL IRON ACQUISITION TACTICS

Although the iron-withholding defenses allied by vertebrate hosts against extracellular pathogens are formidable, microbes have evolved several tactics that surmount a host's iron barriers. The major host iron depots of Fe-transferrin and heme-containing molecules are drawn upon by microbes in at least three ways: (1) the microorganism uses its siderophore to extract iron from Fe-transferrin (or Fe-lactoferrin), (2) the microbe produces a surface receptor for the host's Fe-transferrin (or Fe-lactoferrin) and subsequently removes iron from the protein, and (3) heme is used, either via a microbial heme receptor or by an as yet undefined means. Some bacteria may employ more than one of these mechanisms.[1,23-25]

Siderophore-mediated iron acquisition from Fe-transferrin is an attractive hypothesis; however, in only a few instances have siderophores been shown crucial for virulence of pathogens. Production of the siderophore aerobactin is prominent in some of the pathogenic enterics, such as the invasive *Escherichia coli* strains that infect the bloodstream.[26] Virulence of the marine fish pathogen *Vibrio anguillarum* depends on the siderophore anguibactin.[23] Mutant strains of the plant pathogen *Erwinia chrysanthemi*, which cannot produce or utilize the siderophore chrysobactin, are unable to spread from the point of inoculation into plant tissue.[27] Some siderophores (for example, enterobactin) may be designed to function more efficiently in environments other than the serum of vertebrates.[28] Therefore, not all siderophores are virulence factors. The precise contribution of siderophores to the virulence of organisms that have more than one iron acquisition mechanism is still uncertain.

Some pathogens are able to use Fe-transferrin or Fe-lactoferrin (or both) as a sole iron source without intervention of a siderophore. This ability sometimes correlates with synthesis of a bacterial receptor for the Fe-transferrin molecule. Some pathogens have developed a specificity that permits them to use only the transferrin of their preferred host.[29-33] The biochemical mechanisms by which iron is released from Fe-transferrin by microbial cells is not clear; it could involve a reductive system similar to the membrane-bound reductase identified in some bacteria.[34]

Finally, many pathogenic species have tapped the plentiful supply of host iron held in heme-containing compounds.[12-15,35-37] Access to the iron in heme (which is normally present inside the barrier of the host cell membrane) may require additional microbial elements such as cytolysins and proteases. Regulation of the expression of a number of hemolytic enzymes by iron is not surprising.[35,38,39] Because acquisition of iron from heme may be a common theme in pathogenic bacteria, attention is now directed toward the mechanism(s) of iron removal from heme. The iron-protoporphyrin complex might be taken up through a heme receptor/transport system or the metal might be removed externally from the complex before iron uptake. Iron acquisition from heme is considered to be siderophore independent. In *Plesiomonas shigelloides*, the heme molecule is transported intact into the cell.[37] Outer membrane proteins that either bind heme or which are involved in acquiring iron from heme have been identified in several bacteria;[40-43] however, the heme-binding surface protein in *Shigella* species is not involved in procuring iron from heme.[43]

IRON ACQUISITION IN *AEROMONAS* SPECIES

Aeromonad Infections

Members of the Gram-negative bacterial genus *Aeromonas* are ubiquitous in water and are opportunistic pathogens of humans and aquatic animals, causing human wound, soft tissue, and blood infections as well as acute gastroenteritis.[44] In fish, they cause a fatal hemorrhagic septicemia. Classification within the genus is complex and likely incomplete; DNA-DNA reassociation kinetics suggest at least 14 hybridization groups (genospecies), not all of which can be separated phenotypically.[45] The relative pathogenicity of the hybridization groups may not be identical and most (80%) of the human clinical isolates belong to hybridization groups 1, 4, and 8, while groups 6 and 7 do not appear to cause human infections.[46] The potential virulence mechanism(s) used by these bacteria for obtaining iron from a vertebrate host is the subject of the remainder of this paper.

Aeromonad Siderophore- and Heme-Mediated Iron Uptake

Most of the *Aeromonas* species produce either of two siderophores: amonabactin or enterobactin (the predominant siderophore in most enteric bacteria).[47,48] Both siderophores are phenolates containing 2,3-dihydroxybenzoic acid (2,3-DHB). Amonabactin is produced mainly in two biologically active forms which are composed of 2,3-DHB, lysine, glycine, and either tryptophan or phenylalanine. Studies with 12 of the aeromonad hybridization groups by Zywno et al.[45] showed that the type of siderophore produced sharply divided according to the aeromonad DNA hybridization group. Amonabactin is the predominant siderophore in hybridization groups 1 through 5 and 12. Group 8/10 and group 9 make enterobactin. The few isolates in the other rarely isolated groups which were available for testing made conclusions uncertain, although group 7 may make no detectable siderophore and group 6 may produce a previously unknown siderophore.

Massad et al.[15] determined the ability of isolates of the *Aeromonas* species to use Fe-transferrin as an iron source by measuring growth of the microorganisms in vertebrate serum that was heated to destroy its bactericidal complement activity. Of 54 amonabactin-producing isolates, 50 grew in the serum, whereas none of 30 enterobactin-producing strains were able to grow. Adding iron to the serum at a level sufficient to saturate the transferrin allowed growth of all isolates. Moreover, mutants unable to produce amonabactin were also unable to grow in the serum, although a supplement of amonabactin supported their growth in serum. These data indicate that in *Aeromonas* species amonabactin probably is an iron acquisition virulence factor, and confirm an earlier conclusion[28] that enterobactin is nonfunctional in the host's serum. To determine if the *Aeromonas* species could use various heme compounds as sole sources of iron, Massad et al.[15] assayed the capacity of heme compounds to permit growth in iron-restricted conditions. Regardless of the siderophore produced (or lack thereof), all strains of the *Aeromonas* species used heme and hemoglobin (even when hemoglobin was bound by human haptoglobin). Acquisition of iron from the heme compounds by the *Aeromonas* species is siderophore independent.

Cloning Aeromonad Siderophore (2,3-DHB) Genes

The data summarized above suggest the hypothesis that aeromonad strains producing amonabactin can acquire iron from the host in at least two separate ways (from Fe-transferrin with amonabactin and from heme), while the strains producing enterobactin must rely on heme during an infection. The validity of this thesis can be tested by determining the effects of mutations (and combinations of mutations) in the various iron uptake systems (amonabactin, enterobactin, heme) on the virulence of the microorganism. Because chemical mutagenesis methods simultaneously produce mutations at several chromosomal loci, the experimental approach is to clone genes encoding functions in each of the systems, to then inactivate (mutate) the cloned gene by insertion of a transposon or other genetic element, and finally to return the mutated gene to the aeromonad chromosome, replacing the wild-type gene and generating an isogenic mutant. With the *Aeromonas* species, virulence of the strains then can be conveniently tested in a channel catfish infection model.

A strategy for cloning aeromonad genes involved in biosynthesis of either amonabactin or enterobactin was based on the fact that both siderophores contain 2,3-DHB.[49] In the enterobactin-producing enteric bacteria, 2,3-DHB is synthesized from chorismic acid, the branch point compound in aromatic biosynthesis.[50,51] Four genes *(entC, -E, -B, -A)* are necessary; in *E. coli,* they exist in a single operon. The operon is transcribed during iron-restricted growth. Such regulation is accomplished by a regulatory protein termed Fur (encoded by the *fur* "ferric uptake regulation" gene) which first binds iron and then associates with a region of DNA (called the "iron box") near the promoter site of the iron-regulated operon.[52] Barghouthi et al.[49] identified, in an amonabactin-producing aeromonad, a gene *(amoA)* that complemented a mutated *entC* gene in *E. coli.* The nucleotide sequence of *amoA* revealed a potential Fur binding site and a possible AmoA protein of 389 amino acids.[49] Alignment of the AmoA amino acid sequence with the sequence of EntC (the *E. coli entC* gene product) showed an overall similarity, suggesting functional equivalence of the two proteins but evolutionary divergence of the genes encoding them. The EntC and the AmoA proteins appear to be members of a family of related proteins that bind chorismic acid, directing this compound into the various specific branches of aromatic biosynthesis.[49] Downstream from the *amoA* gene was a gene that complemented a mutation in the *E. coli entE* gene, suggesting that the amonabactin-producing aeromonads have a 2,3-DHB operon that is functionally similar to *entCEBA*.

Massad[53] used a similar strategy to clone the genes *(aebCEBA)* encoding synthesis of 2,3-DHB from an enterobactin-producing strain of the *Aeromonas* species. Based on deletion mapping and complementation studies with *E. coli* mutant strains, it was concluded that the genes were functionally equivalent and in the same order as those *(entCEBA)* on the *E. coli* chromosome.

Possible Divergence of Bacterial 2,3-DHB Genes

The availability of the cloned *entC*-like genes enabled Massad[53] to prepare gene probes and to use them to study (by hybridization technology) the relatedness of the genes encoding synthesis of 2,3-DHB. No hybridization was detected between the functional equivalents: *E. coli entC,* aeromonad amonabactin *amoA,* and aeromonad enterobactin *aebC.* Moreover, the *aebC* probe hybridized only with DNA from aeromonads producing enterobactin; the *amoA* probe was similarly specific. The data suggest at least two distinct, but functionally equivalent, 2,3-DHB operons in the *Aeromonas* species; one is present in the amonabactin producers, the other in the enterobactin producers. Each of these also is functionally similar to, but different from, the 2,3-DHB operon in *E. coli* and some other enteric bacteria.

The critical moiety in the iron-binding centers of the phenolate siderophores is 2,3-DHB. To our knowledge, this structure is not found in other microbial products. In addition to the aeromonads described here, more than 20 other bacterial species produce siderophores (or siderophore-like substances) containing 2,3-DHB conjugated to an amino compound (Table 1). Extrapolation of the comparative studies of the 2,3-DHB operon begun here suggests that many (if not all) of these may have versions of a 2,3-DHB operon which have diverged from a single ancestral gene group. Appropriate studies of the 2,3-DHB genes in these and other organisms might permit construction of the evolutionary history of the phenolate siderophores in bacteria.

PROSPECTUS

It is clear that there are several routes by which pathogenic microorganisms can obtain iron from vertebrate hosts and it is also evident that different pathogenic species may use various combinations of these pathways. There are many unanswered questions about the "iron problems" in host-parasite interactions. The precise contribution of siderophores to acquisition of iron from the host must be clarified. Siderophores probably are designed chemically to function best in specific environments. Most pathogens live a dual existence: partly in a host and partly outside a host. Some siderophores may be used only to support growth outside the host while others may be functional in a host (or even functional only in certain anatomical regions of a host). For those microbes able to obtain iron from Fe-transferrin without intervention of a siderophore, we need to know how they remove iron from the glycoprotein complex. We also must know how widespread is the use of heme as a source of iron. Moreover, the ways that pathogens have devised to procure iron from heme need to be defined. These ways may range from use of a heme transport system with internal release of iron to external release of iron from heme with subsequent transport of the metal. These and other

Table 1.   Microorganisms Producing 2,3-Dihydroxybenzoyl-Amino Conjugates[a]

Microorganism	Ref.
Aeromonas spp.	45,47
Agrobacterium tumefaciens	54
Angiococcus disciformis	55
Azospirillum brasilense	56
Azospirillum lipoferum	57
Azotobacter vinelandii	58,59,60
Bacillus circulans	61
Bacillus subtilis	62
"Bacterium DMS5746"	63
Enterobacter spp.	64
Erwinia carotovora	65
Erwinia chrysanthemi	66
Escherichia coli	64
Klebsiella oxytoca	67
Klebsiella pneumoniae	68,69,70
Paracoccus denitrificans	71
Pseudomonas stutzeri	72
Rhizobium leguminosarum	73
Rhizobium trifolii	74
Rhizobium RA-1	75
Rhizobium spp.	76
Salmonella typhimurium	77
Serratia marcescens	78
Shigella spp.	68,79
Vibrio cholerae	80

[a]   Free (unconjugated) 2,3-DHB is found in various actinomycetes,[81,82] *Acinetobacter calcoaceticus*,[83] and *Brucella abortus*.[84]

questions probably can be answered only through case-by-case studies of each pathogenic species.

Finally, having detailed knowledge of the iron acquisition virulence mechanisms of different pathogens should permit the design of new therapeutic strategies that could complement the host's iron-withholding defenses. Success of these strategies could have significant impact in the treatment of human diseases, as well as diseases of commercially important animals and plants. Beneficial treatment approaches that are based on the microbial iron uptake systems might include use of toxin-siderophore complexes to subvert the siderophore transport process, or vaccines employing components of heme acquisition mechanisms to generate antibodies that would block this process, and others.

ACKNOWLEDGMENTS

The iron research with the *Aeromonas* species described here was supported by U.S. Public Health Service grant AI24535 from the National Institute of Allergy and Infectious Diseases, National Institutes of Health. We appreciate the many contributions, both direct and indirect, made by Shelley M. Payne, Kenneth N. Raymond, Linda Tunstad, and Jason Telford at other institutions. We express our sincere gratitude for dedicated work by our students Sameer Barghouthi (later postdoctoral associate), George Massad, Sabrina Zywno, Charles Lee, and Lisa Williamson, as well as technicians Ronny Young, Jim Rish, and Clinton Bailey. This paper is dedicated to the memory of Lucille Eubanks, with a special thanks for her many years of friendship and service.

REFERENCES

1. Byers, B. R., Pathogenic iron acquisition, *Life Chem. Rep.*, 4, 143, 1987.
2. Weinberg, E. D., Iron withholding: a defense against infection and neoplasia, *Physiol. Rev.*, 64, 65, 1984.
3. Stevens, R. G., Beasley, R. P., and Blumberg, B. S., Iron-binding proteins and risk of cancer in Taiwan, *J. Natl. Cancer Inst.*, 76, 605, 1986.
4. Shelby, J. V. and Friedman, G. V., Epidemiologic evidence of an association between body iron stores and risk of cancer, *Int. J. Cancer*, 41, 677, 1988.
5. Stevens, R. G., Jones, D. Y., Micozzi, M. S., and Taylor, P. R., Body iron stores and risk of cancer, *N. Engl. J. Med.*, 319, 1047, 1988.
6. de Jong, G., van Dijk, J. P., and van Eijk, H. G., The biology of transferrin, *Clin. Chim. Acta*, 190, 1, 1990.
7. Hrkal, Z. and Muller-Eberhard, U., Partial characterization of the heme-binding glycoproteins rabbit and human hemopexin, *Biochemistry*, 10, 1746, 1971.
8. Putnam, F. W., Haptoglobin, *Plasma Proteins: Structure, Function and Genetic Control*, 2nd ed., Putnam, F. W., Ed., Academic Press, New York, 1975, 1.
9. Gutteridge, J. M. C. and Smith, A., Antioxidant protection by haemopexin of haem-stimulated lipid peroxidation, *Biochem. J.*, 156, 861, 1988.
10. Griffiths, E., Rogers, H. J., and Bullen, J. J., Iron, plasmids and infection, *Nature*, 284, 508, 1980.
11. Eaton, J. W., Brandt, P., Mahoney, J. R., and Lee, J. T., Haptoglobin: a natural bacteriostat, *Science*, 215, 691, 1982.
12. Francis, R. T., Booth, J. W., and Becker, R. R., Uptake of iron from hemoglobin and the hemoglobin-haptoglobin complex by hemolytic bacteria, *J. Biochem. (Tokyo)*, 17, 767, 1985.
13. Dyer, D. W., West, E. P., and Sparling, P. F., Effects of serum carrier proteins on the growth of pathogenic *Neisseria* with heme-bound iron, *Infect. Immun.*, 55, 2171, 1987.
14. Stull, T. L., Protein sources of heme for *Haemophilus influenzae*, *Infect. Immun.*, 55, 148, 1987.
15. Massad, G., Arceneaux, J. E. L., and Byers, B. R., Acquisition of iron from host sources by mesophilic *Aeromonas* species, *J. Gen. Microbiol.*, 137, 237, 1991.

16. Weinberg, E. D., Iron and infection, *Microbiol. Rev.,* 42, 45, 1978.

17. Congleton, J. L. and Wagner, E. J., Acute-phase hypoferremic response to lipopolysaccharide in rainbow trout *(Oncorhynchus mykiss), Comp. Biochem. Physiol.,* 98A, 195, 1991.

18. Ellison, R.T., III, LaForce, F. M., Giehl, T. J., Boose, D. S., and Dunn, B. E., Lactoferrin and transferrin damage of the Gram-negative outer membrane is modulated by Ca^{2+} and Mg^{2+}, *J. Gen. Microbiol.,* 136, 1437, 1990.

19. Ellison, R. T., III and Giehl, T. J., Killing of Gram-negative bacteria by lactoferrin and lysozyme, *J. Clin. Invest.,* 88, 1080, 1991.

20. Bellamy, W., Takase, M., Yamauchi, K., Wakabayashi, H., Kawase, K., and Tomita, M., Identification of the bactericidal domain of lactoferrin, *Biochim. Biophys. Acta,* 1121, 130, 1992.

21. Finkelstein, R. A., Sciortino, C. V., and McIntosh, M. A., Role of iron in microbe-host interactions, *Rev. Infect. Dis.,* 5, S759, 1983.

22. Weinberg, E. D., Iron depletion: a defense against intracellular infection and neoplasia, *Life Sci.,* 50, 1289, 1992.

23. Crosa, J. H., Genetics and molecular biology of siderophore-mediated iron transport in bacteria, *Microbiol. Rev.,* 53, 517, 1989.

24. Payne, S. M., Iron and virulence in *Shigella, Mol. Microbiol.,* 3, 1301, 1989.

25. Otto, B. R., Verweij-van Vught, A. M. J. J., and MacLaren, D. M., Transferrins and heme-compounds as iron sources for pathogenic bacteria, *Crit. Rev. Microbiol.,* 18, 217, 1992.

26. Carbonetti, N. H., Boonchai, S., Parry, S. H., Vaisanen-Rhen, V., Korhonen, T. K., and Williams, P. H., Aerobactin-mediated iron uptake by *Escherichia coli* isolates from human extraintestinal infections, *Infect. Immun.,* 51, 966, 1986.

27. Enard, C., Diolez, A., and Expert, D., Systemic virulence of *Erwinia chrysanthemi* 3937 requires a functional iron assimilation system, *J. Bacteriol.,* 170, 2419, 1988.

28. Konopka, K. and Neilands, J. B., Effect of serum albumin on siderophore-mediated utilization of transferrin iron, *Biochemistry,* 23, 2122, 1984.

29. Herrington, D. A. and Sparling, P. F., *Haemophilus influenzae* can use human transferrin as a sole source for required iron, *Infect. Immun.,* 48, 248, 1985.

30. Tsai, J., Dyer, D. W., and Sparling, P. F., Loss of transferrin receptor activity in *Neisseria meningitidis* correlates with inability to use transferrin as an iron source, *Infect. Immun.,* 56, 3132, 1988.

31. Schryvers, A. B. and Morris, L. J., Identification and characterization of the transferrin receptor from *Neisseria meningitidis, Mol. Microbiol.,* 2, 281, 1988.

32. Gonzalez, G. C., Caamano, D. O. L., and Schryvers, A. B., Identification and characterization of a porcine-specific transferrin receptor in *Actinobacillus pleuropneumoniae, Mol. Microbiol.,* 4, 1173, 1990.

33. Gerlach, G.-F., Klashinsky, S., Anderson, C., Potter, A. A., and Willson, P. J., Characterization of two genes encoding distinct transferrin-binding proteins in different *Actinobacillus pleuropneumoniae* isolates, *Infect. Immun.,* 60, 3253, 1992.

34. Evans, S. L., Arceneaux, J. E. L., Byers, B. R., Martin, M. E., and Aranha, H., Ferrous iron transport in *Streptococcus mutans, J. Bacteriol.,* 168, 1096, 1986.

35. Stoebner, J. A. and Payne, S. M., Iron-regulated hemolysin production and utilization of heme and hemoglobin in *Vibrio cholerae, Infect. Immun.,* 56, 2891, 1988.

36. Staggs, T. M. and Perry, R. D., Identification and cloning of a *fur* regulatory gene in *Yersinia pestis, J. Bacteriol.,* 173, 417, 1991.
37. Daskaleros, P. A., Stoebner, J. A., and Payne, S. M., Iron uptake in *Plesiomonas shigelloides*: cloning of the genes for the heme-iron uptake system, *Infect. Immun.,* 59, 2706, 1991.
38. Riddle, L. M., Graham, T. E., and Amborski, R. L., Medium for the accumulation of extracellular hemolysin and protease by *Aeromonas hydrophila, Infect. Immun.,* 33, 728, 1981.
39. Lebeck, G. and Gruenig, H., Relation between the hemolytic property and iron metabolism in *Escherichia coli, Infect. Immun.,* 50, 682, 1985.
40. Hanson, M. S. and Hansen, E. J., Molecular cloning, partial purification, and characterization of a haemin-binding lipoprotein from *Haemophilus influenzae* type b, *Mol. Microbiol.,* 5, 267, 1991.
41. Otto, B. R., Sparrius, M., Verweij-van Vught, A. M. J. J., and McLaren, D. M., Iron regulated outer membrane protein of *Bacteriodes fragilis* involved in heme uptake, *Infect. Immun.,* 58, 3954, 1990.
42. Lee, B. C., Isolation of haemin-binding proteins of *Neisseria gonorrhoeae, J. Med. Microbiol.,* 36, 121, 1992.
43. Daskaleros, P. A. and Payne, S. M., Congo red binding phenotype is associated with hemin binding and increased infectivity of *Shigella flexneri* in the HeLa cell model, *Infect. Immun.,* 55, 1393, 1987.
44. Khardori, N. S. and Fainstein, V., *Aeromonas* and *Plesiomonas* as etiological agents, *Annu. Rev. Microbiol.,* 42, 395, 1988.
45. Zywno, S. R., Arceneaux, J. E. L., Altwegg, M., and Byers, B. R., Siderophore production and DNA hybridization groups of *Aeromonas* spp., *J. Clin. Microbiol.,* 30, 619, 1992.
46. Janda, J. M., Recent advances in the study of the taxonomy, pathogenicity, and infectious syndromes associated with the genus *Aeromonas, Clin. Microbiol. Rev.,* 4, 397, 1991.
47. Barghouthi, S., Young, R., Olson, M. O. J., Arceneaux, J. E. L., Clem, L. W., and Byers, B. R., Amonabactin, a novel tryptophan- or phenylalanine-containing phenolate siderophore in *Aeromonas hydrophila, J. Bacteriol.,* 171, 1811, 1989.
48. Barghouthi, S., Young, R., Arceneaux, J. E. L., and Byers, B. R., Physiological control of amonabactin biosynthesis in *Aeromonas hydrophila, Biol. Metals,* 2, 155, 1989.
49. Barghouthi, S., Payne, S. M., Arceneaux, J. E. L., and Byers, B. R., Cloning, mutagenesis, and nucleotide sequence of a siderophore biosynthetic gene *(amoA)* from *Aeromonas hydrophila, J. Bacteriol.,* 173, 5121, 1991.
50. Shea, C. M. and McIntosh, M., Nucleotide sequence and genetic organization of the ferric enterobactin transport system: homology to other periplasmic binding protein-dependent systems in *Escherichia coli, Mol. Microbiol.,* 5, 1415, 1991.
51. Rusnak, F., Liu, J., Quinn, N., Berchtold, G. A., and Walsh, C. T., Subcloning of the enterobactin biosynthetic gene *entB:* expression, purification, characterization, and substrate specificity of isochorismatase, *Biochemistry,* 29, 1425, 1990.
52. Neilands, J. B., Molecular biology and regulation of iron acquisition by *Escherichia coli* K-12, in *The Bacteria,* Iglewski, B. H. and Clark, V. L., Eds., Academic Press, New York, 1990, 205.

53. Massad, G., Physiology and Genetics of Iron Acquisition By *Aeromonas* species, Ph.D. dissertation, University of Mississippi Medical Center, Jackson, 1991.

54. Ong, S. A., Peterson, T., and Neilands, J. B., Agrobactin, a siderophore from *Agrobacterium tumefaciens*, *J. Biol. Chem.*, 254, 1860, 1979.

55. Kunze, B., Bedorf, N., Kohl, W., Höfle, G., and Reichenbach, H., Myxochelin A, a new iron-chelating compound from *Angiococcus disciformis* (myxobacteriales), *J. Antibiot.*, 42, 14, 1989.

56. Bachawat, A. and Ghosh, S., Iron transport in *Azospirillum brasilense:* role of the siderophore spirilobactin, *J. Gen. Microbiol.*, 133, 1759, 1987.

57. Saxena, B., Modi, M., and Modi, V. V., Isolation and characterization of siderophores from *Azospirillum lipoferum* D2, *J. Gen. Microbiol.*, 132, 2219, 1986.

58. Corbin, J. L. and Bulen, W. A., The isolation and identification of 2,3-dihydroxybenzoic acid and 2-N,6-N-di(2,3-dihydroxybenzoyl)-L-lysine formed by iron-deficient *Azotobacter vinelandii*, *Biochemistry*, 8, 757, 1969.

59. Knosp, O., von Tigerstrom, M., and Page, W. J., Siderophore-mediated uptake of iron in *Azotobacter vinelandii*, *J. Bacteriol.*, 159, 337, 1984.

60. Page, W. J. and von Tigerstrom, M., Aminochelin, a catecholamine siderophore produced by *Azotobacter vinelandii*, *J. Gen. Microbiol.*, 134, 453, 1988.

61. Kobaru, S., Tsunakawa, M., Hanada, M., Konishi, M., Tomita, K., and Kawaguchi, H., Bu-2743E, a leucine aminopeptidase inhibitor, produced by *Bacillus circulans*, *J. Antibiot.*, 36, 1396, 1983.

62. Ito, T. and Neilands, J. B., Products of "low-iron fermentation" with *Bacillus subtilis:* isolation, characterization and synthesis of 2,3-dihydroxybenzoylglycine, *J. Am. Chem. Soc.*, 80, 4645, 1958.

63. Taraz, K., Ehlert, G., Geisen, K., and Budzikiewicz, H., Protochelin, ein Catecholat-Siderophor aus einem Bakterium (DMS Nr. 5746), *Z. Naturforsch.*, 45, 1327, 1990.

64. O'Brien, I. G. and Gibson, F., The structure of enterochelin and related 2,3-dihydroxy-N-benzoylserine conjugates from *Escherichia coli*, *Biochim. Biophys. Acta*, 215, 393, 1970.

65. Bull, C. T. and Loper, J. E., Genetic analysis of catechol siderophore production of *Erwinia carotovora*, *Phytopathology*, 81, 1187, 1991.

66. Persmark, M., Expert, D., and Neilands, J. B., Isolation, characterization, and synthesis of chrysobactin, a compound with siderophore activity from *Erwinia chrysanthemi*, *J. Biol. Chem.*, 264, 3187, 1989.

67. Korth, H., Über das Vorkommen von 2,3-Dihydroxybenzoesäure und ihrer Aminosäurederivate in Kulturmedien von *Klebsiella oxytoca*, *Arch. Mikrobiol.*, 70, 297, 1970.

68. Perry, R. D. and San Clemente, C. L., Siderophore synthesis in *Klebsiella pneumoniae* and *Shigella sonnei* during iron deficiency, *J. Bacteriol.*, 140, 1129, 1979.

69. Podschun, R., Fischer, A., and Ullman, U., Siderophore production of *Klebsiella* species isolated from different sources, *Zbl. Bakt.*, 276, 481, 1992.

70. Tarkkanen, A.-M., Allen, B. L., Williams, P. H., Kauppi, M., Haahtela, K., Siitonen, A., Orskov, I., Orskov, F., Clegg, S., and Korhonen, T. K., Fimbriation, capsulation, and iron-scavenging systems of *Klebsiella* strains associated with human urinary tract infection, *Infect. Immun.*, 60, 1187, 1992.

71. Tait, G. T., The identification and biosynthesis of siderochromes formed by *Micrococcus denitrificans, Biochem. J.,* 146, 191, 1975.

72. Chakraborty, R.N., Patel, H. N., and Desai, S. B., Isolation and partial characterization of a catechol-type siderophore fom *Pseudomonas stutzeri* RC 7, *Curr. Microbiol.,* 20, 283, 1990.

73. Patel, H.N., Chakraborty, R. N., and Desai, S. B., Isolation and partial characterization of a phenolate siderophore from *Rhizobium leguminosarum* IARI 102, *FEMS Microbiol. Lett.,* 56, 31, 1988.

74. Skorupska, A., Choma, A., Derylo, H., and Lorkiewicz, Z., Siderophore containing 2,3-dihydroxybenzoic acid and threonine formed by *Rhizobium trifolii, Acta Biochim. Pol.,* 35, 119, 1988.

75. Modi, H., Shah, K. S., and Modi, V. V., Isolation and characterization of catechol-like siderophore from cowpea *Rhizobium* RA-1, *Arch. Microbiol.,* 141, 156, 1985.

76. Jadhav, R. S. and Desai, A. J., Characterization of siderophore from cowpea *Rhizobium* (peanut isolate), *Curr. Microbiol.,* 24, 137, 1992.

77. Pollack, J. R. and Neilands, J. B., Enterobactin, an iron transport compound from *Salmonella typhimurium, Biochem. Biophys. Res. Commun.,* 38, 989, 1970.

78. Angerer, A., Klupp, B., and Braun, V., Iron transport systems of *Serratia marcescens, J. Bacteriol.,* 174, 1378, 1992.

79. Lawlor, K. M. and Payne, S. M., Aerobactin genes in *Shigella* spp., *J. Bacteriol.,* 160, 266, 1984.

80. Griffiths, G. L., Sigel, S. P., Payne, S. M., and Neilands, J. B., Vibriobactin, a siderophore from *Vibrio cholerae, J. Biol. Chem.,* 259, 383, 1984.

81. Dyer, J. R., Heding, H., and Schaffer, C. P., Phenolic metabolite of "low iron fermentation" of *Streptomyces griseus.* Characterization of 2,3-dihydroxybenzoic acid, *J. Org. Chem.,* 29, 2802, 1964.

82. Ratledge, C. and Chaudrey, M. A., Accumulation of iron-binding phenolic acids by actinomycetales and other organisms related to mycobacteria, *J. Gen. Microbiol.,* 66, 71, 1971.

83. Smith, A. W., Freeman, S., Minnet, W. G., and Lambert, P. A., Characterization of a siderophore from *Acinetobacter calcoaceticus, FEMS Microbiol. Lett.,* 70, 29, 1990.

84. Lopez-Goni, I., Moriyon, I., and Neilands, J. B., Identification of 2,3-dihydroxybenzoic acid as the *Brucella abortus* siderophore, *J. Bacteriol.,* 60, 4496, 1992.

CHAPTER 4

Iron Uptake and Siderophore Formation in the Actinorhizal Symbiont *Frankia*

Gregory L. Boyer and Dallas B. Aronson

INTRODUCTION TO THE
FRANKIA-ACTINORHIZAL SYMBIOSIS

The actinomycete *Frankia* is a Gram-positive, aerobic, filamentous, sporulating, and nitrogen-fixing bacterium. It can exist as a free-living saprophyte in the soil. It can also infect selected actinorhizal plant species to form a root nodule.[1] Once in the nodule, *Frankia* hyphae penetrate and spread throughout the cells of the nodule cortex. They are separated from the host cytoplasm by a plant-synthesized capsule containing plant cell wall materials. Symbiotic and free-living *Frankia* can also differentiate to form vesicles — spherical structures which are the site of nitrogen fixation. This ability to fix atmospheric nitrogen results in a positive mutualistic symbiosis between *Frankia* and selected higher plants similar to that observed for the legume-*Rhizobium* symbioses. *Frankia* show a wide range of cross reactivity with their plant hosts. While one species, *F. alni*, is generally accepted, the considerable variation among strains has inhibited the use of species names until the genus is better understood. Currently, the strain is identified using a three-letter abbreviation indicating the laboratory of origin, followed by up to ten numbers.[2] In an older designation, the strain was identified by the actinorhizal genus/species from which it was isolated, followed by a number; e.g., a strain from *Alnus rubra* would be Ar__ and one isolated from *Comptonia peregrina* would be Cp__.

The actinorhizal plant hosts are a taxonomically diverse group of species found within 8 families and divided further into 21 genera. Actinorhizal plants provide a large fraction of the biologically fixed nitrogen in temperate regions, especially where legumes are not present due to temperature extremes. Their ability to live in nitrogen-poor surroundings means that actinorhizal plants are often pioneer species found early in environmental succession. They have

0-87371-942-5/94/$0.00+$.50

found practical applications in underdeveloped countries that are economically unable to add fertilizers to their soils, or with foresters looking to reclaim clear-cut, burned, or nitrogen-depleted soils for timber growth.[3] Common members of this group include alder trees (e.g., *Alnus glutinosa*), *Ceanothus americanus, Myrica gale,* and several species of *Casuarina.*

In contrast to the *Rhizobium* species, isolated *Frankia* readily grows and fixes dinitrogen in aerobic culture. This genus has been divided into two broad groups, "A" and "B", based on DNA homology and general physiological characteristics. *Frankia* prefers a relatively simple medium for growth, with propionate or succinate as a carbon source.[4] The group "A" *Frankia* can also utilize simple carbohydrates such as fructose, glucose, and sucrose. Static culture conditions are generally preferred, however *Frankia* will not grow under anaerobic conditions. Growth occurs only from the hyphal tips or from branch points along the hyphae. Even under the best conditions, doubling times may be on the order of days. Cultures growing under nitrogen-fixing conditions show even longer doubling times than nitrogen-replete cultures. Presumably, this is due to the high energy requirement associated with the reduction of atmospheric dinitrogen. The need to protect the nitrogenase enzyme complex from oxidation through increased respiration, and the induction of vesicle formation under nitrogen-fixing conditions, also adds to the energetic cost of the process.[4,5]

The ability to fix dinitrogen in aerobic cultures makes *Frankia* an ideal organism for studying the effects of iron limitation on nitrogen fixation in a potentially symbiotic organism. Cells grown without added nitrogen should have a higher requirement for iron. Iron is required both as an essential component of the nitrogenase complex and as part of the electron transport system providing the additional reductants necessary to fix nitrogen. In nodules from *Casuarina* and *Myrica* species, the iron-containing proteins, hemoglobin and leghemoglobin, have also been found at levels similar to those present in the *Rhizobium*/legume symbioses.[5] All these factors would increase the demand for iron under nitrogen-fixing conditions and should provide a strong selective pressure for a high affinity iron-uptake mechanism such as the biosynthesis of siderophores.

Siderophore biosynthesis has been reported for several members of the symbiotic Rhizobiaceae.[6,7] The production of these strong iron chelators is strain specific, and a wide variety of chemical structures and siderophore types exist in species of *Rhizobium* and *Bradyrhizobium*. Siderophores play an important role in iron transport and nitrogen fixation *in planta* by symbiotic *Rhizobium meliloti*.[8] In contrast, siderophore biosynthesis has only recently been reported in the *Frankiaceae*,[9] and even less is known about the importance of siderophores in nitrogen fixation and nodule development in symbiotic actinorhizal associations. Here we will review the production of the siderophore, frankobactin, in free-living cultures of *Frankia* sp. strain 52065 originally isolated from *Ceanothus americanus*. We will also examine growth and siderophore production in five other isolates of *Frankia* grown under both iron- and nitrogen-limiting conditions. Preliminary characterization of the siderophore from strain 52065 will be presented along with its role in iron uptake.

GROWTH AND SIDEROPHORE FORMATION IN *FRANKIA* SP. STRAIN 52065

Growth of *Frankia* sp. Strain 52065 Under Iron-Limiting Conditions

Siderophore production in *Frankia* sp. strain 52065 (DDB 03010210, obtained from Dr. Lawrence Winship, Hampshire College, Amherst, MA) was induced by growing cultures in BAP medium without added iron or nitrogen.[10,11] Strain 52065 grows very slowly in nonstirred free-living culture. Iron-limited cultures showed an extended lag phase and an even slower growth rate when compared to iron-replete (20 μM added iron) controls (Table 1). After day 30, the iron-replete cultures continued to grow whereas the iron-deficient cultures entered a stationary phase. Addition of ferric EDTA to these iron-limited cultures restored their growth and allowed them to reach final protein concentrations similar to those observed in iron-replete cultures. Nitrogen-replete cultures always grew at a higher growth rate than cultures forced to fix dinitrogen, regardless of their iron levels (Table 1).

Siderophore Production in Iron-Limited Cultures

The presence of siderophores can be detected in iron-deficient cultures by using several different techniques. Each approach has advantages and disadvantages. By combining information from several different assays, the type and approximate number of siderophores can be determined. Catechol siderophores, measured using the Arnow assay,[12] were not present in either –N–Fe or –N+Fe cultures of *Frankia* sp. strain 52065. In contrast, hydroxamate compounds, measured using the Csaky assay,[13] appeared starting on day 25 in supernatants from cultures grown under iron-limiting, but not iron-replete, conditions. This suggested that *Frankia* sp. strain 52065 produced a hydroxamate siderophore. *Rhizobium* species are known to produce siderophores that contain neither hydroxamic acid nor catechol functional groups.[14] For this reason, two "general" assays whose response is independent of the chemical nature of the iron-binding ligands were used to measure siderophore formation. The chrome azurol S, or CAS, dye-binding assay[15] measures the ability of a strong iron-chelating compound to remove iron from a colored iron-dye complex. Using this assay, iron-binding activity in supernatants taken from the –N+Fe cultures was not significantly greater than that expected from the residual EDTA in the culture media. In contrast, –N–Fe cultures showed a strong increase in CAS

Table 1. The Increase in Biomass of *Frankia* Strain 52065 in Cultures Grown Under Iron-Deficient, Nitrogen-Deficient, and Nutrient Replete Conditions[a]

Nutrient status[b]	Increase in biomass (μg protein/day)
+N+Fe	1.58
–N+Fe	0.59
+N–Fe	1.03
–N–Fe	0.24

[a] Data adapted from Reference 10.

[b] Cultures were grown in BAP media with 20 μM ferric EDTA and 5 mM ammonium chloride present as indicated. Incubation was at 23°C in the dark and the cultures were shaken once daily. Three independent experiments were run. The growth rates reported here were determined over the linear phase of growth (days 1–19) from a single representative experiment.

activity through the culture cycle. These results again suggested the presence of a siderophore.

While the CAS assay is widely used to detect siderophores in microbial cultures,[16,17] it can give a false positive value when strong reducing compounds are present in the culture medium. For this reason, a second "general" assay, the HPLC iron-binding assay,[18] was used to measure siderophore formation in *Frankia* sp. strain 52065. In this assay the putative siderophore was first labeled using ^{55}Fe, separated by HPLC, and then detected using a flow-through radioactivity detector. The iron-labeling and HPLC conditions for this assay are summarized in Figure 1. This approach has been used for the analysis of hydroxamate siderophores from bacteria, fungi, and cyanobacteria[19] and gives excellent agreement with results from the chemical assays. It is independent of the type of iron-binding ligand and depends only upon the ability of a given chelator to successfully compete with citrate for the radioactive iron. A sample from the iron-replete *Frankia* culture containing no iron-binding activity shows only the presence of the ^{55}Fe citrate complex. This complex eluted in the void volume of the column at 2.2 min. Supernatants from iron-limited cultures of *Frankia* sp. strain 52065 showed the presence of a second strong iron-binding peak eluting at 12.1 min (Figure 1). This peak was first seen *circa* day 12 and increased in size through the late exponential to early stationary phase (day 40). As the amount of radioactive iron eluting in this 12.1 min peak increased, a corresponding decrease was observed in the size of the 2.2 min ferric citrate peak. Only a single new peak was observed, suggesting *Frankia* sp. strain 52065 produces only a single siderophore in response to iron limitation. This compound has been given the trivial name "frankobactin". The evidence supporting its production in cultures of *Frankia* sp. strain 52065 is summarized in Table 2. The HPLC iron-binding assay also allows a quantitative estimate of siderophore concentration.

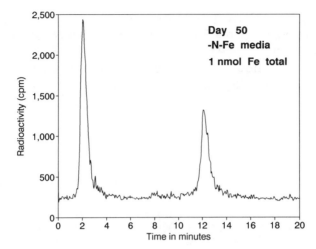

Figure 1. The separation by HPLC of [55]Fe citrate from [55]Fefrankobactin in the supernatant from a 40-day-old –N–Fe culture of *Frankia* sp. strain 52065. Separation conditions consisted of a 10-μm Hamilton PRP-1 column and a 20 min gradient running from water to 100% acetonitrile containing 1% v/v acetic acid. The sample was labeled by mixing an equal volume of the sample and an iron stock containing 107 μ*M* [55]Fe in 0.2 *N* HCl and 5 m*M* citrate, incubated for 1 h at room temperature, and 20 μl injected on column.

The counts per minutes in the 12.1 min peak were converted to micromoles of siderophore assuming a 1:1 ratio of siderophore to iron. When expressed on a volume basis, *Frankia* produced a maximum of 2 μmol siderophore per liter of culture. This is an extremely low level of siderophore formation when compared to the levels reported in culture medium from other bacteria and fungi.[20]

Growth and Siderophore Formation in other *Frankia* Isolates

Five additional strains of *Frankia* were grown under nitrogen-limiting conditions, with and without added iron, to investigate how widespread the occurrence of strong iron-binding compounds was in *Frankia* sp. These experiments are described in detail elsewhere[10,21] and summarized in Table 2. The additional isolates were assayed for growth and siderophore formation throughout the culture cycle and compared to strain 52065. Siderophores were measured using a combination of assays including the Csaky test for hydroxamates, the Arnow test for catechols, the chrome azurol S (CAS) assay, the HPLC iron-binding assay, and the citrate lyase/malate dehydrogenase enzymatic assay[22] for citrate. The growth response to iron and nitrogen limitation varied for the six isolates tested. Two strains (HPFArI3 and ArI5) were unable to grow in culture without added nitrogen. The other strains, 52065, HFPCpI1, HFPCcI3, and AvsI3, showed decreased growth under

Table 2. **A Comparison of Growth Under Iron-Limiting Conditions and Siderophore Production in Six Strains of *Frankia* as Measured Using Five Common Siderophore Assays**

Frankia[a] strain	Growth		Activity in the siderophore assays				
	with N	without N	Csaky	Arnow	CAS	Citrate	HPLC
52065	++[b]	+	+++	−	+++	−	12.1
HFPCpl1	+++	+	−	−	+++	−	−
HFPCcl3	+++	+	−	−	−	−	−
HFPArl3	+++	−	−	−	−	−	−
Arl5	++	−	−	−	−	−	−
Avsl3	++	+	−	−	+++	−	−

[a] Strain 52065 (DDB 03010210) and HFPCcl3 (HFP020203) were obtained from Dr. L. Winship, Hampshire College, Amherst, MA. HFPCPl1 (HFP 070101) was obtained from Dr. D. Benson, University of Connecticut, Storrs, CT. Strains HFPArl3 (HFP013003), Arl5 (DDB 01310310), and Avsl3 (DDB 01360610) were obtained from Dr. D.P. Labeda, USDA Northern Region Research Center, Peoria, Il. Experimental details are described elsewhere.[10,21]

[b] Growth was scored as excellent (+++), moderate (++), poor (+), or none (−). All siderophore assays were conducted on cultures grown without added nitrogen. The Csaky, Arnow, CAS, and citrate activity are reported as +++ (positive reaction) or (−) no activity. For the HPLC assay, a value refers to the retention time (min) of any new iron-binding peaks. Samples which contained only a 2.1 min peak co-eluting with ferric citrate from the labeling solution are reported as no activity.

nitrogen-limiting conditions compared to nitrogen-replete controls. All six strains showed moderate to excellent growth under nitrogen-replete but iron-limiting conditions (Table 2). When cultures were grown under nitrogen-fixing conditions with added iron, only a slight (Avsl3, HFPCpl1, HPFCcl3, 52065) or no improvement (HFPArl3, Arl5) was observed over those cultures grown without added iron or nitrogen.[21] These results indicate that nitrogen limitation places a greater stress on these organisms than iron limitation. The results for the different siderophore assays on these six isolates are shown in Table 2. Strain 52065 was the only isolate to produce a hydroxamate siderophore (frankobactin) under iron-limited conditions. Three strains (HFPCcl3, HFPArl3, and Arl5) showed no siderophore activity in all five assays. Two strains, HFPCpl1 and Avsl3, gave a strongly positive CAS response under iron-limiting conditions, both with and without added nitrogen. Iron-replete cultures of these two isolates were CAS negative. Despite strong CAS activity, the HPLC assay indicated only the presence of a peak eluting in the void volume of the column (2.1 min). These results would suggest that the CAS activity was associated with a small molecular weight organic acid(s) or reducing compound(s) excreted into the medium. Further work on the characterization of these compounds is in progress.

IMPORTANCE OF FRANKOBACTIN IN IRON UPTAKE

Production of a strong iron-binding compound by *Frankia* under iron-limiting conditions does not necessarily establish its function as a siderophore. The compound must also promote iron uptake in *Frankia* sp. strain 52065 under iron-limiting conditions. Cultures were grown either with or without added iron, and their ability to absorb ^{55}Fe chelated by frankobactin measured. Cells grown without added iron (–N–Fe cells) exhibited a rapid period of uptake within the first minute of the assay, followed by a linear period of uptake of from 1 to 15 min. This curve reached saturation after 30 to 45 min. Nonspecific uptake was measured using cells (–N–Fe) incubated with ^{55}Fe chelated by nitrilotriacetic acid (NTA). The same rapid uptake during the first minute was followed by no further increase over time. Iron-replete cells grown with 20 μM added iron showed a much slower linear uptake of ^{55}Fe-labeled ferrifrankobactin after the first minute than iron-limited cultures (Table 3). When cells were first treated with 2,4-dinitrophenol to block active transport, the uptake rate for ^{55}Fe in the iron-limited cultures was comparable to their iron-replete controls. These results indicate that iron is absorbed using an inducible active transport system. Uptake rates measured through the linear region between 1 and 15 min (Table 3) are comparable with siderophore-mediated iron uptake rates reported for other iron-limited bacteria and fungi.[23] It is unknown at this time if frankobactin itself is transported into the cell. The low rate of ^{55}Fe uptake observed in iron-replete cultures could represent specific transport or breakdown of the frankobactin iron complex followed by nonspecific iron uptake.

CHARACTERIZATION OF FRANKOBACTIN

Isolation and Purification of Frankobactin

Frankobactin was purified using a modification of procedures reported for schizokinen and pseudobactin.[24,25] The siderophore was extracted from lyophilized cell-free media using 1:1 phenol:chloroform and then back-extracted into water by adding 5:1 ether:water to the chloroform layer. The aqueous fraction was further purified by chromatography on Biogel P-2 using 10% aqueous methanol. Biogel P-2 is an excellent resin for the separation of a wide number of siderophores and is routinely used in our laboratory to separate siderophores such as ferrichrome, schizokinen, and frankobactin from impurities and breakdown products. The pooled frankobactin fractions from Biogel P-2 were purified to homogeneity using a semipreparative Hamilton PRP-1 HPLC column and a gradient running from water to 75% aqueous acetonitrile containing 1% acetic acid. While iron was not deliberately added to the frankobactin preparation prior to purification, ferrifrankobactin did show up as

Table 3.   Uptake Rates For Iron Chelated By Frankobactin in *Frankia* sp. Strain 52065 Grown Under Iron-Replete and Iron-Limiting Conditions

Culture condition	Uptake rate
–N–Fe (exp 1)	193 pmol Fe/min/mg
–N–Fe (exp 2)	198 pmol Fe/min/mg
–N+Fe (exp 1)	57 pmol Fe/min/mg
–N+Fe (exp 2)	50 pmol Fe/min/mg
–N–Fe +DNP (exp 1)	52 pmol Fe/min/mg

Note: *Frankia* sp. strain 52065 was grown under nitrogen-fixing conditions with (–N+Fe) and without (–N–Fe) added iron. Actively growing cells (ca. day 10) were harvested, washed once with uptake medium, resuspended at a protein concentration of 20 μg/ml, and incubated with 40 nM [55]Fe chelated with 80 nM frankobactin. [55]Fe uptake was measured at 5 min intervals and the rate calculated between 1 and 16 min. The nonspecific uptake rate during this time period, measured using [55]Fe NTA, was 5 pmol Fe/min/mg. To test for active transport, dinitrophenol (DNP) was added to the incubation culture at a concentration of 0.2 mM.

a separate peak on the HPLC due to iron contaminants in the solvent and glassware. Frankobactin eluted slightly behind the ferrifrankobactin complex. This procedure yielded 5 to 10 mg of siderophore, sufficiently pure for NMR spectroscopy, from 8 l of a late log phase culture.

Characterization of Frankobactin

The UV absorbance spectrum of deferrated frankobactin, given in Figure 2, showed a major peak at 301 nm ($\varepsilon = 7640$) along with a characteristic shoulder at 240 and 247 nm. This shoulder was indicative of a phenyl oxazoline ring similar to that reported in agrobactin.[26] Ferrifrankobactin showed strong absorbance in the ultraviolet and the characteristic hydroxamate-iron band at 440 nm. The calculated extinction coefficient ($\varepsilon = 8062$) was approximately twice that reported for other siderophores.[27] The absorbance band at 440 nm showed a slight bathochromic (red) shift as the pH was lowered to 2, suggesting that frankobactin was a dihydroxamate compound.[28]

Amino acid analysis following acid hydrolysis indicated the presence of equimolar amounts of glycine, serine, and ornithine. A fourth amino acid, eluting between histidine and arginine, was tentatively identified as

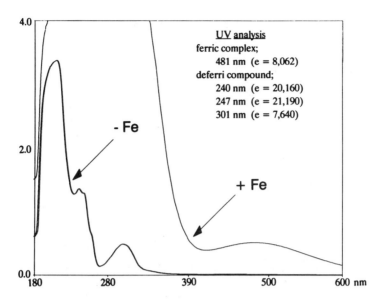

Figure 2. The ultraviolet and visible spectrum of ferri- and deferrifrankobactin in water.

hydroxyaspartate. Hydroxyaspartate has been reported as a component of the oxazoline ring system of several of the mycobactin siderophores produced by species of mycobacteria.[29] Hydroxylysine, a modified amino acid present in all known mycobactin siderophores, was not present in the hydrolyzed frankobactin sample.

The molecular weight of frankobactin was determined using fast atom bombardment (FAB) mass spectrometry (Figure 3). The glycerol matrix peaks were present as the series 277.1, 369.2, 461.5, 553.3, 645.3, 737.5, and 829.5. A peak at m/z 782.5 [M+1] was observed and the fragment ions at m/z 766.5 and 752.4, corresponding to [M–16] and [M–30], can be ascribed to the loss of oxygen from the hydroxamate functional groups as described by Dell and co-workers.[30] The molecular ion of ferrifrankobactin at m/z 835 was not observed in this spectrum. The FAB spectrum was run a second time using a thioglycerol matrix to confirm that m/z 782.5 represented the [M+1] ion for frankobactin and was not due to the glycerol matrix. Again, a strong peak at m/z 782.6 was observed along with a fragment ion at m/z 752.4 [M–30]. No other characteristic fragmentation of frankobactin was observed. A high resolution +FAB mass spectrum gave an average mass for the [M+H] peak of 782.3685. The best fit for this molecular mass would be $C_{33}H_{52}O_{13}N_9$ (expected 782.3686) although numerous other possibilities fell within a 2- to 3-milli mass unit deviation window. This best fit would correspond to a molecular formula of $C_{33}H_{51}O_{13}N_9$ (molecular weight: 781) for the frankobactin parent compound.

One- and two-dimensional proton and carbon NMR spectroscopy was used to characterize structural fragments of frankobactin. There were 33 carbon

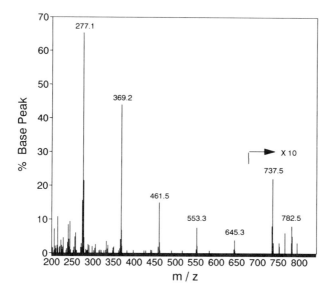

Figure 3.   The positive fast atom bombardment mass spectrum for frankobactin run in a glycerol matrix. The region above m/z 675 has been expanded tenfold.

signals present in the natural abundance carbon-13 NMR spectrum. Their attached proton signals were identified by an inverse-detected H,C COSY spectroscopy (Table 4). The presence of an *o*-hydroxyphenyloxazoline ring system is evident from the salicylyl resonances at 7.40, 7.13, 6.69, and 6.65 ppm in the aromatic region of the proton NMR[31] (Figure 4) and the corresponding signals in the carbon spectrum. Upon standing, a second series of peaks in the aromatic regions of proton and carbon spectra was observed. The chemical nature of this minor form of frankobactin is unknown. However, work with the siderophore agrobactin has shown that the oxazoline ring structure is readily hydrolyzed to its open ring form, agrobactin A.[26] A similar hydrolysis would explain the shift in the aromatic region of the NMR spectra observed with frankobactin.

These results indicate that frankobactin is a novel siderophore. Oxazoline ring systems have been previously reported in the parabactin and mycobactin siderophores. However parabactin is a catechol-based siderophore, lacking any hydroxamic acid functionalities. The known mycobactin siderophores are generally lipophilic and contain the modified amino acid, hydroxylysine. In contrast, frankobactin lacks both a catechol functionality and hydroxylysine. It is a highly water-soluble dihydroxamate siderophore and has an apparent molecular weight of 781 Da. This would correspond to a mid-sized siderophore containing three to four amino acids and the oxazoline functionality (Figure 5). Further studies on the structure of this compound using high field carbon and proton NMR are currently in progress.

Table 4. The 75 MHz Carbon NMR Spectrum
of Frankobactin

Chemical shift	Attached proton (multiplicity)
178.92	—
174.92	—
172.40	—
171.54	—
171.29	—
170.10	—
168.50	—
164.32	—
161.12	7.85 s
158.79	7.51 s
135.28	7.13 t
129.48	7.38 d
120.64	6.64 t
117.20	6.69 d
110.96	—
70.37	
68.25	4.71 t
67.38	3.70 m
61.87	
55.65	3.76 m
54.15	4.04 m
50.92	3.23 m
48.24	3.33
43.05	
39.61	2.61 m
32.41	2.10 s
29.13	
28.65	1.2–1.5
28.06	1.2–1.5
23.81	1.2–1.5
23.64	
21.88	
18.69	0.78 d

Note: All values are reported in ppm downfield from
TMS. Multiplicity: (s)inglet, (d)oublet, (t)riplet,
(m)ultiplet.

ACKNOWLEDGMENTS

The authors would like to acknowledge the Cornell University Mass Spectroscopy Facilities for the high and low resolution FAB mass spectra, Mr. Robert Sherwood at the Cornell Biotechnology center for the amino acid analysis, Mr. David Kiemle at SUNY College of Environmental Science and

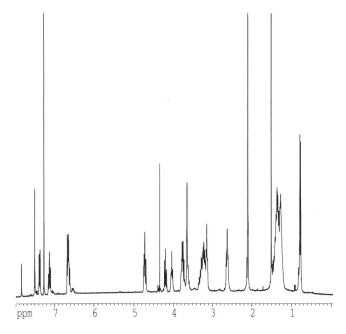

Figure 4. The 300 mHz proton NMR spectrum of frankobactin in deuterium oxide.

Serine
Glycine
Ornithine
HO-Aspartate

1 Phenyl oxazoline ring +   4 Amino acids +   2 Hydroxamates

Figure 5. Structural components of frankobactin.

Forestry for the high resolution NMR studies, and Professor James Nakas at SUNY-CESF for many fruitful discussions. This work was supported by the USDA McIntire Stennis cooperative forestry research program.

REFERENCES

1. Schwintzer, C. R. and Tjepkema, J. D., Eds., *The Biology of* Frankia *and Actinorhizal Plants,* Academic Press, New York, 1990, 408.
2. Lechevalier, M. P. and Lechevalier, H. A., Systematics, isolation and culture of *Frankia*, in *The Biology of* Frankia *and Actinorhizal Plants,* Schwintzer, C. R. and Tjepkema, J. D., Eds., Academic Press, New York, 1990, 35.

3. Benoit, L. F. and Berry, A. M., Methods for production and use of actinorhizal plants in forestry, low maintenance landscapes, and revegetation, in *The Biology of* Frankia *and Actinorhizal Plants,* Schwintzer, C. R. and Tjepkema, J. D., Eds., Academic Press, New York, 1990, 281.

4. Benson, D. R. and Schultz, N. A., Physiology and biochemistry of *Frankia* in culture, in *The Biology of* Frankia *and Actinorhizal Plants,* Schwintzer, C. R. and Tjepkema, J. D. Eds., Academic Press, New York, 1990, 107.

5. Silvester, W. B., Harris, S. L., and Tjepkema, J. D., Oxygen regulation and hemoglobin, in *The Biology of* Frankia *and Actinorhizal Plants,* Schwintzer, C. R. and Tjepkema, J. D., Eds., Academic Press, New York, 1990, 157.

6. Guerinot, M. L., Iron uptake and metabolism in the *Rhizobia*/legume symbiosis, *Plant and Soil,* 130, 199, 1991.

7. Barton, L. L, Vester, C. R., Gill, P. R., Jr., and Neilands, J. B., The role of siderophores in symbiotic nitrogen-fixation by root nodule bacteria, in *The Biochemistry of Metal Micronutrients in the Rhizosphere,* Manthey, J. A., Ed., Lewis Publishers, Chelsea, MI, 1993, (Chapter 5 in this volume).

8. Barton, L. L., Fekete, R. A., Vester, C. R., Gill, P. R., Jr., and Neilands, J. B., Physiological characteristics of *Rhizobium meliloti* 1021 TN5 mutants with altered rhizobactin activities, *J. Plant Nutr.,* 15, 2145, 1992.

9. Aronson, D. B. and Boyer, G. L., *Frankia* produces a hydroxamate siderophore under iron limitation, *J. Plant Nutr.,* 15, 2193, 1992.

10. Aronson, D. B., Growth, Nitrogen Fixation, and Siderophore Production: The Physiological Role of Iron in the Genus *Frankia,* State University of New York, College of Environmental Science and Forestry, MS thesis, 1992.

11. Murry, M. A., Fontaine, M. S., and Torrey, G. J., Growth kinetics and nitrogenase induction in *Frankia* sp. HFP Ar15 grown in batch culture, *Plant Soil,* 78, 61, 1984.

12. Arnow, L. E., Colorimetric determination of the components of 3,4-didroxy-phenylalanine-tyrosine mixtures., *J. Biol. Chem.,* 118, 531, 1937.

13. Gilman, A. H., Lewis, A. G., and Andersen, R. J., Quantitative determination of hydroxamic acids, *Anal Chem.,* 53, 841, 1981.

14. Smith, M. J., Shoolery, J. N., Schwyn, B., Holden, I., and Neilands, J. B., Rhizobactin, a structurally novel siderophore from *Rhizobium meliloti, J. Am. Chem. Soc.,* 107, 1739, 1985.

15. Schwyn, B. and Neilands, J. B., Universal chemical assay for the detection and determination of siderophores., *Anal. Biochem.,* 160, 47, 1987.

16. Alexander, D. B. and Zuberer, D. A., Use of chrome azurol S reagents to evaluate siderophore production by rhizosphere bacteria, *Biol. Fertil. Soils,* 12, 39, 1991.

17. Ames-Gottfred, N. P., Christie, B. R., and Jorden, D. C., Use of the chrome azurol S agar plate technique to differentiate strains and field isolates of *Rhizobium leguminosarum* bovar *trifolii, Appl. Environ. Microbiol.,* 55, 707, 1989.

18. Speirs, R. J. and Boyer, G. L., Analysis of [55]Fe labeled hydroxamate siderophores by high-performance liquid chromatography, *J. Chromatogr.,* 537, 259, 1991.

19. Boyer, G. L., Speirs, R. J., Morse, P. D., and Anderson, W. B., The use of high performance liquid chromatography for the detection of iron chelates in culture, *Appl. Environ, Microbiol.,* submitted.

20. Neilands, J. B., Methodology of siderophores, *Struct. Bonding,* 58, 1, 1984.

21. Aronson, D. B. and Boyer, G. L., Growth and siderophore formation in six iron-limited strains of *Frankia, Soil Biol. Biochem.,* in press.
22. Moellering, H. and Gruber, W., Determination of citrate with citrate lyase, *Anal. Biochem.,* 17, 369, 1966.
23. Winkelmann, G., Specificity of iron transport in bacteria and fungi, in *Handbook of Microbial Iron Chelates,* Winkelmann, G., Ed., CRC Press, Boca Raton, FL, 1991, 65.
24. Mullis, K. B., Pollack, J. R., and Neilands, J. B., Structure of schizokinen, an iron-transport compound from *Bacillus megaterium, Biochemistry,* 10, 4894, 1971.
25. Yang, C.-C. and Leong, J., Structure of pseudobactin 7SR1, a siderophore from a plant-deleterious *Pseudomonas, Biochemistry,* 23, 3534, 1984.
26. Ong, S. A., Peterson, T., and Neilands, J. B., Agrobactin, a siderophore from *Agrobacterium tumefaciens, J. Biol. Chem.,* 254, 1860, 1979.
27. Matzanke, B. F., Structures, coordination chemistry and functions of microbial iron chelates, in *Handbook of Microbial Iron Chelates,* Winkelmann, G., Ed., CRC Press, Boca Raton, FL, 1991, 15.
28. Neilands, J. B. and Nakamura, K., Detection, determination, isolation, characterization and regulation of microbial iron chelates, in *Handbook of Microbial Iron Chelates,* Winkelmann, G., Ed., CRC Press, Boca Raton, FL, 1991, 1.
29. Snow, G. A., Mycobacteria and mycobactins. A review, *Bacteriol. Rev.,* 34, 99, 1970.
30. Dell, A., Hider, R. C., Barber, M., Bordoli, R. S., Sedgwick, R. D., Tyler, A. N., and Neilands, J. B., Field desorption and fast atom bombardment mass spectrometry of hydroxamate containing siderophores, *Biomed. Mass Spectrosc.,* 9, 158, 1982.
31. Greatbanks, D. and Bedford, G. R., Identification of mycobactins by nuclear magnetic resonance spectroscopy, *Biochem. J.,* 115, 1047, 1969.

CHAPTER 5

The Role of Siderophores in Symbiotic Nitrogen Fixation by Legume Root Nodule Bacteria

Larry L. Barton, Craig R. Vester, Paul R. Gill, Jr., and J.B. Neilands

INTRODUCTION

Biological nitrogen fixation is a phenotypic character displayed by only a few genera of bacteria. The reduction of dinitrogen to ammonia requires about the same amount of energy, on a mole basis, as fixation of carbon dioxide. Cells committed to nitrogen fixation spend a significant quantity of cellular energy for this metabolism. While nitrogen fixation is supported in cyanobacteria with energy derived from the photosynthesis reactions, free-living heterotrophic bacteria obtain energy from various soluble energy sources in the environment. Root nodule bacteria have optimized cellular energetics through use of nutrients derived from the plant in a close symbiotic association. Nodules develop on lateral root hairs and contain differentiated bacterial cells (bacteroids) inside a peribacteroid membrane. The level of oxygen in the nodule is reduced by leghemoglobin to enable nitrogenase, an oxygen-sensitive enzyme localized within the bacteroids, to function.

Root nodule bacteria of the genus *Rhizobium* and *Bradyrhizobium* require a specific legume species for the establishment of a successful symbiotic association. Over the years, the biochemical components involved in the various stages of nodulation have been examined and numerous molecular biological approaches have provided useful new information. With respect to micronutrient requirements of the bacteria in the nodules, or of the plant in general, trace metals may function to activate molecules by providing appropriate structural conformation or may serve as signals for nodular initiation. For example, cobalt is present in vitamin B_{12} (cobalamin) and nickel is needed for optimal urease and hydrogenase activities.[1] Copper has been considered important for nodule function and molybdenum is needed for nitrogenase production;[2] however, the nature of these requirements are not at present clearly defined for the alfalfa root nodule system.

0-87371-942-5/94/$0.00+$.50
© 1994 by CRC Press, Inc.

Alternately, mineral elements may be used as regulators to provide signals for nodulation or development. Calcium has long been known to be essential for successful nodulation[3] and the calcium-binding protein isolated from *Rhizobium fredii*[4] may be part of the elaborate signal mechanism essential for the regulation of nodule initiation. The synthesis of rhicadhesin by root hairs of pea is influenced by calcium[5] and this molecule serves to attach cells of *R. leguminosarum* bv. *viciae*. An extracellular protein produced by the *nodO* gene in bacteria binds calcium and plays a putative role in nodule initiation.[6]

Iron is an essential nutrient for the root nodule bacteria because it is used for synthesis of nonheme iron proteins, nitrogenase, heme-containing enzymes, and cytochromes. Additionally, iron has been shown to have an important role in the initiation of nodulation;[7] however, this regulatory role for iron is not as well defined as for calcium. In bacterial iron nutrition, siderophore production has been demonstrated for numerous strains and this activity was recently reviewed by Winkelmann.[8] A prerequisite for siderophore synthesis is the absence of available iron in an aerobic environment. Since iron is one of the most abundant elements in the soil, the formation of iron-insoluble complexes in aerobic environments contributes to the lack of iron availability for bacteria. This review will focus on the role that siderophores can play in biochemical events culminating with nitrogen fixation by *Rhizobium* spp. growing *in planta*.

HIGH AFFINITY IRON TRANSPORT

Siderophore Production

Systems for high affinity assimilation of iron have been found in almost all aerobic bacteria surveyed[8] and siderophores have been reported for several *Rhizobia* spp. No one single class of siderophore is produced by the root nodule bacteria, but rather a series of different molecular structures which range from moderate to strong binding of Fe^{3+}. While the specificity of siderophores in *Rhizobium* has been previously reviewed,[9,10] a current listing of siderophore production and distribution in *Rhizobium* is summarized in Table 1. Molecules with moderately high affinity for iron would include rhizobactin M, which is an amino polycarboxylic acid with ethylenediaminedicarboxylic plus hydroxycarboxylic moieties.[11] Some root nodule bacteria produce citric acid[12,13] and malic acid;[13] however, considering malic acid as a siderophore is unjustified since the production of malic acid is not related to the level of iron in the medium.[13] In a number of bacterial strains citric acid production is not influenced by iron levels, and in these bacteria it would be difficult to justify citric acid as a siderophore. The limited effectiveness of citric acid as an iron chelator as compared to citrate substituted on carboxyls of amino acids has been discussed by Neilands.[14] Anthranilic acid, *o*-aminobenzoic acid, is produced by a strain of *R. leguminosarum* bv. *viciae* and is considered to function as a siderophore.[15-17]

Table 1.   Siderophore Production by Various Isolates of *Rhizobium*
             and *Bradyrhizobium*

Organism	Type of siderophore	Common name	Ref.
R. meliloti DM4	Carboxylate	Rhizobactin M	11
B. japonicum 61A152	Carboxylate	Citric acid	12
B. japonicum[a]	Carboxylate	Citric acid	13
Bradyrhizobium TAL100	Carboxylate	Citric acid	13
B. lupini WUB	Carboxylate	Citric acid	13
R. leguminosarum MNF 300	Carboxylate	Citric acid	13
R. leguminosarum bv. *viviae*	Carboxylate	Anthranilic acid	15–17
R. trifolii	Phenolate	—	18
R. leguminosarum	Phenolate	—	19
Rhizobium RA-1	Phenolate	—	51
Bradyrhizobium NC92	Catechol	—	52
R. leguminosarum[b]	Hydroxamate	—	13,53
R. meliloti 1021	Hydroxamate	Rhizobactin 1021	14
Bradyrhizobium sp.[c]	Unclassified	—	54

[a] Strains UDSA 110 and 61A76.
[b] Biovar *trifolii* Tl and biovar *viciae* WU235, WSM1130, WSM709, and MNF710.
[c] Symbiont from *Acacia manfium,* a woody legume.

Phenolate siderophores that contain 2,3 dihydroxybenzoic acid (DHBA) conjugated to threonine are produced by *R. trifolii*[18] and *R. leguminosarum,*[19] while 2,3 DHBA conjugated to threonine and glycine are produced by cowpea *Bradyrhizobium* RA-1.[19] Additionally, 2,3-DHBA and 3,4-DHBA appear conjugated to lysine and alanine in the siderophore produced by a strain of peanut *Rhizobium.*[20] Hydroxamic acid-containing siderophores have been reported for strains of *R. leguminosarum.*[13] Since the determinants for siderophores have the potential to be plasmid encoded, mobility of plasmids in terrestrial environments could account for the numerous molecular types of siderophores produced by *rhizobia* and concern has been raised over the validity of some of the high affinity systems of iron transport by root nodule bacteria.[12]

Mutants Defective in Rhizobactin 1021 Synthesis and Transport

R. meliloti 1021, an alfalfa symbiont, produces a siderophore that is unlike rhizobactin M from *R. meliloti* DM4 and has been designated rhizobactin 1021. A 35-kb segment of genomic DNA has been isolated that encoded the biosynthesis and regulation of this high affinity iron uptake system.[21] Through use of the Tn5 insertion mutagenesis system, Gill and Neilands obtained numerous mutants defective in iron transport and grouped them into the following three classes:[21] (1) Rzb⁻, mutants unable to synthesize rhizobactin 1021; (2) Rbu⁻, mutants that produce rhizobactin 1021 but are unable to take up this molecule; and (3) Rbr⁻, mutants unable to repress synthesis of siderophore in the presence of available iron. While there may be some difficulties interpreting phenotypic characteristics in mutants derived by transposon mutagenesis resulting from secondary transpositions or rearrangement of bacterial chromosome,[22] the *R. meliloti* strains defective in rhizobactin 1021 activities displayed stable characteristics of nodulation and nitrogen fixation[23,24] as well as ferric siderophore reductase.[24] When the growth response of 11 Tn5 mutants of *R. meliloti* 1021 was examined, 10 of the mutant cultures displayed growth comparable to the parent culture.[24] One strain, PRR22, grew slower than the other mutant strains and this may reflect a genotype that is not characteristic of a change at a single site. Evidence for single-site transposon insertion in these mutants must come from transductional analysis or probing with a Tn5 sequence, which thus far has not been conducted.

NITROGEN-FIXATION STUDIES

Plants Grown in Potting Mixture

When mutants of *R. meliloti* 1021 defective in high affinity iron uptake were a source of inoculum for alfalfa *(Medicago sativa)* cultivated in a sterilized potting mixture with adequate iron levels but deficient in soluble nitrogen, the amount of shoot wet weight was less than the parent strain (Table 2), thereby indicating the reduced effectiveness of nitrogen fixation. The inefficient nitrogen-fixation system with bacteria deficient in rhizobactin 1021 synthesis (Rzb⁻) and uptake (Rbu⁻) supports the hypothesis that siderophore function is required by *rhizobia* that fix nitrogen *in planta*. It would be expected that both the parent *R. meliloti* 1021 and Rbr⁻ strains would produce rhizobactin 1021 when grown in plants stressed for iron; but plant growth resulting from inoculations with bacteria classified as Rbr⁻ was significantly less than with the parent strain. A possible explanation is that the regulatory domain of the siderophore operon for rhizobactin 1021 may influence phenotypic activities to such an extent that nitrogen fixation is severely affected. An alternate explanation is that excess rhizobactin may interfere with plant iron metabolism. The peribacteroid region in nodules with a Rbr⁻ strain would be expected to contain high quantities of rhizobactin 1021, but it remains to be demonstrated that this siderophore can traverse the peribacteroid membrane into the root vascular region.

Table 2.   Alfalfa Growth as Influenced by Mutants of *R.*
           meliloti Defective in High Affinity Iron Transport[23]

Phenotype of R. meliloti 1021	Shoot weight (g)[b]		Number of nodules/plant
	Mean	SE	
Parent	0.332	0.025	3.3
Rzb[−a]	0.160	0.025	2.9
Rbu[−]	0.117	0.018	1.9
Rbr[−]	0.088	0.020	6.1

[a] Values are an average based on nine mutant strains.
[b] Weight of uninoculated plants was 0.113 g with 0.025 SE.

Cultivation of Plants on Mineral Salts Agar Medium

Another system used to follow nodulation has been to inoculate the plants growing on an agar support medium with the desired bacterial culture.[25] The advantages of this system are that nodule development is readily observed, the plants are exposed to only the bacterial culture used as inoculum, and acetylene reduction to measure nitrogen fixation can be done on the undisturbed plant-nodule culture. When Tn5-derived mutants deficient in the rhizobactin 1021 uptake system were tested, the alfalfa plants nodulated with these mutants produced levels of plant growth significantly lower than with the parent *R. meliloti* 1021 culture. Parameters of symbiosis can be expressed in various ways, i.e., nodule number, nodule size, plant weight, amount of nitrogen fixed per nodule, or amount of nitrogen fixed per plant. However, we believe that the most meaningful index of the effectiveness of bacterial strains in symbiotic nitrogen fixation is the activity as to the amount of ethylene produced (nitrogen fixed) per plant mass. The activity with the strain mutagenized in regulation (Rbr[−]) is comparable to the plants inoculated with the parent strain (Table 3). Mutants in uptake (Rbu[−]) or biosynthesis (Rzb[−]) resulted in levels of nitrogen fixed markedly less than the parent strain. These data support the hypothesis that iron is limiting *in planta* and that siderophore production enhances nitrogen fixation in the alfalfa-*R. meliloti* system. Caution should be used in extrapolating the results found with the *R. meliloti*-alfalfa system to other legumes. Many root bacteria do not produce siderophores but remain effective in establishing nodules and conducting nitrogen fixation.

Hydroponic Culture of Inoculated Plants

Plants are frequently cultivated in hydroponic systems to study growth because nutrients can be readily controlled.[26] Studies of the iron nutrition of plants grown in hydroponic culture have been recently reviewed by Johnson and Barton.[1] In experiments using Fe-limited hydroponic culture of alfalfa

Table 3.   Nitrogenase Activity (Acetylene Reduction) for Alfalfa Inoculated with Wild-Type and Rhizobactin Mutants of Rhizobium meliloti 1021[23]

Mutant class of *R. meliloti* 1021	Acetylene reduction rate (mmol ethylene g^{-1} ow h^{-1})	Activity (%)
Parent	4.48	100
Rzb^{-a}	1.44	32
Rbu^{-b}	1.49	33
Rbr$^-$	3.94	88

[a] Data are averages of results from seven mutant strains.
[b] Results averaged from two mutant strains.

inoculated with *R. meliloti* 1021 (parent or mutant strains of the Rzb$^-$ class) Barton et al.[24] found that the nitrogenase activity of the symbiotic bacteria was greater with the parent than with the mutant (Table 4).

While the number of bacteria in colony forming units (CFU) is over seven times greater with the strains defective in rhizobactin 1021 biosynthesis, the nitrogenase activity/CFU of bacteria was greater with the parent strain of *R. meliloti* 1021. Only 13% of the activity present in the parent strain occurred with Rzb$^-$ strains. While bacterial cells in the nodule with nitrogenase activity may be incapable of initiating colonies on nutrient plates, the assumption is made that the ratio of nitrogen-fixing bacteria to colony-initiating bacteria in nodules of the same age would be constant. The hydroponic cultures of alfalfa inoculated with mutants of *R. meliloti* 1021 hold considerable promise for future experimentation.

IRON FOR LEGUME-RHIZOBIA ASSOCIATION

Siderophores and the Initiation of Nodulation

Considerable attention has been given to the physiological activities of nodular development involving both the host legume plant and the specific bacterial strain. The major steps identified by Gallon and Chaplin[27] that are important in this symbiotic process are listed in Table 5, and several of these steps merit exploration. In some of the steps the requirement for iron is well established and we would address the putative role of siderophore production in the rhizobial metabolic process. It is readily apparent that siderophore production is not an obligatory requirement for root nodule bacteria because studies devoted to examination of nodulating strains of bacteria failed to produce siderophores.[12,13] The persistence of *rhizobia* in soil reflects the capacity of these bacteria to satisfy their nutritional requirements.

Calcareous soils with limited available iron are found in numerous agricultural regions of the world[28] and survival of *rhizobia* in this environment may

**Table 4. Nitrogenase Activity (Acetylene Reduction) by
Nodules from Alfalfa Expressed as Activity of
Cells Cultivated from the Nodule[24]**

Bacterial inoculum	CFU bacteria/nodule[a]	Nitrogen fixed/CFU bacteria (μmol ethylene produced/10^6 cells)
Parent	0.8×10^6	9.44
Rzb[-b]	5.9×10^6	1.21

[a] CFU = colony forming unit.
[b] Results are an average of two strains: PRR20 and PRR30.

be enhanced through siderophore activity. Indeed, siderophores of microbial origin have been readily demonstrated in numerous soils,[29-31] and the subject has been reviewed by Crowley et al.[32] Through the use of Tn5-generated mutants of *R. meliloti,* motility toward root hairs of alfalfa has been shown to be important in nodule formation.[33] Positive chemotaxis of *Rhizobium* spp. to root exudates has been thoroughly examined and is important in efficiency of legume nodulation.[34-40] Citric acid, long considered to be a siderophore,[41] could function as a root-secreted chemoattractant as well as function as a mediator of highly efficient iron transport. It is conceivable that ferric citrate may be a chemoattractant for *Rhizobium* spp. and, thus, iron may play an indispensible role for rhizobial migration to roots. While other soil bacteria may also be attracted to the root zone, the competitive feature of *rhizobia* in nodulation would be attributed to recognition of bacterial cell for the plant roots. Plant lectins are important in recognizing surface polysaccharides of *Rhizobium* spp.[42] Other recognition activities may play supporting roles[4-6] and experiments should be designed to test if iron-binding proteins on the outer membranes of *rhizobia* could be receptors for nodule initiation. Precedence for plant recognition of bacterial proteins comes from fimbriae (protein appendages) on *rhizobia,* leading to polar attachment of bacteria to hair roots.[43]

Siderophores and Development of the Nodule

Cellular differentiation is important in nodule maturation and symbiosis requires high levels of iron (Table 5). Expert and Gill[44] have reviewed the literature concerning the effect of iron deficiency on functional nodule development. They suggest that siderophores can be important for symbiotic iron metabolism by contributing to the bacterial nutrition. However, Expert and Gill remind us that the ability of specific root nodule bacteria to use siderophores produced by other rhizobial strains is highly variable. Plant species vary in their response to iron deficiency, with the peanut symbiosis considered to be sensitive to iron deficiency for early nodule development and not for nodule initiation.[45] With lupins, iron limitation influences nodule initiation more than nodule development.[46] Translocation of iron from leaves to the nodules or from

Table 5.   Physiological Processes in Successful Infection of Legumes by *Rhizobia*

Steps of infection[a]	Iron requirement
Survival and multiplication in the soil	Yes
Migration to rhizosphere	No
Recognition of bacterial cell and plant roots	No
Attachment of *Rhizobium* to host	No
Successful infection (root hair curling, penetration of plant, formation of infection thread)	Yes
Nodule formation	Yes (?)
Leghemoglobin synthesis	Yes
Bacteroid formation	?
Peribacteroid membrane synthesis	?
Nitrogenase production	Yes

[a] The sequence of steps is modified from the review by Gallon and Chaplin, 1987.[27]

one area of the root to another, as determined by split-root experiments, can not supply iron for the developing lupin nodule.[47] The role of ferritin in providing iron to the plant for nodule development, leghemoglobin, and nitrogenase production is unresolved; however, several reviews have addressed these topics.[48,49] In developing nodules a microaerophilic condition is produced, and in this environment 25% of the soluble protein is leghemoglobin.[50] If one considers that heme is supplied by the bacterial cells and the apoprotein from the plant,[27,48] a compelling argument can be made for siderophore involvement in nodule development to provide the high levels of iron needed for successful symbiosis. There is evidence from some systems that the plant nodule makes the heme, and in this case iron needs of the plants would be greater than iron needs of plants without nodules. It may be that before we can understand the iron nutrition of the nodule, we may have to first characterize the subtilties of iron metabolism coupled to iron translocation in the host plant and explore the bacterial responsibilities for heme synthesis in various root bacteria-legume systems.

PERSPECTIVES

It is now well established that iron is required for the legume symbiosis. From our results with Rzb⁻ mutants, and from numerous results in the literature, *Rhizobium* spp. can fix nitrogen without the apparent production of a high affinity iron binding molecule. Additionally, siderophore production may not be required for initiation of nodulation processes. However, bacteria capable of siderophore production would be more robust in terms of iron metabolism *in planta* if iron is indeed limiting in the nodule. We do not know if Tn5 mutants of *R. meliloti* with modified production of rhizobactin 1021 can use citrate as an alternate siderophore; this will be important to know in establishing the iron nutrition of root nodule bacteria. Through the use of these and additional mutants of *R. meliloti* deficient in iron transport, we will seek to determine if rhizobactin 1021 synthesis does occur *in planta* and address the role that this rhizobial siderophore can play in nodule development.

REFERENCES

1. Johnson, G.V. and Barton, L.L., Selected physiological responses associated with Fe(III) and Fe(II) metabolism, in *Iron Chelation in Plants and Soil Microorganisms,* Academic Press, San Diego, 1993, 371.
2. O'Hara, G.W., Boonkerd, N., and Dilworth, M.J., Mineral constraints to nitrogen fixation, *Plant Soil,* 108, 93, 1988.
3. Albrecht, W.A. and Davis, F.L., Physiological importance of Ca^{2+} in legume inoculation, *Bot. Gaz.,* 88, 310, 1929.
4. Krishnan, H.B. and Pueppke, S.G., Purification, partial characterization, and subcellular localization of a 38 kilodalton, calcium-regulated protein of *Rhizobium fredii* USDA 208, *Arch. Microbiol.,* 159, 250, 1993.
5. Smit, G., Lonman, T.J.J., Boerrigter, M.E.T.I., Kijne, J.W., and Lugtenberg, G.J.J., Purification and characterization of the *Rhizobium leguminosarum* bv. *viciae* Ca^{2+}-dependent adhesion, which modifies the first step in attachment of cells of the family Rhizobiaceae to plant root hair tips, *J. Bacteriol.,* 171, 4054, 1989.
6. Economou, A., Hamilton, W.D.O., Johnston, A.W.B., and Downie, J.A., The *Rhizobium* nodulation gene *nodO* encodes a Ca^{2+}-binding protein that is exported without N-terminal cleavage and is homologous to haemolysis and related proteins, *EMBO J.,* 9, 349, 1990.
7. Lie, T.A. and Van Egeraat, A.W.S.M., Iron-ethylenediaminetetraacetic acid, a specific inhibitor for root-nodule formation in the legume *Rhizobium* symposis, *J. Plant Nutr.,* 11, 1025, 1988.
8. Winkelmann, G., Specificity of iron transport in bacteria and fungi, in *Handbook of Microbial Iron Chelates,* Winkelmann, G., Ed., CRC Press, Boca Raton, FL, 1991, 65.
9. Guerinot, M.L., Iron uptake and metabolism in the rhizobia/legume symbiosis, *Plant Soil,* 130, 199, 1991.
10. Schwyn, B. and Neilands, J.B., Siderophores from agronomically important species of the Rhizobiaceae, *Comments Agric. Food Chem.,* 1, 95, 1987.
11. Smith, M.J., Schoolerg, J.N., Schwyn, B., Holden, I., and Neilands, J.B., Rhizobactin, a structurally novel siderophore from *Rhizobium meliloti, J. Am. Chem. Soc.,* 107, 1739, 1985.
12. Guerinot, M.L., Meidl, E.J., and Plessner, O., Citrate as a siderophore in *Bradyrhizobium japonicum, J. Bacteriol.,* 172, 3298, 1990.
13. Carson, K.C., Holliday, S., Glenn, A.R., and Dilworth, M.J., Siderophore and organic acid production in root nodule bacteria, *Arch. Microbiol.,* 157, 264, 1992.

14. Neilands, J.B., Overview of bacterial iron transport and siderophore systems in *Rhizobia*, in *Iron Chelation in Plants and Soil Microorganisms*, Barton, L.L. and Hemming, B.C., Eds., Academic Press, San Diego, 1993, 180.

15. Rioux, C.R., Jordan, D.C., and Rattray, J.B.M., Iron requirement of *Rhizobium leguminosarum* and secretions of anthranilic acid during growth on an iron-deficient medium, *Arch. Biochem. Biophys.*, 248, 175, 1986.

16. Rioux, C.R., Jordan, D.C., and Rattray, J.B.M., Anthranilate-promoted iron uptake in *Rhizobium leguminosarum*, *Arch. Biochem. Biophys.*, 248, 183, 1986.

17. Nadler, K.D., Johnston, A.W.B., Chen, J.-W., and John, T.R., A *Rhizobium leguminosarum* mutant defective in symbiotic iron acquisition, *J. Bacteriol.*, 172, 670, 1990.

18. Skorupska, A., Choma, A., Derylo, H., and Lorkiewicz, Z., Siderophore containing 2,3-dihydroxybenzoic acid and threonine formed by *Rhizobium trifolii*, *Acta Biochim. Pol.*, 35, 119, 1988.

19. Patel, H. N., Chakraborty, R.N., and Desai, S.B., Isolation and partial characterization of a phenolate siderophore from *Rhizobium leguminosarum* IARI 102, *FEMS Microbiol. Lett.*, 56, 131, 1988.

20. Jadhav, R.S. and Desai, A.J., Isolation and characterization of siderophore from cowpea *Rhizobium* (peanut isolate), *Curr. Microbiol.*, 24, 137, 1992.

21. Gill, P.R. and Neilands, J.B., Cloning of genomic region required for a high-affinity iron-uptake system in *Rhizobium meliloti* 1021, *Mol. Microbiol.*, 3, 1183, 1989.

22. Ishimaru, C.A. and Loper, J.E., Biochemical and genetic analysis of siderophores produced by plant-associated *Pseudomonas* spp. and *Erwinia* spp., in *Iron Chelation in Plants and Soil Microorganisms*, Barton, L.L. and Hemming, B.C., Eds., Academic Press, San Diego, 1993, 28.

23. Gill, P.R., Jr., Barton, L.L., Scoble, M.D., and Neilands, J.B., A high-affinity iron transport system of *Rhizobium meliloti* may be required for efficient nitrogen fixation *in planta*, *Plant Soil*, 130, 211, 1991.

24. Barton, L.L., Fekete, F.A., Vester, C.R., Gill, P.R., Jr., and Neilands, J.B., Physiological characteristics of *Rhizobium meliloti* 1021 Tn5 mutants with altered rhizobactin activities, *J. Plant Nutr.*, 15, 2145, 1992.

25. Michiels, K.W., Vanderleyden, J., Van Gool, A.P., and Singer, E.R., Isolation and characterization of *Azospirillum brasilense* loci that correct *Rhizobium meliloti exoB* and *exoC* mutations, *J. Bacteriol.*, 170, 5401, 1988.

26. Treeby, M., Marschner, H., and Römheld, V., Mobilization of iron and other micronutrient cations from a calcareous soil by plant-borne, microbial, and synthetic metal chelators, *Plant Soil*, 114, 217, 1989.

27. Gallon, J.R. and Chaplin, A.E., *An Introduction to Nitrogen Fixation*, Cassel Educational, London, 1987, 276.

28. Vose, B.P., Iron nutrition in plants: a world overview, *J. Plant Nutr.*, 5, 233, 1982.

29. Powell, P.E., Szaniszlo, P.J., Cline, G.R., and Reid, C.P.P., Hydroxamate siderophores in the iron nutrition of plants, *J. Plant Nutr.*, 5, 653, 1982.

30. Powell, P.E., Cline, G.R., Reid, C.P.P., and Szaniszlo, P.J., Occurrence of hydroxamic siderophore iron chelators in soils, *Nature (London)*, 287, 833, 1980.

31. Powell, P.E., Szaniszlo, P.J., and Reid, C.P.P., Confirmation of occurrence of hydroxamate siderophores in soil by a novel *Escherichia coli* assay, *Appl. Environ. Microbiol.,* 46, 1080, 1983.

32. Crowley, D.E., Reid, C.P.P., and Szaniszlo, P.J., Microbial siderophores as iron sources for plants, in *Iron Transport in Microbes, Plants and Animals,* Winkelmann, G., van der Helm, D., and Neilands, J.B., Eds., VCH, Weinheim, Germany, 1987, chap. 20.

33. Malek, W., The role of motility in the efficiency of nodulation by *Rhizobium meliloti, Arch. Microbiol.,* 158, 26, 1992.

34. Malek, W., Chemotaxis in *Rhizobium meliloti* strain L5.30, *Arch. Microbiol.,* 152, 611, 1989.

35. Caetano-Anolles, G., Wall, L.G., DeMicheli, A.T., Macchi, E.M., Bauer, W.D., and Favelukes, G., Role of motility and chemotaxis in efficiency of nodulation by *Rhizobium meliloti, Plant Physiol.,* 86, 1228, 1988.

36. Mellor, H.Y., Glenn, A.R., Arwas, R., and Dilworth, M.J., Symbiotic and competitive properties of motility mutants of *Rhizobium trifolii* TA1, *Arch. Microbiol.,* 148, 34, 1987.

37. Gaworzewska, E.T. and Carlisle, M.J., Positive chemotaxis of *Rhizobium leguminosarum* and other bacteria towards root exudates from legumes and other plants, *J. Gen. Microbiol.,* 128, 1179, 1982.

38. Ames, P. and Bergman, K., Competitive advantage provided by bacterial motility in the formation of nodules by *Rhizobium meliloti, J. Bacteriol.,* 148, 728, 1981.

39. Kusch, A.K. and Dadarwall, K. R., Root exudates as preinvasive factors in the nodulation of chick pea varieties, *Soil Biol. Biochem.,* 13, 51, 1981.

40. Currier, W.W. and Strobel, G.A., Chemotaxis of *Rhizobium* spp. to plant root exudates, *Plant Physiol.,* 57, 820, 1976.

41. Neilands, J.B., Microbial iron compounds, *Annu. Rev. Microbiol.,* 50, 715, 1981.

42. Sprent, J.I. and Sprent, P., *Nitrogen Fixing Organisms,* Chapman and Hall, New York, 1990, 256.

43. Vesper, S.J. and Bauer, W.D., Role of pili (fimbriae) in attachment of *Bradyrhizobium japonicum* to soybean roots, *Appl. Environ. Microbiol.,* 52, 134, 1986.

44. Expert, D. and Gill, P.R., Jr., Iron: a modulator in bacterial virulence and symbiotic nitrogen-fixation, in *Signals in Plant-Microbe Communication,* Verma, D.P., Ed., CRC Press, Boca Raton, Fl, 1992, 229.

45. O'Hara, G.W., Dilworth, M.J., Boonkerd, N., and Parkpian, P., Iron deficiency specifically limits nodule development in peanut inoculated with *Bradyrhizobium* sp., *New Phytol.,* 108, 51, 1988.

46. Tang, C., Robson, A.D., and Dilworth, M.J., The role of iron in nodulation and nitrogen fixation in *Lupinus angustifolius* L., *New Phytol.,* 114, 173, 1990.

47. Tang, C., Robson, A.D., and Dilworth, M.J., A split-root experiment shows that iron is required for nodule initiation in *Lupinus angustifolius* L., *New Phytol.,* 115, 61, 1990.

48. Guerinot, M.L., Iron and the nodule, in *Iron Chelation in Plants and Soil Microorganisms,* Barton, L.L. and Hemming, B.C., Eds., Academic Press, San Diego, 1993, 197.

49. Theil, E.C. and Hase, T., Plant and microbial ferritins, in *Iron Chelation in Plants and Soil Microorganisms,* Barton, L.L. and Hemming, B.C., Eds., Academic Press, San Diego, 1993, 134.
50. Verma, D.P.S. and Long, S., The molecular biology of *Rhizobium*-legume symbiosis, *Int. Rev. Cytol. Suppl.,* 14, 211, 1983.
51. Modi, M., Shah, K.S., and Modi, V.V., Isolation and characterization of catechol-like siderophore from cowpea *Rhizobium* RA-1, *Arch. Microbiol.,* 141, 156, 1985.
52. Nambiar, P.T.C. and Sivaramakrishnan, S., Detection and assay of siderophores in cowpea rhizobium *(Bradyrhizobium)* using radioactive Fe (^{59}Fe), *Appl. Microbiol. Lett.,* 4, 37, 1987.
53. Carson, K.C., Dilworth, M.J., and Glenn, A.R., Siderophore production and iron transport in *Rhizobium leguminosarum* bv. viciaeMNF710, *J. Plant Nutr.,* 15, 2203, 1992.
54. Lesueur, D., Diem, H. G., and Meyer, J.M., Iron requirement and siderophore production in *Bradyrhizobium* strains isolated form *Acacia mangium, J. Appl. Bacteriol.,* 74, 675, 1993.

CHAPTER 6

Microbial Siderophores and Rhizosphere Ecology

Jeffrey S. Buyer, Marian G. Kratzke, and Lawrence J. Sikora

INTRODUCTION

Siderophores are low molecular weight iron-chelating agents produced by almost all bacteria and fungi under iron-limiting conditions.[1] Siderophores are excreted in order to bind and solubilize extracellular Fe(III), which is then available to the microorganism. The chemistry, biochemistry, and molecular biology of siderophores have been intensively studied and many excellent reviews have been published.[2-4] However, relatively little is known about the importance of siderophores in microbial ecology.

We are particularly interested in the role of siderophores in rhizosphere ecology. The rhizosphere is commonly defined as the zone where root activity significantly influences biological properties.[5] Competition for iron between microorganisms could be an important factor in determining species distribution in the rhizosphere. This could lead to biological control of plant pathogens or, if the plant pathogen was a successful competitor, to increased plant disease. There are several worthwhile reviews on the role of siderophores in plant-microbe interactions.[6-8]

Our research has taken several distinct paths. First, we developed a model for siderophore-mediated microbial interactions. Next, we developed a growth medium modeling the inorganic components of the rhizosphere and used it for *in vitro* studies. Finally, we developed an immunoassay for a siderophore and used it to determine siderophore concentration *in vivo*. In this article we will summarize our work in these three areas.

MODEL

If iron is the limiting nutrient, and if ferric siderophores are the only source of iron, then ferric siderophore concentration should determine the growth rate of microorganisms. We have previously published a model for siderophore-mediated iron competition between two microorganisms[9] (Figure 1).

Each organism produces a siderophore which chelates iron, at a rate determined by the kinetics of the chelation process and the concentrations of siderophore and iron, to produce the ferric siderophore. Because all siderophores have very high formation constants for Fe(III), and because soil contains large amounts of iron, at equilibrium virtually all the siderophore would form ferric siderophore. However, much of the available iron is present as insoluble ferric oxyhydroxide polymers which react slowly, so the reaction is limited by kinetics and by the amount of siderophore. An organism which produced more siderophore, or a siderophore which chelated iron more rapidly, would have a competitive advantage. There have been no reports, to our knowledge, on the chelation kinetics of siderophores and soil iron. Clearly, soil organic matter and rhizosphere pH would strongly affect the rate of chelation and solubilization.

Each organism takes up the iron from its own ferric siderophore at a rate determined by the concentration of the siderophore and the intrinsic rate of the transport system (K_M and V_{max}). An organism which required less iron or had a more efficient transport system would have an advantage. An organism which could use another organism's siderophore would have a large advantage. Indeed, among fluorescent pseudomonads the ability to use another pseudomonad's siderophore for iron transport has been shown to be a critical factor in competition for iron.[10] This may partially explain why microorganisms produce multiple siderophores and have transport systems for siderophores they do not produce. Bacteria and fungi, if competing with other organisms for iron, may have evolved multiple siderophores in an attempt to produce siderophores that other organisms cannot use. At the same time, each organism may have evolved multiple transport systems in order to use siderophores produced by other organisms.

Siderophores directly compete with each other for iron through ligand exchange. A siderophore with a higher equilibrium constant for iron, at the rhizosphere pH, will be favored, as will a siderophore present in higher concentration. Ligand exchange will only be significant if it occurs at a rate faster than other processes affecting siderophore concentration. Ligand exchange between trihydroxamate siderophores is extremely slow,[11] but it has not been studied for most siderophores.

Siderophores and ferric siderophores may be removed from the rhizosphere by adsorption onto surfaces, biological or chemical degradation, or by leaching. A siderophore less prone to adsorption or degradation might be of advantage to the producing microorganism. We know of no quantitative studies of adsorption or degradation of siderophores in soil.

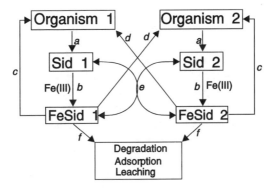

Figure 1. A model for siderophore-mediated microbial interactions — a: siderophore production; b: iron chelation and solubilization; c: uptake of ferric siderophore; d: uptake of ferric siderophore produced by other organism; e: ligand exchange; f: removal of siderophore and ferric siderophore.

The concentrations of siderophores in the rhizosphere at any given point in time and space should depend on all of these factors. The siderophore concentration plays a key role in this model, since it is central to competition between microorganisms and to the rate of iron uptake by each organism. Given the complexity of the system, we decided to use an *in vitro* model in order to study siderophore-mediated antagonism between two organisms.

IN VITRO STUDIES

Introduction

Variations in media components are known to result in production of different siderophores. For example, *Pseudomonas* B10 produces pseudobactin in rich media and pseudobactin A in minimal media,[12] while *Pseudomonas* A214 produces a different siderophore when grown on glucose and ammonium ion than when given glycerol and casamino acids.[34] This suggests that the C and N sources available to biocontrol agents and rhizosphere pathogens might have an important influence on siderophore-mediated antagonism and the resulting biocontrol of plant disease.

Another factor which may affect siderophore-mediated antagonism is pH. Low pH should make iron more available to microorganisms and might favor fungal growth over bacteria, while high pH should induce iron starvation and promote siderophore production. Variations in pH might also affect the specific siderophores produced, since *Gaeumannomyces graminis* var. *tritici* produces predominantly coprogen B at pH 4 to 5 while dimerum acid is preferentially produced at pH 6 to 7.[13] Temperature may be an important environmental variable, as higher temperatures have been reported to reduce biosynthesis of siderophores in several species, including pseudomonads.[14]

We have introduced a microbial growth medium, named Rhizosphere Medium (RSM), for the study of siderophore-mediated interactions.[15] Because the inorganic composition of RSM is similar to the rhizosphere it should provide a useful medium for examining the effects of potential rhizosphere nutrients and pH on inhibition of rhizosphere pathogens by rhizobacteria. We therefore studied the effects of C sources, pH, and temperature on the *in vitro* inhibition of *Gaeumannomyces graminis* var. *tritici,* the take-all pathogen of wheat, by plant growth-promoting *Pseudomonas* B10. *Pseudomonas* B10 produces the siderophore pseudobactin,[16] while *G. graminis* produces the siderophores coprogen B and dimerum acid in liquid culture.[13] However, *G. graminis* apparently produces little or no siderophore on RSM plates at pH 6.8.[15] While certain other pseudomonads antagonize *G. graminis* through the production of antibiotics,[17] these antibiotics have not been observed in experiments involving *Pseudomonas* B10.

Methods

RSM was prepared as follows. For 1 l of medium, 0.75 g Ca(NO$_3$)$_2$·4H$_2$O (3.18 mM), 0.246 g MgSO$_4$·7H$_2$O (1.00 mM), 18.22 g 2-[(2-amino-2-oxoethyl)-amino]ethanesulfonic acid (ACES) (0.1 M), and 2 g NaOH (0.05 M) were dissolved in 853 ml deionized H$_2$O. For plates, 15 g agar was added. After autoclaving, the following sterile stock solutions were added: 1 ml 1 M KH$_2$PO$_4$ pH 7 (1 mM), 1 ml of 7 × 10^{-4} M ZnSO$_4$·7H$_2$O (7 × 10^{-7} M) and 9 × 10^{-4} M MnSO$_4$·4H$_2$O (9 × 10^{-7} M), 1 ml of 20 mg/l thiamine·HCl (5.9 × 10^{-5} M) and 1 mg/l biotin (4.1 × 10^{-6} M), 100 ml 10% casamino acids, and 33.3 ml 30% sucrose (29.2 mM). The phosphate, casamino acids, and sucrose were auto-claved separately, while the metal mixture and vitamins were filter-sterilized. These solutions were added after autoclaving in order to prevent precipitation of insoluble salts and avoid degradation of the vitamins.

For variations in carbon source the casamino acids and sucrose were re-placed by a selected C source and 1 g NH$_4$Cl per liter was added. Stock solutions of various C sources were autoclaved separately and added to pro-duce a final C content of 0.421%. The C:N ratio was 13:1.

Variations in pH were accomplished by replacing the ACES with other buffers, also at 0.1 M concentration, and titrating to the desired pH with HCl or NaOH. The citrate buffers were filter-sterilized and added after autoclaving to prevent precipitation of calcium citrate.

All cultures were grown at 22°C. A disk of *G. graminis* grown on RSM agar was inoculated into 25 ml of liquid RSM and grown for 6 to 8 days without shaking. The contents of the flasks were homogenized and the suspension was diluted with sterile RSM salts (liquid RSM without casamino acids, sucrose, minerals, or vitamins) to a final absorbance of 0.60 at 650 nm. A 2 ml aliquot was mixed into 100 ml of molten RSM agar at 45 to 50°C and plates were poured. After hardening, 5 µl of an overnight culture of *Pseudomonas* B10 grown in liquid RSM was added to the center of the plate and incubated. Antagonism was observed as a less dense zone of *G. graminis* growth around the bacterial colony. Experiments were performed in duplicate and repeated three to five times. High iron controls contained 200 μM $FeCl_3$. In order to study the effect of temperature on antibiosis, RSM plates were incubated at 15, 22, 27, and 30°C.

Results and Discussion

No antagonism was observed in high iron controls, as expected (data not shown). Results of varying C sources are presented in Table 1. Due to variations in growth rates and fungal density between various C sources, no attempt was made to quantify antagonism. There was a general correlation between pseudobactin production and antagonism, as expected. Casamino acids plus sucrose (RSM), dextrose, mannose, and xylose resulted in highly reproducible antagonism, while galactose resulted in occasional antagonism. Glycerol and mannitol supported good growth of *Pseudomonas* B10 and detectable pseudobactin production, but too little *G. graminis* growth to measure antagonism. Arabinose, cellobiose, lactose, rhamnose, and sucrose resulted in little or no pseudobactin production, good growth of *G. graminis,* and no inhibition of *G. graminis* by *Pseudomonas* B10.

These results suggest that the C sources present in the rhizosphere may be crucial to biocontrol of *G. graminis* by *Pseudomonas* B10. Carbon sources which either result in pseudobactin production or better growth of *Pseudomonas* B10 than *G. graminis* might be favorable to biocontrol, while C sources which do not support pseudobactin production or favor *G. graminis* growth would presumably lessen biocontrol.

Our results on *Pseudomonas* B10 growth and pseudobactin production are in partial agreement with those of Loper and Schroth.[14] They found *Pseudomonas* B10 to produce pseudobactin on a wider variety of carbon sources than we did. Differences may be due to a different basal medium or to their use of fluorescence rather than visible color to detect siderophore production. We find trace amounts of pseudobactin to produce large fluorescent halos, making it difficult to distinguish between levels of production. They did not measure antibiosis.

Results of varying pH on *Pseudomonas* B10 growth, *G. graminis* growth, and antibiosis are summarized in Table 2. Because the specific absorbance of pseudobactin increases with increasing pH, it was impossible to assay pseudobactin production by visual inspection in this experiment. Instead, the distance between the edge of the bacterial colony and the outer edge of the inhibition zone was measured. At pH 7.3 and 7.9 *G. graminis* grew more slowly and inhibition zones were visually clearer than at pH 6.8, although the size of the inhibition zones was not significantly larger. The relatively less dense growth within the inhibition zone at higher pH could be due to lower availability of iron, less efficient iron transport by *G. graminis,* more efficient iron transport by *Pseudomonas* B10, or due simply to less growth of *G. graminis*

Table 1. Effect of Carbon Sources on *Pseudomonas* Growth, Pseudobactin Production, *G. graminis* Growth, and Antagonism (Each Number Represents the Mean of Six to Ten Replicates)

Carbon Source	B10[a]	Ps[b]	Ggt[c]	Antagonism[d]
RSM	3.0	3.0	3.0	+
Arabinose	0.2	0.0	2.2	−
Cellobiose	0.0	0.0	3.0	−
Citrate	0.2	0.0	0.0	ND
Dextrose	2.5	0.9	3.0	+
Dulcitol	0.3	0.0	1.5	ND
Galactose	3.0	1.5	2.0	V
Glycerol	2.5	2.4	1.3	ND
Lactose	0.2	0.0	2.3	−
Mannitol	2.8	0.9	1.0	ND
Mannose	2.3	1.3	2.9	+
Rhamnose	0.2	0	2.3	−
Sucrose	1.3	0.2	2.3	−
Xylose	1.4	1.3	2.8	+

[a] 0: no growth; 1: very small colony; 2: medium colony; 3: large colony.
[b] 0: white or tan; 1: pale yellow; 2: yellow; 3: intense yellow.
[c] 0: no growth; 1: slow growth of individual spots; 2: thin field of hyphae; 3: thick field of hyphae.
[d] +: zone of diminished fungal growth around bacterial colony; −: no zone of inhibition; V: inhibition occurred on some replicates but not all; ND: not determined because fungal growth too uneven to observe less dense zone.

Table 2. Effect of pH on *Pseudomonas* B10 Growth, *G. graminis* Growth, and Antagonism (Each Number Represents the Mean of Six to Ten Replicates)

Buffer	pH	B10[a]	Ggt[b]	Antagonism[c] (cm)
Citrate	4.5	0.0	2.4	0.0
Citrate	5.5	2.4	3.0	0.2
ACES	6.7	2.9	3.0	0.8
Tricine[d]	7.3	3.0	2.8	1.0
Tricine[d]	7.9	3.0	3.0	0.9

[a] Rating system same as Table 1.
[b] Rating system same as Table 1.
[c] Width of inhibition zone.
[d] Growth of *G. graminis* was slower than at pH 6.8, and hyphae within the inhibition zone were less dense than at pH 6.8.

at higher pH values.[18,19] Changes in efficiency of iron transport with pH could reflect variations in amount of siderophore produced, greater stability of the ferric siderophore complex at higher pH,[20] or alterations in K_M or V_{max} of the ferric siderophore transport system.

In a previous study, Misaghi et al.,[21] found that a partially purified siderophore from a *Pseudomonas fluorescens* strain inhibited *Pythium aphanidermatum* more at pH 8.0 than at pH 6.0. While our results are in qualitative agreement, we do not necessarily agree with their conclusion that biological control by siderophores will only occur at alkaline pH. Both *in vitro* assays indicated inhibition of the test fungus at pH levels less than 7, although less inhibition than at pH values greater than 7. Since we have no idea what minimum value of *in vitro* antibiosis corresponds to a threshold above which biocontrol in the field will occur, we can only conclude that biocontrol may be more effective at higher pH levels. We also feel that rhizosphere pH, rather than bulk soil pH, may be the crucial factor, and in many cases rhizosphere pH will be quite different from the surrounding soil.[22,23] A further complication may arise from variations in pH along a root due to localization of ion uptake.[24]

Both organisms grew and antagonism was observed at all temperatures except 30°C, at which temperature *G. graminis* did not grow. We conclude from these experiments that soil temperature is not likely to be a direct factor in biocontrol of take-all disease by *Pseudomonas* B10, although indirect effects are certainly possible. For example, the growth rate of other organisms interacting with *Pseudomonas* B10 could vary with temperature, thus affecting the population density of the biocontrol agent.

The most important factor in siderophore-mediated interactions is probably the availability of iron, since this regulates siderophore production. Our work suggests that the C sources available to the microorganisms may also be a significant factor. The rhizosphere certainly contains a wide variety of C-containing compounds, many of which may be utilized by any particular strain. Schwab et al.[25] found that soluble sugars and amino acids in root exudates varied with plant species, but generally glucose, fructose, and arabinose made up more than 10% of the carbohydrate fraction, while glutamate, glycine, lysine, and serine constitute greater than 10% of the total amino acids.

It is conceivable that selection of cultivar or crop management techniques could be used to maximize certain C sources in the rhizosphere, resulting in greater biocontrol. Our results suggest that pH is less important, while temperature may play no role at all for biocontrol of *G. graminis* by *Pseudomonas* B10.

While models and *in vitro* results are interesting, they are primarily useful in suggesting *in vivo* experiments. In order to test the conclusions of our model and the relevance of our *in vitro* system we need methods for determining C sources and siderophore concentration in rhizosphere samples. We decided to first determine the concentration of a particular siderophore in the rhizosphere. In the next section we describe a method for the determination of ferric pseudobactin, the siderophore of *Pseudomonas* B10, from rhizosphere samples, and give some preliminary data. Eventually, we plan to use this method to determine what factors, such as soil pH, temperature, water potential, and plant species, affect the production of rhizosphere siderophores.

SIDEROPHORE IMMUNOASSAY

Introduction

The detection of siderophores has most often involved the use of iron-deficient media and growth rates of auxotrophic mutants. Most of the assays using microorganisms test for the presence of a class of siderophores, such as hydroxamates or catechols, as opposed to a specific siderophore. An enterobactin defective *Escherichia coli* K-12 strain was used to detect ferrichrome-type siderophores in soil,[26] while a panel of *E. coli* auxotrophic mutants were used to determine enterobactin, ferrichromes, a group comprised of coprogen, ferrioxamines, and rhodotorulic acid, and unidentified siderophores.[27] No bio-assays have been developed for specific siderophores produced by fluorescent pseudomonads, which is not surprising considering the ability of pseudomonads to use each other's siderophores for iron transport.

In order to measure the concentration of a specific siderophore in the rhizosphere we have developed an immunoassay for ferric pseudobactin (FePs), the siderophore of plant growth-promoting *Pseudomonas* B10. This assay is highly sensitive and specific for ferric pseudobactin, but applying it to rhizo-sphere samples has proven difficult. This section will describe our general methods, with emphasis on problems and pitfalls. Experimental details have been reported elsewhere.[28,29]

Methods

Antibody

The production of monoclonal antibody (MAb) 5D4-C3 using mice immu-nized with chromic pseudobactin conjugated to bovine serum albumin has been previously reported.[28] We currently purify the antibody by ammonium sulfate precipitation followed by chromatography on immobilized Protein G. The purified antibody is stored at $-80°C$.

Immunoassay

The competitive enzyme-linked immunoassay (ELISA) for FePs was previ-ously described.[28] Standards or samples were combined with MAb and diluted to 250 µl in each well of a 96-well plate previously blocked with nonfat dry milk to prevent protein adsorption. The plate was incubated for 1 h at 37°C to allow pseudobactin-antibody complexes to form. Aliquots were transferred to a 96-well plate containing immobilized FePs and incubated for 18 h at 4°C to allow Mab that was not complexed to solution-phase pseudobactin to bind with immobilized FePs. The plate was washed to remove solution-phase Mab. Anti-mouse immunoglobulin conjugated to alkaline phosphatase was added and incubated for 1 h at 37°C. The plate was washed to remove solution-phase antibody-alkaline phosphatase conjugate and the alkaline phosphatase substrate, *p*-nitrophenyl phosphate, was added. The plate was incubated for 1 to 4 h to allow the reaction product, *p*-nitrophenol, to form. Plates were read at 405 nm.

As we reported,[28] immobilization of the siderophore was difficult. We are currently immobilizing pseudobactin (Ps) by binding it covalently to Nunc CovaLink plates using the homobifunctional crosslinker, disuccinimidyl suberate, and then converting it to FePs using FeEDTA. This method, combined with using very low concentrations of Mab, has given us lower detection limits than those previously reported.[29]

We were particularly concerned with identifying interferences in our samples. We therefore constructed 14-point standard addition curves by combining 25 μl of FePs standard, typically ranging from 10^{-8} to 10^{-13} mol per well, with 200 μl of sample and 25 μl of Mab. These curves were compared to ones containing identical amounts of FePs standard and Mab, but with buffer instead of sample.

Extraction

A number of authors have found that siderophores are difficult to extract from soil, with yields of less than 1% reported.[30] A novel bioassay using soil suspension was developed to circumvent this problem.[31] We have not been able to apply our immunoassay to samples containing solids, presumably due to adsorption to the solid surfaces. In order to develop an extraction method we added Ps or FePs to sand or soil and extracted with a wide variety of inorganic buffers, salt solutions, and organic solvents. A partial list of the conditions employed is given in Table 3. While FePs extracts from sand with aqueous buffer at pH 6.8, we found it impossible to directly extract Ps. We therefore shake the sand with 1 mM FeEDTA in buffer to convert it to FePs, which then extracts with a yield of 50%. In soil, the only extractant we have found effective for Ps and FePs is liquefied phenol.[29]

Sample Preparation

The soil and sand extracts are not directly usable in our ELISA. We previously showed that culture filtrates of certain pseudomonads contain high molecular weight compounds that gave false positives and speculated that these were proteases.[28] Similar results have been obtained with rhizosphere samples, so we routinely use ultrafiltration with 3000-molecular-weight cutoff filters to eliminate these compounds (Centricon 3 or Centriprep 3, Amicon).

Table 3. Partial List of Attempted Extractants for Pseudobactin and Ferric Pseudobactin from Soil and Sand

| | % Yield | | | |
| | Pseudobactin | | Ferric Pseudobactin | |
Extractant	Sand	Soil	Sand	Soil
Buffer pH 6.8	0	NA[a]	100	NA
1 mM FeEDTA	50	24	100	NA
Phenol	0	100	25	100
1 M NaCl	0	8	NA	NA
10% MeOH	NA	18	NA	NA
1 M NH$_4$OAc	0	8	NA	NA
2 M Pyridine-acetic acid	NA	0	NA	NA
Pyridine	NA	0	NA	NA
1:1 Chloroform:Phenol	NA	0	NA	NA

[a] NA, not attempted.

While the FeEDTA and EDTA in sand extracts do not interfere, ELISAs of the sand extract are extremely variable with high noise levels as indicated by jagged, nonreproducible standard addition curves. We found that heating the samples for 40 min at 80°C and/or going through repeated freeze-thaw cycles, combined with filtration and ultrafiltration, helped but did not completely eliminate these difficulties. We are currently working on a solid-phase extraction method for sand extracts.

For soil samples, the liquefied phenol is removed by freezing the sample and subliming off the water and phenol. The extract is suspended in buffer, FeEDTA is added to convert any free Ps to FePs, and the extract is then ultrafiltered. This method has resulted in samples which routinely give good analyses, presumably because the phenol does not extract interfering compounds. We found that similar results were obtained either by adding varying levels of FePs to the unknown samples and looking for a plateau (the standard addition curve described above), or by adding FePs to a control plant extract to produce a standard curve and comparing the absorbance of unknowns to that standard curve. The second method is simpler, more economical, and easier to quantitate.[29]

Colonization

We chose *Pseudomonas* B10 because the structure of FePs was previously determined with a high degree of certainty[16] and because the organism was known to inhibit several plant pathogens.[32,33] In order to study colonization, a double mutant resistant to rifampicin and nalidixic acid was obtained by spreading B10 on RSM plates containing 100 μg/ml each of rifampicin and nalidixic acid. We found that *Pseudomonas* B10 rif[r] nal[r] was a poor colonizer, so the antibiotic-resistant strain was inoculated onto barley plants grown in sand, reisolated after several weeks of growth, and then inoculated onto barley plants again. This process was repeated several times in order to maximize colonization. The final isolate was named *Pseudomonas* Sm 1-3.[29]

Siderophore Production In Vivo

Barley seeds were surface-sterilized and planted in either acid-washed sand or soil limed to pH 7 and autoclaved. The plants were grown at 22°C. *Pseudomonas* Sm 1-3 was grown overnight at 22°C. The culture was centrifuged down and the cells resuspended in plant nutrient solution at a density of 4×10^8 cfu/ml. Each plant was watered with 2.5 ml. One day after inoculation certain plants received casamino acids in buffer. Three days after inoculation plants were harvested and the roots cut up. Samples were extracted as described above and analyzed by ELISA.

Results and Discussion

The ELISA results for sand-grown plants are shown in Figure 2. The untreated plant (Control) gives a response similar to that of B10 only, and casamino acids only, treated plants. The plant treated with B10 and casamino acids clearly contains FePs, but due to the limitations of the assay these data are qualitative rather than quantitative. It must be stated that these curves are atypical, and generally ELISAs of sand extracts are much noisier and difficult to interpret.

Since casamino acids could either increase Ps production by an existing B10 population or support a larger B10 population, which would then result in increased Ps production (or both), we performed colonization studies. The addition of casamino acids caused a 10- to 100-fold increase in B10 population, suggesting that higher colonization levels might result in higher FePs concentration in the rhizosphere. We therefore boosted colonization, as described above, and finally isolated *Pseudomonas* Sm 1-3. *Pseudomonas* Sm 1-3 typically colonized plants tenfold higher than the parent strain *Pseudomonas* B10.[29]

We were unable to detect FePs in the rhizosphere of barley grown in sand and inoculated with *Pseudomonas* Sm 1-3, but this may be due primarily to the poor quality of the ELISA curves. We therefore began to work with soil-grown plants while we continued trying to improve our sand extraction procedure.

Small quantities of FePs could be detected in rhizosphere samples of soil-grown barley inoculated with *Pseudomonas* Sm 1-3. We typically find 10^{-10} mol FePs per gram sample in these experiments, with populations of 10^7 cfu/g.[29]

There are a very limited number of tools available for measuring siderophores in natural ecosystems. Other authors in this volume have developed HPLC methodology using radiometric detection. Our immunoassay

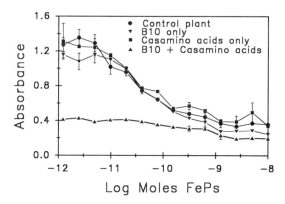

Figure 2. Results of a competitive ELISA on rhizosphere extracts of sand-grown barley plants. Serial dilutions of FePs were added to aliquots of extract. The curves plateau when the concentration of added FePs is less than that already in the extracts.

method appears to be more sensitive than the HPLC method but is restricted to one siderophore. The high specificity of our antibodies is advantageous for the specific measurement of FePs from *Pseudomonas* B10, but does not permit us to assay for any other siderophore with these antibodies. Antibodies to the fluorescent group might provide an immunoassay for all pseudobactin-type siderophores.

SUMMARY

Competition for iron in the rhizosphere may well affect microbial ecology, community structure, and provide a mechanism for biological control of plant pathogens. The problem is extremely complex and ultimately requires methods for studying iron availability in the rhizosphere, microbial physiology *in vivo*, and quantitation of all competing siderophores.

We have designed a simple model of siderophore-mediated interactions which suggests that the concentration of siderophore is crucial in these interactions. Kinetic factors, including rates of chelation of insoluble iron and uptake by the microorganisms, also appear to be important. Agar plate experiments suggest that the C sources available for microbial growth may affect the growth of the microorganisms, the amount of siderophore produced by each microorganism, and the resulting siderophore-mediated antagonism.

One particular siderophore, ferric pseudobactin, has been detected in the rhizosphere of barley plants inoculated with *Pseudomonas* B10. This experiment was performed in soil limed to pH 7 and autoclaved, so while somewhat artificial, it still demonstrates that siderophores are produced in soil.

We are currently determining the effects of the inoculation method, inoculum density, and soil pH on ferric pseudobactin concentration in rhizosphere samples. This will give us a starting place in attempting to apply our model to real systems. Eventually, we hope to measure C sources in rhizosphere samples so that our *in vitro* experiments reported here may be compared to *in vivo* systems.

ACKNOWLEDGMENTS

This work was supported in part by the U.S.-Israel Binational Agricultural Research and Development Fund.

REFERENCES

1. Neilands, J. B., Microbial iron compounds, *Annu. Rev. Biochem.,* 50, 715, 1981.
2. Braun, V. and Winkelmann, G., Microbial iron transport structure and function of siderophores, *Prog. Clin. Biochem. Med.,* 5, 67, 1987.
3. Neilands, J. B., Molecular aspects of regulation of high affinity iron absorption in microorganisms, *Adv. Inorg. Biochem.,* 8, 63, 1990.
4. Winkelmann, G., Structural and stereochemical aspects of iron transport in fungi, *Biotech. Adv.,* 8, 207, 1990.
5. Youssef, R. A., Kanazawa, S., and Chino, M., Distribution of microbial biomass across the rhizosphere of barley (*Hordeum vulgare* L.) in soils, *Biol. Fertil. Soils,* 7, 341, 1989.
6. Leong, J., Siderophores: their biochemistry and possible role in the biocontrol of plant pathogens, *Annu. Rev. Phytopathol.,* 24, 187, 1986.
7. Leong, S. A. and Expert, D., Siderophores in plant-pathogen interactions, in *Plant Microbe Interactions,* Vol. 3., Kosuge, T. and Nester, E., Eds., Academic Press, New York, 1990, 62.
8. Loper, J. E. and Buyer, J. S., Siderophores in microbial interactions on plant surfaces, *Mol. Plant-Microbe Interact.,* 4, 5, 1991.
9. Buyer, J. S. and Sikora, L. J., Rhizosphere interactions and siderophores, *Plant Soil,* 129, 101, 1990.
10. Buyer, J. S. and Leong, J., Iron transport-mediated antagonism between plant growth-promoting and plant-deleterious *Pseudomonas* strains, *J. Biol. Chem.,* 261, 791, 1986.
11. Tufano, T. P. and Raymond, K. N., Coordination chemistry of microbial iron transport compounds. 21. Kinetics and mechanisms of iron exchange in hydroxamate siderophore complexes, *J. Am. Chem. Soc.,* 103, 6617, 1981.
12. Teintze, M. and Leong, J., Structure of pseudobactin A, a second siderophore from plant growth promoting *Pseudomonas* B10, *Biochemistry,* 20, 6457, 1981.
13. Dori, S., Solel, Z., Kashman, Y., and Barash, I., Characterization of hydroxamate siderophores and siderophore-mediated iron uptake in *Gaeumannomyces graminis* var. *tritici, Physiol. Mol. Plant Pathol.,* 37, 95, 1990.
14. Loper, J. E. and Schroth, M. N., Importance of siderophores in microbial interactions in the rhizosphere, *NATO ASI Series,* 117, 85, 1986.
15. Buyer, J. S., Sikora, L. J., and Chaney, R. L., A new growth medium for the study of siderophore-mediated interactions, *Biol. Fertil. Soils,* 8, 97, 1989.

16. Teintze, M., Hossain, M. B., Barnes, C. L., Leong, J., and van der Helm, D., Structure of ferric pseudobactin, a siderophore from a plant growth promoting *Pseudomonas, Biochemistry,* 20, 6446, 1981.

17. Thomashow, L. S. and Weller, D. M., Role of antibiotics and siderophores in biocontrol of take-all disease of wheat, *Plant Soil,* 129, 93, 1990.

18. Balis, C., A comparative study of *Phialophora radicicola,* an avirulent fungal root parasite of grasses and cereals, *Ann. Appl. Biol.,* 66, 59, 1970.

19. Gindrat, P. D., Recherches sur la nutrition et le développement de *Gaeumannomyces graminis* (Sacc.) von Arx et Olivier, agent du piétin-échaudage des céréales, *Schweiz. Landwirtsch. Forsch.,* 7, 197, 1968.

20. Hider, R. C., Siderophore mediated absorption of iron, *Struct. Bond.,* 58, 25, 1984.

21. Misaghi, I. J., Olsen, M. W., Cotty, P. J., and Donndelinger, C. R., Fluorescent siderophore-mediated iron deprivation — a contingent biological control mechanism, *Soil Biol. Biochem.,* 20, 573, 1988.

22. Nye, P. H., Acid-base changes in the rhizosphere, *Adv. Plant Nutr.,* 2, 129, 1986.

23. Römheld, V. and Marschner, H., Mobilization of iron in the rhizosphere of different plant species, *Adv. Plant Nutr.,* 2, 155, 1986.

24. Harrison-Murray, R. S. and Clarkson, D. T., Relationships between structural development and absorption of ions by the root system of *Cucurbita pepo, Planta,* 114, 1, 1973.

25. Schwab, S. M., Leonard, R. T., and Menge, J. A., Quantitative and qualitative comparison of root exudates of mycorrhizal and nonmycorrhizal plant species, *Can. J. Bot.,* 62, 1227, 1984.

26. Powell, P. E., Szaniszlo, P. J., and Reid, C. P. P., Confirmation of occurrence of hydroxamate siderophores in soil by a novel *Escherichia coli* bioassay, *Appl. Environ. Microbiol.,* 46, 1080, 1983.

27. Nelson, M., Cooper, C. R., Crowley, D. E., Reid, C. P. P., and Szaniszlo, P. J., An *Escherichia coli* bioassay of individual siderophores in soil, *J. Plant Nutr.,* 11, 915, 1988.

28. Buyer, J. S., Sikora, L. J., and Kratzke, M. G., Monoclonal antibodies to ferric pseudobactin, the siderophore of plant growth-promoting *Pseudomonas putida* B10, *Appl. Environ. Microbiol.,* 56, 419, 1990.

29. Buyer, J. S., Kratzke, M. G., and Sikora, L. J., A method for detection of pseudobactin, the siderophore produced by a plant-growth-promoting *Pseudomonas* strain, in the barley rhizosphere, *Appl. Environ. Microbiol.,* 59, 677, 1993.

30. Powell, P. E., Cline G. R., Reid, C. P. P., and Szaniszlo, P. J., Occurrence of hydroxamate siderophores in soil, *Nature,* 287, 833, 1980.

31. Bossier, P. and Verstraete, W., Detection of siderophores in soil by a direct bioassay, *Soil Biol. Biochem.,* 18, 481, 1986.

32. Kloepper, J. W., Leong, J., Teintze, M., and Schroth, M. N., *Pseudomonas* siderophores: a mechanism explaining disease-suppressive soils, *Curr. Microbiol.,* 4, 317, 1980

33. Kloepper, J. W., Leong, J., Teintze, M., and Schroth, M. N., Enhanced plant growth by siderophores produced by plant growth-promoting rhizobacteria, *Nature,* 286, 885, 1980.

34. Buyer, J. S., unpublished results.

CHAPTER 7

Role of Siderophores in Plant Growth Promotion and Plant Protection by Fluorescent Pseudomonads

Monica Höfte, Marleen Vande Woestyne, and Willy Verstraete

INTRODUCTION

Plant roots continuously release nutrients such as sugars, amino acids, organic acids, and vitamins which serve as food for microorganisms. As a consequence, a number of soil bacteria and fungi become closely associated with the developing rhizosphere during plant growth. Most of these microorganisms are harmless saprophytes. Others, however, are pathogenic and can cause serious diseases. A third group may have a beneficial effect on the plant; these are mainly rhizobacteria that actively colonize the root. This means that they can keep pace with the growing root in soil.

In 1978, Kloepper and Schroth[1] proposed the term "PGPR" — Plant Growth Promoting Rhizobacteria — for beneficial rhizobacteria. The best-studied PGPR are members of the fluorescent *Pseudomonas* group, mainly *P. putida* and *P. fluorescens*. Growth promotion is evidenced by increases in seedling emergence, vigor, seedling weight, root system development, and yield.[2] Recently, it was stated by Kloepper[3] that there is no clear separation of growth promotion and biological control induced by bacterial inoculants. Indeed, most PGPR enhance plant growth indirectly by reductions in populations of deleterious microorganisms, or the so-called minor pathogens.[4-6] Most PGPR strains also show biological control activity and are able to reduce crop damage caused by major plant pathogens, especially soilborne pathogenic fungi.[3,5] Several mechanisms have been described by which PGPR exhibit growth promotion and biological control, including antibiosis, competition for nutrients, and siderophore production. Kloepper et al. were the first to show that siderophores can be involved in PGPR activity.[7]

0-87371-942-5/94/$0.00+$.50
© 1994 by CRC Press, Inc.

P. AERUGINOSA 7NSK2 IS A PGPR

Our interest in PGPR started with the work of Iswandi. Iswandi[8] tested 274 pseudomonads isolated from the roots of barley, wheat, and tomato for plant growth promotive capacities and found that 17% of these pseudomonads stimulated the growth of barley or tomato in nonsterile soil. One of the most promising strains was the fluorescent *Pseudomonas* strain 7NSK2. This strain increased the dry weight of maize, barley and wheat by 15 to 25% under greenhouse conditions. Iswandi et al.[9,10] evaluated the influence of soil microbial density and activity on the beneficial effect of seed inoculation with *Pseudomonas* strain 7NSK2 (Table 1). In a first series of experiments,[9] microbial activity in the soil was stimulated through growing barley plants. In a second series of experiments,[10] sugars and amino acids were added to the soil. Pretreated soils were subsequently used in seed bacterization experiments with maize and barley. The growth of maize and barley decreased with increasing microbial activity and density in soil, probably due to the development of deleterious microorganisms. The beneficial effect of the *Pseudomonas* strain 7NSK2 was apparently due to the protection of the plant against these deleterious microorganisms rather than to direct growth promotion. Iswandi made no attempts to unravel the mechanisms by which *Pseudomonas* 7NSK2 stimulates plant growth. There were some indications, however, that siderophore production played a role. Indeed, the *in vitro* inhibitory effect of the 7NSK2 strain against five soil fungi (*Gaeumannomyces graminis, Fusarium oxysporum, Ustilago hordei, Fusarium culmorum,* and *Trichoderma* sp.) was more pronounced in the absence of iron.[11]

Additional interest in siderophores came from the work of Bossier and Verstraete.[12,13] They developed a bioassay to detect hydroxamate siderophores in soil. It was observed that siderophore concentrations found in soil depended upon availability of substrates, water activity in soil, the adsorption capacity for siderophores, the humus content, and the concentration of extractable iron.

Strain identification revealed that *Pseudomonas* 7NSK2 is a *Pseudomonas aeruginosa.* This restricts its application in practice since *P. aeruginosa* is known to be a human pathogen. Cornelis et al.[14] were able to cluster clinical *P. aeruginosa* strains into three different groups according to the amino acid composition of their pyoverdins and the specificity of ferripyoverdin uptake. *P. aeruginosa* 7NSK2 and other *P. aeruginosa* strains from soil, however, are quite distinct from medical isolates and can be classified separately according to their pyoverdin pattern as visualized by isoelectric focusing.[15] Another drawback of the 7NSK2 strain is that it does not grow at temperatures below 12°C, which restricts its use in temperate climates. This is one of the reasons why we also focused on a second beneficial *Pseudomonas* strain isolated by Iswandi — *P. fluorescens* ANP15 — which grows readily at low temperatures.

Interestingly, however, certain *P. aeruginosa* strains produce two siderophores, a yellow-green fluorescent siderophore of the pyoverdin type,[16]

Table 1. **Effect of Seed Inoculation with *Pseudomonas* Strain 7NSK2 on the Growth of Barley (cv. *Iban*) in Soils with Different Levels of Microbial Activity[9,10]**

Soil pretreatment	Soil resp. (mg CO_2–C kg^{-1} soil day^{-1})	Plant growth (g DW/pot)		Increase (%)
		Control	Inoculated	
Addition of C and N sources (per kg soil)				
None	3.00a	0.223	0.249	11.6
66 mg C + 6.6 mg N[a]	3.19b	0.208	0.236	13.5[c]
660 mg C + 66 mg N[a]	4.20c	0.193	0.238	23.3[d]
66 mg C + 6.6 mg N[b]	3.18b	0.215	0.240	11.6[c]
660 mg C + 66 mg N[b]	4.82d	0.208	0.242	16.3[d]
Precropping period with barley				
None	1.67a	0.437	0.459	5.0
1 month	1.97b	0.394	0.414	5.1
2 months	2.95c	0.360	0.410	16.9
3 months	3.49d	0.350	0.410	18.0[c]
4 months	2.50c	0.335	0.389	16.0

Note: Data followed by the same letter are not significantly different for $p = 0.05$.
[a] C sources: galactose, glucose, arabinose, xylose (each sugar 25% of the total C). N source: NH_4NO_3.
[b] C sources: sugars as in footnote a. N source: L-asparagine and aspartic acid.
[c] Significantly different from the control for $p = 0.05$.
[d] Significantly different from the control for $p = 0.01$.

and pyochelin, a thiazolin derivate.[17] The physiological role of pyochelin is not very clear since it has a much weaker affinity for iron(III) than pyoverdin and is produced in much lower quantities *in vitro*.

PYOVERDIN PRODUCTION IS INVOLVED IN PLANT GROWTH PROMOTION

Via transposon mutagenesis several siderophore-negative mutants of both ANP15 and 7NSK2 were isolated. Plant experiments carried out with wild-type strains and pyoverdin-negative mutants indicated that pyoverdin production is important in plant growth promotion by ANP15 and 7NSK2 (Table 2).

Several cereals and vegetables were grown in sugar-pretreated soil, with and without seed bacterization with 7NSK2. A pyoverdin mutant MPFM1 was unable to promote plant growth, in contrast with the wild-type strain 7NSK2.[18] Mutant MPFM1 carries a single chromosomal Tn5 insertion and is nonfluorescent. This mutant is also negative in the Csáky hydroxamate assay. Seong et al.[19] likewise showed that pyoverdin mutants of ANP15 are unable to

Table 2.   Effect of Wild Type Strains 7NSK2 and ANP15 and Pyoverdin-Negative Mutants MPFM1 and JBFM1 on the Growth of Various Crops in Sugar-Treated Soil[18,19]

Crop	Strains	Plant dry weight (mg DW/pot)	Increase (%)
Maize	Control	713	0.0a
	7NSK2	794	11.3b
	MPFM1	740	3.7a
Cucumber	Control	1604	0.0a
	7NSK2	1764	10.0b
	MPFM1	1664	3.7a
Spinach	Control	650	0.0a
	7NSK2	751	15.4b
	MPFM1	692	6.5a
Maize	Control	1393	0.0a
	ANP15	1653	18.6b
	JBFM1	1442	3.5a

Note: Values are the mean of five replicates. For each crop, and within columns, values followed by the same letter are not significantly different for $p = 0.05$.

promote the growth of maize in a sugar-pretreated soil. It was also shown that the beneficial effect of 7NSK2 on the growth of maize, spinach, and corn is more pronounced when the strain was acclimatized by adding it to the soil 2 to 4 weeks prior to sowing.[20] During this acclimatization period 7NSK2 population densities in soil decreased significantly, but the remaining bacteria colonized the roots actively, resulting in a significant decrease in the fungal populations on the roots and in the rhizosphere. Although the root colonizing capabilities of mutant MPFM1 were not impaired, the presence of this strain did not influence the fungal population or the plant growth.

ECO-PHYSIOLOGY OF 7NSK2 AND ANP15

To study the behavior of both 7NSK2 and ANP15 in more detail, Mud(*lac*)-marked strains were constructed. A *lacZ* marker and a kanamycin resistance gene were introduced into the chromosome of ANP15 and 7NSK2.[21] This allowed very selective and sensitive recovery of both strains from nonsterile environments. The *lacZ*-marked 7NSK2 strain, called MPB1, could be recovered with an efficiency of about 100% from soils on minimal X-gal medium containing sebacic acid as a selective carbon source and kanamycin. The limit of detection is about 10 CFU/g soil. With the aid of these marked strains it could be demonstrated that soil inoculation resulted in a far better root colonization than seed inoculation.[21,22]

Pyoverdin production, survival in soil, and root colonization of 7NSK2 and ANP15 at various temperatures were compared in a series of experiments conducted by Seong et al.[19] 7NSK2 and ANP15 clearly have different optimum temperature ranges for growth and siderophore production. 7NSK2 is a mesophilic strain with a maximum pyoverdin production around 20°C, while ANP15 shows its maximum pyoverdin production around 12°C. ANP15 survived better at subzero temperatures, while 7NSK2 survived better at 28°C. 7NSK2 colonized maize roots more effectively at 30°C, while at 18°C ANP15 was the better root colonizer (Figure 1).

Boelens and Verstraete[22] studied the role of motility in root colonization by *P. fluorescens* ANP15. They found that a nonmotile Tn5 mutant of ANP15 colonized roots at population densities similar to the wild-type strain. Moreover, the absence of the motility trait had no negative effect on survival and dispersal of the strain in soil. Capillary assays revealed, however, that motility might be vital in sites where a distinctive nutrient gradient is present.

PROMOTION OF SEEDLING EMERGENCE

Both ANP15 and 7NSK2 are able to increase the germination rate of seeds in soil subjected to unfavorable conditions. Both strains significantly increased germination of maize seeds which had been subjected to cold for 10 days by 30 to 60%.[23] When maize seeds were subjected to cold for 21 days ANP15, but not 7NSK2, was able to increase germination (Table 3).[24] This is probably due to the fact that at 4°C ANP15 is biochemically active, while 7NSK2 only becomes active at temperatures above 12°C. The mechanisms by which both strains protect the seeds are not clear. During the cold period, 7NSK2 probably protects the seeds merely by its physical presence. Once conditions become favorable for germination, 7NSK2 may actively promote seedling emergence, probably by protecting the seedlings against other microorganisms. Other mechanisms besides the production of pyoverdin seem to play a role in this protection since the pyoverdin mutant MPFM1 still had emergence-promoting capacities.[23]

BIOLOGICAL CONTROL OF *PYTHIUM*

P. aeruginosa 7NSK2 shows an iron-regulated antagonism against *Pythium* sp. on agar plates. Interestingly, however, the pyoverdin mutant MPFM1 also showed iron-regulated antagonism.[25] This observation prompted us to investigate the antagonistic role of the second siderophore of 7NSK2, pyochelin. By chemical mutagenesis of pyoverdin mutant MPFM1, a mutant was isolated that did not produce orange halos on CAS agar plates. The growth of this mutant was completely inhibited by 5 mg/l EDDHA. Pyochelin was not found in the

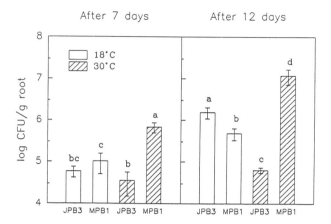

Figure 1. Root colonization capability of *P. fluorescens* JPB3 (Mud*(lac)* marked strain of ANP15) and *P. aeruginosa* 7NSK2 (Mud*(lac)* marked strain of 7NSK2) at 18°C and 30°C. Within a panel, bars followed by the same letter are not statistically different for P = 0.01. The initial inoculum density of JPB3 was 2.7 × 10⁴ CFU/g soil, and 8.0 × 10⁴ CFU/g soil for MPB1.[19] (Reprinted from *Soil Biol. Biochem.*, Vol. 23, Seong, K.Y. et al., p. 423, 1991, with kind permission from Pergamon Press Ltd., Headington Hill Hall, Oxford OX3 OBW, U.K.)

Table 3. Effect of Inoculation with *Pseudomonas* Strains 7NSK2 and ANP15 on the Germination of Maize Seeds after a Cold Period of 11 and 21 Days[24]

Treatment	Number of seeds germinated per pot (10 seeds sown)	Emergence (%)
Cold period of 11 days		
Control	5.17a	51.7
ANP15	8.50b	85.0
7NSK2	8.17b	81.7
LSD (*p* = 0.05)	2.45	
Cold period of 21 days		
Control	1.00a	10.0
ANP15	4.17b	41.7
7NSK2	1.83a	18.3
LSD (*p* = 0.05)	1.04	

Note: Values are the mean of six replicates. Values followed by the same letter are not significantly different for *p* = 0.05.

supernatant of low-iron cultures of the mutant by using the ethyl-acetate extraction method. This mutant, which apparently was no longer able to produce pyochelin, was called KMPCH. The siderophore double mutant KMPCH did not show antagonism against *Pythium* on agar plates. *In vivo* antagonistic tests with tomato as a test plant revealed that 7NSK2, MPFM1,

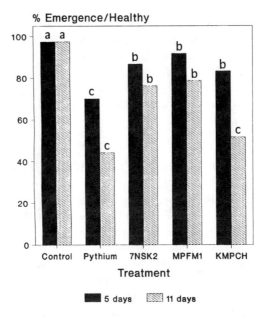

Figure 2. Average amount of seed emergence after 5 days and average amount of healthy tomato seedlings after 11 days in an artificially *Pythium*-infested pot soil. Tomato seeds were untreated (control) or dipped in a suspension of 7NSK2 (wild type), MPFM1 (pyoverdin-negative mutant), or KMPCH (pyoverdin- and pyochelin-negative mutant). Within each time/data set, treatments with the same letter are not significantly different for $p = 0.01$.[25]

and KMPCH protected tomato seeds against *Pythium* preemergence damping-off (Figure 2, results after 5 days). The presence of the bacteria around the seed possibly protects the seed against *Pythium* by site exclusion and competition for nutrients. In addition, sporangial germination and chemotaxic *Pythium* attraction is probably prevented because the seed exudates are used by the numerous bacteria around the seed.

Siderophore production does not seem to play a role in the protection against preemergence damping-off since wild-type and siderophore mutants gave good protection. In fact most fluorescent pseudomonads give good protection against preemergence damping-off. But simple site exclusion and deprivation of leaking nutrients does not suffice for controlling postemergence damping-off. At this point, *in situ* siderophore production seems to be important since the siderophore-negative mutant KMPCH was no longer able to give protection against postemergence damping-off. The pyoverdin-negative mutant MPFM1 protected very well against postemergence damping-off (Figure 2, results after 11 days). These results suggest that pyochelin, rather than pyoverdin, is important in protection against postemergence damping-off. It is possible that pyochelin has an antibiotic action against *Pythium* sp. Alternatively, mutant KMPCH might be unable to take up sufficient iron to carry out normal secondary metabolism, which could result in a reduced production of antibiotics or other relevant factors. Currently, we are trying to isolate a pyoverdin-positive, pyochelin-negative mutant of 7NSK2 in order to unravel the exact role of pyochelin in *Pythium* antagonism.

The role of siderophores in the suppression of plant diseases seems to be pathogen specific since 7NSK2, but not mutant MPFM1, was antagonistic against *Fusarium oxysporum* on agar plates. This suggests a role for pyoverdin in the control of *Fusarium* wilt.

EFFECT OF ZN(II) ON SIDEROPHORE PRODUCTION

It was observed that high concentrations of Zn(II) and various other metals induce pyoverdin production by *Pseudomonas aeruginosa* 7NSK2 on noniron-limiting Luria Bertani (LB) medium, and increase pyoverdin production on iron-limiting media. This phenomenon was investigated in more detail.[26] Concentrations of Zn(II) above 0.2 mM significantly increased the production of pyoverdin in iron-limiting CAA medium (Table 4). Remarkably, pyoverdin production in the presence of Zn(II) concentrations above 0.2 mM could not be repressed by addition of iron(III) concentrations as high as 100 μM, while 5 to 10 μM iron(III) is sufficient to completely repress pyoverdin production in CAA medium in the absence of Zn(II). The growth of the pyoverdin mutant MPFM1 and especially the siderophore double mutant KMPCH was seriously impaired in Zn(II)-containing CAA medium in comparison with the wild type. Fe-pyoverdin, but not FeCl$_3$, could restore the growth of KMPCH to wild-type levels, suggesting that Zn(II) interferes with low affinity iron uptake. Zn(II) specifically induced an iron-regulated 85-kDa outer membrane protein in 7NSK2 which is probably the pyoverdin receptor. In MPFM1, however, pyochelin production and a 78-kDa receptor, most likely the pyochelin receptor, were strongly induced while the 85-kDa protein was repressed. In KMPCH the 78-kDa receptor was completely absent, but a 90-kDa iron-regulated outer membrane protein was induced in the presence of Zn(II).[26]

These results suggest that besides the general regulation of iron uptake systems, which most likely involves a Fur-like repressor protein like the one in *Escherichia coli*,[27,28] a second regulation level may exist. The second regulation level may be specific for each siderophore uptake system, possibly involving siderophore-dependent activation of biosynthesis and uptake genes. Recently, it was shown by Gensberg et al.[29] that *Pseudomonas aeruginosa* PAO1 can respond specifically to the presence of either pyoverdin or pyochelin in the growth medium.

Via Tn5 transposon mutagenesis two mutants of *P. aeruginosa* 7NSK2 were isolated, which suggests that Zn(II) not only affects low affinity iron uptake, but directly interferes with siderophore regulation. Mutant SSS still produces pyoverdin on iron-limiting media, but is unable to produce pyoverdin on LB agar plates supplied with Zn(II) or Ca(II), in contrast with the wild type strain. The *sss* mutation also affects survival in soils with a high microbial

Table 4. **Pyoverdin Production (μM Pvd/OD$_{600}$) by *P. aeruginosa* 7NSK2 in Iron-Limiting CAA Medium With Increasing Concentrations of Zn(II) — With and Without Added Fe(III) (50 μM)[26]**

Zn(II) (mM)	−Fe(III)		+Fe(III)	
	Mean	SD	Mean	SD
0	127a	4	7.7a	0.3
0.1	133a	19	9.1a	1.4
0.2	121a	27	13.9a	0.5
0.5	196b	18	34.1b	8.6
1	213b	20	40.5b	0.6
LSD				
(p = 0.05)	52		15.6	

Note: Values followed by the same letter are not significantly different for p = 0.05. Values of one representative experiment are shown and are the mean of three repetitions.

activity[30] or in soils contaminated with heavy metals. Mutant SSS is still plant growth-promoting and antagonistic against *Pythium,* although its root coloniz-ing capabilities are slightly affected (Table 5).[30] The corresponding *sss* gene was cloned and sequenced and shows strong homology with the *xerC* gene of *Escherichia coli.*[31] The *xerC* gene product is a member of the lambda integrase family of site-specific recombinases. One attractive hypothesis is that the *sss* gene product may introduce gene rearrangements, e.g., DNA inversions, in promoter regions of genes involved in the biosynthesis or regulation of pyoverdin production. These rearrangements might occur as a response to various envi-ronmental stimuli including Zn^{2+} and Ca^{2+}.

Mutant M2 is unable to produce both pyoverdin and pyochelin on iron-limiting media, but readily produces pyoverdin on media supplied with Zn(II). Growth of this mutant in iron-limiting media is extremely poor and cannot be restored by addition of $FeCl_3$, ferripyochelin, or ferripyoverdin. Mutant M2 is apparently unable to switch on the genes for both siderophore biosynthesis and uptake in normal conditions of iron limitation. Mutant M2 has impaired root colonizing capabilities (Table 5),[30] probably because the strain is unable to produce siderophores and to take up exogenous siderophores. It is thought that mutant M2 is damaged in a general activator that regulates siderophore production and uptake. According to O'Sullivan and O'Gara,[28] a *trans*-acting factor for *Pseudomonas* sp. strain M114 is needed to express an iron-regulated promoter from strain M114 in *E. coli.* It is possible that the gene coding for such a *trans*-acting factor is damaged in mutant M2. Zn(II) induction of pyoverdin synthesis may be independent from the normal iron regulation, since the presence of Zn(II) is able to induce pyoverdin production by mutant M2. It is hypothesized that besides the activator for siderophore produc-tion, which is apparently damaged in mutant M2, another regulator might exist which induces the high affinity iron uptake system in the presence of Zn(II).

Table 5. Root Colonization By Various
Mutants of *P. aeruginosa* 7NSK2.[30]

Strain	Root colonization log CFU/g dry root	
	Mean	SD
7NKS2	6.90a	0.12
MPFM1	6.67a	0.13
KMPCH	6.71a	0.05
SSS	6.33b	0.12
M2	6.06c	0.21
LSD ($p = 0.05$)	0.24	

Note: MPFM1: pyoverdin-negative mutant; KMPCH: pyoverdin- and pyochelin-negative mutant; SSS and M2: regulatory mutants. Values followed by the same letter are not significantly different for $p = 0.05$.

REFERENCES

1. Kloepper, J.W. and Schroth, M.N., Plant growth-promoting rhizobacteria on radishes, in Proc. Fourth Int. Conf. Plant Path. Bacteria, Station de Pathologie Végétale et Phytobactériologie, INRA, Angers, Gibert-Clarey, Tours, France, 1978, 879.
2. Kloepper, J.W., Lifshitz, R., and Zablotowicz, R.M., Free-living bacterial inocula for enhancing crop productivity, *Trends Biotechnol.*, 7, 39, 1989.
3. Kloepper, J.W., Plant growth-promoting rhizobacteria as biological control agents, in *Soil Microbial Ecology,* Blaine-Metting, F., Ed., Marcel Dekker, New York, 1993, 255.
4. Leong, J., Siderophores: their biochemistry and possible role in the biocontrol of plant pathogens, *Annu. Rev. Phytopathol.*, 24, 187, 1986.
5. Weller, D.M., Biological control of soil-borne plant pathogens in the rhizosphere with bacteria, *Annu. Rev. Phytopathol.*, 26, 379, 1988.
6. Schippers, B., Bakker, A.W., and Bakker, P.A.H.M., Interactions of deleterious and beneficial rhizosphere microorganisms and the effect of cropping practices, *Annu. Rev. Phytopathol.*, 25, 339, 1987.
7. Kloepper, J.W., Leong, J., Teintze, M., and Schroth, M.N., Enhanced plant growth by siderophores produced by plant growth-promoting rhizobacteria, *Nature,* 286, 885, 1980.
8. Iswandi, A., unpublished data, 1986.
9. Iswandi, A., Bossier, P., Vandenabeele, J., and Verstraete, W., Relation between soil microbial activity and the effect of seed inoculation with the rhizopseudomonad strain 7NSK2 on plant growth, *Biol. Fertil. Soils,* 3, 147, 1987.
10. Iswandi, A., Bossier, P., Vandenabeele, J., and Verstraete, W., Effect of seed inoculation with the rhizopseudomonad strain 7NSK2 on the root microbiota of maize *(Zea mays)* and barley *(Hordeum vulgare), Biol. Fert. Soils,* 3, 153, 1987.

11. Bossier, P., Höfte, M., and Verstraete, W., Ecological significance of siderophores in soil, *Adv. Microb. Ecol.*, 10, 385, 1988.

12. Bossier, P. and Verstraete, W., Detection of siderophores in soil by a direct bioassay, *Soil Biol. Biochem.*, 18, 481, 1986.

13. Bossier, P. and Verstraete, W., Ecology of *Arthrobacter* JG-9-detectable hydroxamate siderophores in soil, *Soil Biol. Biochem.*, 18, 487, 1986.

14. Cornelis, P., Hohnadel, D., and Meyer, J.M., Evidence for different pyoverdine-mediated iron uptake systems among *Pseudomonas aeruginosa* strains, *Infect. Immun.*, 57, 3491, 1989.

15. Cornelis, P., unpublished data, 1992.

16. Wendenbaum, S., Demange, P., Dell, A., Meyer, J.M., and Abdallah, M.A., The structure of pyoverdine Pa, the siderophore of *Pseudomonas aeruginosa*, Tetrahedron Lett., 24, 4877, 1983.

17. Cox, C.D., Rinehart, K.L., Moore, M.L., and Cook, J.C., Pyochelin: novel structure of an iron chelating growth promoter for *Pseudomonas aeruginosa*, *Proc. Natl. Acad. Sci. U.S.A.*, 78, 4256, 1981.

18. Höfte, M., Seong, K.Y., Jurkevitch, E., and Verstraete, W., Pyoverdin production by the plant growth promoting *Pseudomonas* strain 7NSK2. Ecological significance in soil, *Plant Soil,* 130, 249, 1991.

19. Seong, K.Y., Höfte, M., Boelens, J., and Verstraete, W., Growth, survival and root colonization of plant growth beneficial *Pseudomonas fluorescens* ANP15 and *Pseudomonas aeruginosa* 7NSK2 at different temperatures, *Soil Biol. Biochem.*, 23, 423, 1991.

20. Seong, K.Y., Höfte, M., and Verstraete, W., Acclimatization of plant growth promoting *Pseudomonas* strain 7NSK2 in soil: effect on population dynamics and plant growth, *Soil Biol. Biochem.*, 24, 751, 1992.

21. Höfte, M., Mergeay, M., and Verstraete, W., Marking the rhizopseudomonas strain 7NSK2 with a Mud*(lac)* element for ecological studies, *Appl. Environ. Microbiol.*, 56, 1046, 1990.

22. Boelens, J., Vande Woestyne, and Verstraete, W., Ecological importance of motility for the plant growth-promoting rhizopseudomonas strain ANP15, *Soil Biol. Biochem.*, in press.

23. Höfte, M., Boelens, J., and Verstraete, W., Seed protection and promotion of seedling emergence by the plant growth beneficial *Pseudomonas* strains 7NSK2 and ANP15, *Soil Biol. Biochem.*, 23, 407, 1991.

24. Boelens, J., unpublished data, 1992.

25. Heungens, K., Höfte, M., Buysens, S., and Poppe, J., Role of siderophores in biological control of *Pythium spp.* by *Pseudomonas* strain 7NSK2, *Meded. Fac. Landbouwwet. Rijksuniv. Gent,* 57, 365, 1992.

26. Höfte, M., Buysens, S., Koedam, N., and Cornelis, P., Zinc affects siderophore-mediated high affinity iron uptake systems in the rhizosphere *Pseudomonas aeruginosa* 7NSK2, *BioMetals,* 6, 85, 1993.

27. Bagg, A. and Neilands, J.B., Ferric uptake regulation protein acts as a repressor, employing iron(II) as a cofactor to bind the operator of an iron transport operon in *Escherichia coli, Biochemistry,* 26, 5471, 1987.

28. O'Sullivan, D.J. and O'Gara, F., Regulation of iron assimilation: nucleotide sequence analysis of an iron-regulated promoter from a fluorescent pseudomonad, *Mol. Gen. Genet.*, 228, 1, 1991.

29. Gensberg, K., Hughes, K., and Smith, A.W., Siderophore-specific induction of iron uptake in *Pseudomonas aeruginosa*, *J. Gen. Microbiol.*, 138, 2381, 1992.
30. Höfte, M., Boelens, J., and Verstraete, W., Survival and root colonization of mutants of plant growth-promoting pseudomonads affected in siderophore bio-synthesis or regulation of siderophore production, *J. Plant Nutr.*, 15, 2253, 1992.
31. Höfte, M., Dong, Q., and Mergeay, M., unpublished data, 1992.

CHAPTER 8

Contribution of Mycorrhizal Fungi to Micronutrient Uptake by Plants

Eckhard George, Volker Römheld, and Horst Marschner

INTRODUCTION

Most vascular plant species are colonized by vesicular-arbuscular mycorrhizal (VAM), ectomycorrhizal (ECM), or other mycorrhizal fungi. These associations occur in almost all ecosystems.[1] As a rule, plant uptake of phosphate (VAM, ECM) and also often of nitrogen (ECM) is considerably increased by mycorrhizas. In addition, mycorrhizas also affect the plant micronutrient uptake. However, until recently most studies on micronutrient uptake of plants have been carried out using plants with nonmycorrhizal root systems or root systems with unknown mycorrhizal status, or in nutrient solutions where mycorrhizal fungi are either absent or not active. Because of the ubiquity of mycorrhizal associations in nature, and because of the potential influence of mycorrhizas on metal uptake by plants and the effects on (mycor)rhizosphere ecology, metal ion uptake studies need to account for mycorrhizal effects. Furthermore, it has become clear that metal ion translocation within the plant can also be affected by mycorrhizal colonization. Recent studies have been directed towards an investigation of the specific role of the extraradical fungal mycelium in soil, and on mycorrhizal effects on plant metal utilization. These studies are summarized in the following sections, to highlight the importance of mycorrhizal fungi in plant metal ion uptake and utilization.

PROBLEMS IN COMPARISONS OF MYCORRHIZAL AND NONMYCORRHIZAL PLANTS

Using sterilized soil in pot experiments, nonmycorrhizal plants can be compared to mycorrhizal (inoculated) plants in growth and nutrient uptake.

0-87371-942-5/94/$0.00+$.50

Mycorrhizal plants often take up more phosphate (P) from soil and grow better than nonmycorrhizal plants. In many cases, the uptake of other elements also differs between mycorrhizal and nonmycorrhizal plants. However, it is not possible to calculate the direct contribution of the fungi to plant micronutrient uptake by simply subtracting the uptake of nonmycorrhizal plants from the nutrient uptake of mycorrhizal plants. There are several direct or indirect effects of mycorrhizal colonization on plant growth. Many of these are related to better P uptake of mycorrhizal plants from low P soils, leading to greater shoot growth, while root growth and root length, in particular, are less increased.[2]

Even in soils with adequate P levels, VAM and nonVAM plants can differ in their shoot and root morphology.[3] For example, total root length and specific root length were reduced in mycorrhizal compared to nonmycorrhizal maize plants by approximately 40%, although shoot dry weight of these plants was not affected by mycorrhizal colonization (Table 1). In addition, the root branching patterns differ between mycorrhizal and nonmycorrhizal plants. Both more and less intense branching can occur in mycorrhizal roots. These observations may partly be explained by direct (nonnutritional) effects of mycorrhizal colonization on root tip activity and cortical cell membrane potential.[4] In the presence or absence of mycorrhiza, plants can differ in so many aspects that differences in the micronutrient content of plants does not necessarily reflect, for example, micronutrient uptake and transport via mycorrhizal hyphae.

MICRONUTRIENT UPTAKE BY MYCORRHIZAL AND NONMYCORRHIZAL PLANTS

Despite the problems in comparing mycorrhizal and nonmycorrhizal plants, most of the present information about mycorrhizal effects on plant micronutrient uptake is derived from such comparisons. In VAM plants, (apart from P) the concentration and total content of zinc (Zn) and copper (Cu) is often increased.[5,6] This becomes especially clear when mycorrhizal plants are compared to nonmycorrhizal plants which were fertilized with additional P, in order to achieve similar P uptake in both treatments (Table 2).[7] While plant Zn and Cu concentrations and total content increased, the uptake of iron (Fe) and manganese (Mn) decreased in mycorrhizal soybean plants (Table 2).

However, there are many other studies showing that plant micronutrient uptake was not affected by VAM colonization. Plant species and cultivar,[8] fungal type, soil pH,[9] soil physical conditions,[10] soil temperature,[11] and the levels of nutrient supply[12] all influence the mycorrhizal effect on micronutrient uptake. Thus, broad generalizations are not possible for plants colonized by VAM fungi. For ECM-colonized plants, a higher uptake of micronutrients at low supply is less frequently reported than for VAM plants. Because of the combination of direct and indirect effects of mycorrhizal colonization, it is

Table 1.   Effect of VAM *(Glomus mosseae)* on Shoot and Root Growth and Morphology of Adequately P-Fertilized Six-Week-Old Maize Plants[3]

	Shoot dry wt. (g plant^{-1})	Root dry wt. (g plant^{-1})	Root length (m plant^{-1})	Specific root length (m g^{-1} root dry wt.)
−VAM	20.0	4.8	619	130
+VAM[a]	22.8^{ns}	4.6^{ns}	367*	81*

[a] Comparison of −VAM and +VAM is ns, not significant; *, significantly different at $p < 0.05$.

Table 2.   Effect of VAM *(Glomus fasciculatum)* on Plant Growth and Leaf Nutrient Concentration in Nine-Week-Old Soybean; Additional P was Supplied to Nonmycorrhizal Plants to Compensate for Higher P Uptake of Mycorrhizal Plants[7]

	Dry wt. (g plant^{-1})	Nutrient concentration in leaf				
		P (mg g^{-1})	Fe	Mn	Zn	Cu
			(μg g^{-1})			
High P −VAM	18.42	1.4	97.0	161	32.4	20.5
Low P +VAM[a]	19.38^{ns}	1.3^{ns}	80.3*	99*	56.9*	64.1*

[a] Comparison of −VAM and +VAM is ns, not significant; *, significantly different at $p < 0.05$.

necessary to investigate in more detail the uptake properties of extraradical hyphae in order to predict changes in micronutrient uptake resulting from mycorrhizal colonization.

NUTRIENT MOBILIZATION AND UPTAKE BY HYPHAE

The quantity and the biochemical mechanisms of fungal micronutrient uptake can be better studied when extraradical hyphae of the fungi are grown spatially separated from plant root systems. This avoids confounding the root and hyphal effects on nutrient mobilization and uptake. Cultures of VAM fungi cannot be maintained without host plants, but ECM fungi (similar to saprophytic fungi) can be grown host-free. Although the uptake of micronutrients by ECM fungi can therefore be conveniently studied on agar, fungal activity may be quite different when the fungi grow in association with plants. For example, hyphae of *Scleroderma citrinum* are very sensitive to a high supply of Zn when grown axenically, but are much more tolerant when grown in association with *Pinus sylvestris*.[13] For both types of mycorrhizas, therefore, hyphal uptake must be investigated using mycelia interconnected with living plant roots. The density of the fungal mycelium in soil can be very high (up to 25m of hyphae per cubic centimeter of soil).[14] The mycelium of ECM fungi can proliferate in soil over large distances, and VAM hyphae also can extend more than 10 cm away from the root surface.[15]

To differentiate between hyphal and root uptake, screens with a sufficiently fine mesh size to allow passage of hyphae, but not of roots, may be used to separate the soil into zones with and without roots.[3] Nutrients can be supplied to the soil zones explored by the hyphae only, and hyphal uptake may be quantified by measuring soil nutrient depletion in these zones or by comparing nutrient uptake of mycorrhizal and nonmycorrhizal plants. Such systems have been used to examine ^{32}P or ^{15}N uptake, and also in studies of micronutrient uptake.[16-18] In an experiment using a subsoil of a loess (Luvisol), tubes (diameter 3.2 cm) with fertilized soil were buried into soil in pots planted with ryegrass grown in association with the VAM fungus, *Glomus mosseae,* or without mycorrhizal colonization. Each tube had two windows closed by nylon screens excluding the roots from the soil from the tube, but not hyphae. After 40 days, hyphae had considerably depleted the soil of extractable P, Zn, and potassium (K), but not of magnesium (Mg), Mn, or Cu (Table 3). The importance of VAM in Zn nutrition of plants is also reflected in reports of a decline in Zn concentrations in plants grown on low P soils with increasing rates of P fertilizer. This depressed the VAM colonization of roots.[19] Similar experimental systems with nylon screens in soil were set up to measure hyphal nutrient uptake in the field[20] and to show that VAM hyphae transport only very small amounts of water from the soil to the plant.[21] These systems can also be used to study uptake mechanisms of mycorrhizal hyphae and their effect on the "hyphosphere" (or "mycosphere"[22]) soil — the soil immediately surrounding an individual hypha.[14] For example, depending on the form of nitrogen supply, the pH in the soil adjacent to the root can be lower or higher than in the bulk soil. As with roots, NH_4^+-supplied hyphae of *Glomus mosseae* associated with white clover decreased the pH in the hyphosphere,[14] and may thereby have enhanced the uptake of P and of certain micronutrients.

It has not yet been tested specifically whether extraradical hyphae of VAM or ECM fungi, under conditions of micronutrient deficiency, can enhance the release of compounds like siderophores or other chelators and organic acids into the hyphosphere to mobilize micronutrients. Such a release of mobilizing substances is highly likely, at least for some ECM fungi when growing in association with plant roots. Similar to other fungi,[23,24] these fungi when grown host-free produce hydroxamate siderophores,[25] reductants,[26] and oxalic acid,[27] and possibly phenolic compounds.[28] Ectomycorrhizal pine roots can have higher Fe contents than nonmycorrhizal roots, although it is not clear whether the higher Fe contents are a property of the host or the endophyte.[29] In undisturbed forest soils, concentrations of hydroxamate siderophores are high,[30] especially in rhizosphere soil.[31]

Table 3. **Effect of VAM *(Glomus mosseae)* Hyphae Growing in Association With Ryegrass on Depletion of Nutrients in Soil (P, K: Calcium Acetate Lactate Extractable; Mg: CaCl$_2$ [0.0125M] Extractable; Mn, Zn, Cu: DTPA Extractable) During a Period of 40 Days in Root-Free Zones of a Luvisol**

	Soil nutrient concentration					
	P	K	Mg	Mn	Zn	Cu
	(mg g^{-1})				(μg g^{-1})	
Without hyphae (−VAM)	27.5	124	108	6.8	4.5	4.5
With hyphae (+VAM)[a]	10.6*	85*	100[ns]	7.4[ns]	3.0*	4.0[ns]

[a] Comparison of −VAM and +VAM is ns, not significant; *, significantly different at $p < 0.05$.

Hydroxamate siderophores can be produced by ericoid mycorrhizal fungi.[32,33] However, for VAM-colonized roots the only report so far suggesting siderophore production comes from an experiment with nonaxenic plants growing in nutrient solution,[34] and no equivalent observations have been made for extraradical hyphae. Nutrient mobilization by exudation of organic solutes appears to be frequent in ECM fungi, but may not be common in VAM fungi. Nutrient mobilization by ECM hyphae does not necessarily imply that nutrients are also taken up by the hyphae and translocated to the plant. The mobilized nutrients may also accumulate in the hyphae and not be translocated to the host root (see below).

In general, the contribution of mycorrhizal fungi to plant nutrient acquisition at low nutrient supply depends on the spatial distribution and the chemical status of the nutrients in soil. In pot experiments where the soil is separated by screens to restrict root growth to part of the pot volume, between 25 and 60% of the shoot Zn and Cu content was taken up via the VAM fungal hyphae (Figure 1). In undisturbed soils, the contribution of mycorrhizal hyphae to plant nutrient uptake will be high when (1) extraradical hyphae proliferate abundantly, (2) hyphae have access to nutrients chemically not available to roots, or (3) roots do not fully exploit the soil volume. This could be the case, for example, in plant species with coarse, less branched roots (e.g., citrus or cassava), in problem soils where root growth is restricted, and for nutrients with low mobility in soil.

TRANSLOCATION WITHIN THE FUNGUS-PLANT ASSOCIATION

To affect host plant nutrition and shoot growth, the mineral nutrients taken up by the fungal hyphae must be translocated within the mycelium to the

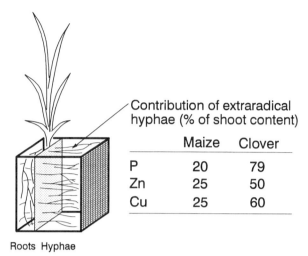

Contribution of extraradical hyphae (% of shoot content)		
	Maize	Clover
P	20	79
Zn	25	50
Cu	25	60

Roots Hyphae

Figure 1.   Estimated contribution of extraradical hyphae of the VAM fungus *Glomus mosseae* to uptake of P, Zn, and Cu by six-week-old maize and by seven-week-old white clover plants grown in a Luvisol in compartmented boxes.[17,18]

host-plant root, pass the fungus-root interface (Hartig net in ECM, intraradical hyphae, and hyphae in VAM), and then move from root to shoot. In many studies only the shoots are analyzed for nutrient concentrations, and this may partly explain the discrepancies between the results of different experiments on mycorrhizal involvement in micronutrient uptake. As an example of this, hyphae of ECM fungi can contain large amounts of metal ions such as Zn,[35,36] thereby reducing plant Zn concentrations. At high Zn supply, ECM fungal hyphae contained up to 17,400 µg Zn g^{-1} dry weight.[13] Similarly, fungi can also accumulate cesium (Cs) in their mycelia.[37] Moreover, concentrations of Zn and Cu can be increased in mycorrhizal plants in roots, but not in shoots.[17,38] This may be regarded as micronutrient storage in fungal tissue within the root, or in root tissue.[38]

Root-to-shoot translocation of micronutrients can be drastically affected by the amount of P transported by the extraradical VAM fungal hyphae to the root. To study this effect, white clover plants were grown in pots with soil divided into compartments by screens to allow hyphae, but not roots, to pass. While the root compartments were fertilized with similar quantities of nutrients in all treatments, the outer (hyphal) compartments were fertilized at three different rates of P (0, 20, or 50 mg kg^{-1}).[18] By comparison to nonmycorrhizal plants, or plants without access to the outer compartments, the quantity of nutrients in plants taken up from the outer compartments by hyphae of mycorrhizal plants was calculated.[18] For example, at the lowest rate of P supply in the outer compartment, mycorrhizal plants with access to the outer (hyphal) compartments contained in shoots approximately 1.4 mg P per pot, and mycorrhizal plants without access to the outer

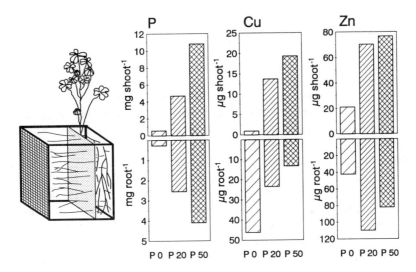

Figure 2. Distribution of the P, Cu, and Zn taken up by VAM hyphae *(Glomus mosseae)* into shoots (upper) and roots (lower) of white clover grown in a Luvisol in compartmented boxes. For details of calculation see Reference 18. Soil in the outer (hyphal) compartments was fertilized with three different levels of P (0, 20, and 50 mg kg⁻¹ soil = P 0, P 20, P 50) as $Ca(H_2PO_4)_2$.[18]

compartments contained in shoots 0.8 mg P per pot.[18] At the lowest rate of P supply, the fungal hyphal contribution to shoot P content was therefore estimated as 0.6 mg P per pot (Figure 2). The quantity of P taken up by hyphae and translocated to the plant was thus very low when the soil in the outer compartments was not P fertilized (Figure 2), while at this low P supply fungal hyphae transported relatively large amounts of Cu (approximately 46 µg Cu per pot) into the roots of mycorrhizal plants (Figure 2). As expected, a higher supply of soil P to hyphae drastically increased P transport by hyphae to the plants, but the P supply did not clearly increase total hyphal Cu uptake.[18] However, when more P was supplied to the hyphae much more Cu was translocated from the root to the shoot in the host plant (Figure 2). For Zn, hyphal uptake was increased by the lower rate of P supply (20 mg P kg⁻¹), and root-to-shoot translocation was increased at the higher rate of P supply (Figure 2).

For Cu, this example demonstrates that the role of VAM on micronutrient uptake and host plant nutrition is insufficiently described when only the shoot contents are analyzed. Mycorrhizal colonization may alter element concentrations in roots, in shoots, or in both, and mycorrhizal colonization as well as P supply can affect root-to-shoot translocation of microelements. Furthermore, in view of the importance of polyphosphates in mycorrhizal hyphal transport processes, and the subsequent polyphosphate hydrolysis at the fungus-root cell interface, metal cations adsorbed to the polyphosphates may be delivered and released to the host plant cells.

MYCORRHIZOSPHERE EFFECTS

Mycorrhizal colonization not only alters the morphology and physiology of the host root and delivers mineral elements via the extraradical hyphae to the host root, but colonization also changes the conditions in the "mycorrhizosphere", the soil zone immediately surrounding the mycorrhizal root.[39,40] Because root exudation may differ considerably between VAM and nonmycorrhizal roots,[41] the energy sources for microorganisms may also differ, affecting the microorganism populations (see Chapters 6 and 14). Conditions in the mycorrhizosphere probably influence, for example, the mycorrhizal effect on plant Mn uptake. Manganese concentrations in VAM plants may be either decreased or increased compared to nonVAM plants.[42] In a range of soils tested, VAM colonization significantly increased leaf Mn concentration in *Poncirus trifoliata* in some (low Mn) soils, but in other soils leaf Mn concentration was significantly decreased.[43] Similarly, for ECM fungi, both lower Mn concentrations in inoculated plants[44] and higher Mn uptake in ECM plants[28] have been found.

The mechanisms responsible for this contrasting behavior are most probably related to different (and often lower) root exudation in the presence of mycorrhizas. It has been shown, for example, that lower root exudation of VAM-colonized roots caused a lower number of bacterial Mn reducers in the mycorrhizosphere, lower soil Mn availability, and lower Mn uptake.[45] Alternatively, the population of Mn-oxidizing bacteria can be higher in the mycorrhizosphere, also contributing to lower soil Mn availability.[46] Also, the exudation of Mn-chelating exudates may be decreased in VAM-colonized plants.[47] On the other hand, ECM fungi can release reducing substances mobilizing soil Mn.[28]

In a pot experiment, soybean plants were inoculated with either a VAM fungus or fertilized with additional P. After 9 weeks of growth, shoot weight and P concentration in the leaves did not differ in both treatments because the additional P supply to nonmycorrhizal plants balanced the effect of mycorrhizal colonization on plant P uptake (Table 2).[7] However, in leaves of VAM plants both Mn and Fe concentrations were decreased compared to nonmycorrhizal plants (Table 2).[7] Thus, in a manner similar to Mn, colonization with VAM fungi can also lead to lower plant Fe uptake.[7,17] However, the effects of VAM fungi on plant Fe absorption and utilization have not yet been thoroughly investigated. In graminaceous species, decreased root exudation by VAM plants may result in lower release of plantborne Fe(III) chelators (phytosiderophores). Decreased root exudation may also lead to reductions in the populations of siderophore-producing bacteria, thereby reducing their role in plant Fe supply.[48] Alternatively, fungal siderophores could compete with the plant for soil Fe, or fungi could decrease direct plant Fe uptake by degradation of plantborne Fe(III) chelators as do bacteria.[49] For boron (B), experimental evidence for a role of mycorrhiza in B nutrition of host plants is at present inconclusive. *Pinus echinata* trees responded to B fertilization by increased mycorrhizal formation and growth, but B uptake was not affected by ECM colonization, indicating a neglible contribution of mycorrhiza to B nutrition of trees in forest soils.[50]

Some of the VAM effects on plant nutrient uptake are summarized schematically in Figure 3. In addition to P, substantial quantities of other elements, including metals, can be absorbed by VAM fungal hyphae and be transported to the root (Figure 3). Also, uptake and delivery of NH_4^+–N, for example, may be considerable.[21] On the other hand, concentrations of calcium (Ca) or Mg are in most cases not increased in VAM plants, so that a substantial hyphal uptake of these elements is unlikely. Use of labeled Ca has shown some Ca uptake by VAM hyphae,[51] but the quantity delivered to the plant by hyphae is probably very small compared to the quantity absorbed directly by the root. Besides this direct nutrient uptake by hyphae, VAM effects may also be related to changes in root exudation in the presence of VAM. While ECM hyphae may release organic solutes such as organic acids, phenolics, and siderophores into the hyphosphere, such release has not been convincingly reported for VAM fungal hyphae. In spite of these particularities of ECM, not much is known about the importance of these fungi in the host-plant micronutrient status.

DEFICIENCY OR TOXICITY

Mycorrhizas appear to affect plant uptake of certain micronutrients to the benefit of the plant, both under conditions of deficiency and toxicity and when a toxic metal is supplied.[32,38,47] Under conditions of deficiency (at least in VAM plants) the uptake of Zn and Cu can be improved (see above), while under conditions of toxicity (at least in ECM plants) shoot concentrations of the respective elements can be decreased. In many cases decreased shoot concentrations are caused by growth enhancement only ("dilution effect" due to better growth of mycorrhizal plants), although there is also experimental evidence showing a decrease in the amounts of the respective elements in the shoots. For example, under high Zn supply, inoculation with *Paxillus involutus* not only decreased Zn concentration in shoots of *Pinus sylvestris,* but also shoot Zn total content (Table 4).[13] A decreased Zn uptake by ECM-colonized plants was related to lower Zn supply to roots due to binding of the metal by hyphal cell-wall polymers or extrahyphal slime.[35] A similar mechanism can also explain better growth of ericoid mycorrhizal plants at high supply of Cu or Zn.[52] Thus, in general, fungi with large mycelia may have the greatest effect in overcoming metal toxicity.[53] Furthermore, metals could be deposited, at least temporarily, in polyphosphate granules within the fungus[54] or on pectic substances at the interface between the fungus and host root. In some fungi, when exposed to high levels of cadmium (Cd), specific metal binding

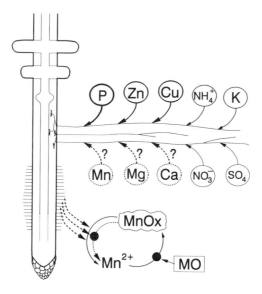

Figure 3. Examples for effects of root colonization with VAM fungi on plant nutrient uptake;
MO = microorganisms.

Table 4. Effect of ECM *(Paxillus involutus)* on Shoot Zn Content and Shoot and Root (Short Lateral Root) Zn Concentration in Seedlings of *Pinus sylvestris* Grown for Nine Months in Vermiculite Watered With Nutrient Solution — Seedlings Were Supplied With a High Rate (14mM) of Zn[13]

	Fungal biomass (% of short root biomass)	Shoot Zn content (mg plant⁻¹)	Shoot Zn concentration ($\mu g\ g^{-1}$ dry wt.)	Root Zn concentration ($\mu g\ g^{-1}$ dry wt.)
−ECM	—	3.19	197	273
+ECM[a]	54	1.52*	106*	708*

[a] Comparison of −VAM and +VAM is ns, not significant; *, significantly different at $p < 0.05$.

proteins such as phosphoglycoproteins[55] or metallothioneins[38,56] can be present, although it is not yet clear whether these proteins commonly occur in mycorrhizal plants under metal toxicity.

Deposition of potentially toxic elements in mycorrhizal fungal structures in the root, or within root tissue itself, could lead to higher concentrations in roots of mycorrhizal plants, but lower concentrations in shoots, as shown for Zn in ECM-colonized pine seedlings (Table 4).[13] Aluminum toxicity can also be overcome by ECM colonization due to the higher P uptake of mycorrhizal plants[57] and chelation of Al by fungal organic compounds.[58] To date it is not known to what extent the changes in root morphology and physiology by mycorrhizal colonization contribute to the lower metal uptake in mycorrhizal plants at conditions of high external supply.

However, at least in VAM plants, increased uptake of micronutrients with depressing effects on growth also has been reported when toxic levels were supplied.[9] In conclusion, for both VAM and ECM plants a protective role of mycorrhizal fungi at high metal supply is possible, but mycorrhizas may also have no effects, or even be detrimental.

EFFECTS OF MICRONUTRIENTS AND HEAVY METALS IN SOIL ON MYCORRHIZA FORMATION

A possible beneficial role of mycorrhizas in micronutrient nutrition and in host plant growth depends on the ability of the fungus to colonize plant roots, both under conditions of very low or high micronutrient supply. Low micronutrient supply normally reduces mycorrhizal formation no more than plant growth. One notable exception is B. Foliar application of B increased mycorrhizal colonization of both VAM[59] and ECM[42] plants. Possibly, additional B supply decreases root exudation of phenolics or formation of phenolics in response to root colonization, thus generally enhancing fungal activity in the rhizosphere and, presumably, also mycorrhizal root colonization.[60]

On the other hand, a high micronutrient or metal supply can drastically reduce mycorrhizal colonization[61] and extraradical hyphal growth.[13] Fungal species and isolates differ considerably in their effect on plant growth under high metal supply.[13] These differences between species or isolates can be related more to differences in compatibility between the host and fungal type than to a specific tolerance of metals by individual fungal isolates.[62] The reduction of mycorrhizal activity at high metal concentrations restricts, of course, the possible significance of mycorrhizas in the adaptation of plants to soils with toxic concentrations of mineral elements.

CONTRIBUTIONS TO PLANT GROWTH

Almost all evidence for mycorrhizal effects comes from pot experiments. In field or forest soils mycorrhizas are ubiquitous, and therefore it is difficult to compare mycorrhizal and nonmycorrhizal plants. When several plant species growing in a grassland were harvested at different times over a 2-year period, Zn, Cu, or Mn concentrations in shoots were not related to differences in VAM colonization.[63] Phenomena other than VAM or ECM colonization may often be more important for regulation of micronutrient uptake by plants. It must also be kept in mind that mycorrhizas derive their carbon requirement from the host plant, using 10% or more of the carbon transported to the roots.[64-66] Hyphal nutrient uptake or toxic metal immobilization can be most effective in fungi with large, proliferating mycelia (see above). However, because of the correspondingly high carbon costs of a large mycelium and also (in ECM fungi) high nitrogen storage in hyphae, these fungi can also be the most inhibitory to plant growth.[67,68]

Therefore, mycorrhizal associations may not always be advantageous to the plant.[69] In addition, other adaptive plant mechanisms to low micronutrient supply, such as root exudation of phytosiderophores, can respond faster and more specifically to temporal changes in supply, so that their activity (and thus carbon demand) can be better regulated when the nutrient supply is adequate after a period of deficiency. In ecological terms, this suggests that benefit/cost ratios of mycorrhizas are highest in slow-growing plant species, but that release of metal mobilizing root exudates is more effective in fast growing species with a low carbon/nutrient balance.

CONCLUSIONS

For an understanding of the micronutrient acquisition of soil-grown plants, studies have to include mycorrhizal plants because mycorrhizal colonization can affect micronutrient uptake by the plant and utilization within the plant. At a low supply of micronutrients, mycorrhizal plants may acquire increased quantities of micronutrients, either by direct element uptake from the soil by extraradical hyphae and subsequent translocation to the plant, or by mycorrhizal effects on root and microbial activity. At a high supply of metals, at least ECM fungi may protect the plant from excess elements by storage of these elements in fungal tissue or by alteration of rhizosphere conditions with an effect on other microorganisms. However, general predictions about mycorrhizal effects on plant microelement balances are not possible because the effects depend on the specific element, soil conditions, and plant and fungal type. Moreover, mycorrhizal fungi obtain their carbon requirement for growth and maintenance from their host plants. Although mycorrhizal fungi are ubiquitous in agricultural fields and forests, at present it is not possible on a large scale to manipulate mycorrhizal associations in order to raise plant productivity on soils with deficient or excessive microelement supply.

ACKNOWLEDGMENTS

We thank Dr. Jan Colpaert for comments on the manuscript.

REFERENCES

1. Read, D.J., Mycorrhizas in ecosystems, *Experientia*, 47, 376, 1991.
2. Gnekow, M.A. and Marschner, H., Influence of the fungicide pentachloronitrobenzene on VA-mycorrhizal and total root length and phosphorus uptake of oats *(Avena sativa)*, *Plant Soil*, 114, 91, 1989.
3. Kothari, S.K., Marschner, H., and George, E., Effect of VA mycorrhizal fungi and rhizosphere microorganisms on root and shoot morphology, growth and water relations in maize, *New Phytol.*, 116, 303, 1990.
4. Fieschi, M., Alloatti, G., Sacco, S., and Berta, G., Membrane potential hyperpolarisation in vesicular arbuscular mycorrhizae of *Allium porrum* L.: a non-nutritional long-distance effect of the fungus, *Protoplasma*, 168, 136, 1992.
5. Tinker, P.B. and Gildon, A., Mycorrhizal fungi and ion uptake, in *Metals and Micronutrients: Uptake and Utilization by Plants*, Robb, D.A. and Pierpoint, W.S., Eds., Academic Press, New York, 1983, chap. 2.
6. Sharma, A.K. and Srivastava, P.C., Effect of vesicular-arbuscular mycorrhizae and zinc application on dry matter and zinc uptake of greengram *(Vigna radiata* L. Wilczek), *Biol. Fertil. Soils*, 11, 52, 1991.
7. Pacovsky, R.S. and Fuller, G., Mineral and lipid composition of *Glycine-Glomus-Bradyrhizobium* symbioses, *Physiol. Plant.*, 72, 733, 1988.
8. Lambert, D.H., Cole, H., Jr., and Baker, D.E., Variation in the reponse of alfalfa clones and cultivars to mycorrhizae and phosphorus, *Crop Sci.*, 20, 615, 1980.
9. Killham, K. and Firestone, M.K., Vesicular arbuscular mycorrhizal mediation of grass response to acidic and heavy metal depositions, *Plant Soil*, 72, 39, 1983.
10. Kucey, R.M.N. and Janzen, H.H., Effects of VAM and reduced nutrient availability on growth and phosphorus and micronutrient uptake of wheat and field beans under greenhouse conditions, *Plant Soil*, 104, 71, 1987.
11. Raju, P.S., Clark, R.B., Ellis, J.R., and Maranville, J.W., Effects of species of VA-mycorrhizal fungi on growth and mineral uptake of sorghum at different temperatures, *Plant Soil*, 121, 165, 1990.
12. Raju, P.S., Clark, R.B., Ellis, J.R., and Maranville, J.W., Mineral uptake and growth of sorghum colonized with VA mycorrhiza at varied soil phosphorus levels, *J. Plant Nutr.*, 13, 843, 1990.
13. Colpaert, J.V. and Van Assche, J.A., Zinc toxicity in ectomycorrhizal *Pinus sylvestris, Plant Soil*, 143, 201, 1992.
14. Li, X.-L., George, E., and Marschner, H., Phosphorus depletion and pH decrease at the root-soil and hyphae-soil interfaces of VA mycorrhizal white clover fertilized with ammonium, *New Phytol.*, 119, 397, 1991.
15. Li, X.-L., George, E., and Marschner, H., Extension of the phosphorus depletion zone in VA-mycorrhizal white clover in a calcareous soil, *Plant Soil*, 136, 41, 1991.
16. Kothari, S.K., Marschner, H., and Römheld, V., Direct and indirect effects of VA mycorrhizal fungi and rhizosphere microorganisms on acquisition of mineral nutrients by maize *(Zea mays* L.) in a calcareous soil, *New Phytol.*, 116, 637, 1990.

17. Kothari, S.K., Marschner, H., and Römheld, V., Contribution of the VA mycorrhizal hyphae in acquisition of phosphorus and zinc by maize grown in a calcareous soil, *Plant Soil,* 131, 177, 1991.

18. Li, X.-L., Marschner, H., and George, E., Acquisition of phosphorus and copper by VA-mycorrhizal hyphae and root-to-shoot transport in white clover, *Plant Soil,* 136, 49, 1991.

19. Lambert, D.H., Baker, D.E., and Cole, H., Jr., The role of mycorrhizae in the interactions of phosphorus with zinc, copper, and other elements, *Soil Sci. Soc. Am. J.,* 43, 976, 1979.

20. Jakobsen, I., Phosphorus transport by external hyphae of vesicular-arbuscular mycorrhizas, in *Mycorrhizas in Ecosystems,* Read, D.J., Lewis, D.H., Fitter, A.H., and Alexander, I.J., Eds., CAB International, Wallingford, U.K., 1992, chap. 6.

21. George, E., Häussler, K.-U., Vetterlein, D., Gorgus, E., and Marschner, H., Water and nutrient uptake by hyphae of *Glomus mosseae, Can. J. Bot.,* 70, 2130, 1992.

22. Gilbert, R.G. and Linderman, R.G., Increased activity of soil microorganisms near sclerotia of *Sclerotium rolfsii* in soil, *Can. J. Microbiol.,* 17, 557, 1971.

23. Winkelmann, G., Structures and functions of fungal siderophores containing hydroxamate and complexone type iron binding ligands, *Mycol. Res.,* 96, 529, 1992.

24. Thieken, A. and Winkelmann, G., Rhizoferrin: a complexone type siderophore of the Mucorales and Entomophtorales (Zygomycetes), *FEMS Microbiol. Lett.,* 94, 37, 1992.

25. Szaniszlo, P.J., Powell, P.E., Reid, C.P.P., and Cline, G.R., Production of hydroxamate siderophore iron chelators by ectomycorrhizal fungi, *Mycologia,* 73, 1158, 1981.

26. Cairney, J.W.G. and Ashford, A.E., Reducing activity at the root surface in *Eucalyptus pilularis-Pisolithus tinctorius* ectomycorrhizas, *Aust. J. Plant Physiol.,* 16, 99, 1989.

27. Lapeyrie, F., Picatto, C., Gerard, J., and Dexheimer, J., T.E.M. study of intracellular and extracellular calcium oxalate accumulation by ectomycorrhizal fungi in pure culture or in association with *Eucalyptus* seedlings, *Symbiosis,* 9, 163, 1990.

28. Cairney, J.W.G. and Ashford, A.E., Release of a reducing substance by the ectomycorrhizal fungi *Pisolithus tinctorius* and *Paxillus involutus, Plant Soil,* 135, 147, 1991.

29. Leyval, C. and Reid, C.P.P., Utilization of microbial siderophores by mycorrhizal and non-mycorrhizal pine roots, *New Phytol.,* 119, 93, 1991.

30. Perry, D.A., Rose, S.L., Pilz, D., and Schoenberger, M.M., Reduction of natural ferric iron chelators in disturbed forest soils, *Soil Sci. Soc. Am. J.,* 48, 379, 1984.

31. Reid, R.K., Reid, C.P.P., Powell, P.E., and Szaniszlo, P.J., Comparison of siderophore concentrations in aqueous extracts of rhizosphere and adjacent bulk soils, *Pedobiologia,* 26, 263, 1984.

32. Shaw, G., Leake, J.R., Baker, A.J.M., and Read, D.J., The biology of mycorrhiza in the Ericaceae. XVII. The role of mycorrhizal infection in the regulation of iron uptake by ericaceous plants, *New Phytol.,* 115, 251, 1990.

33. Haselwandter, K., Dobernigg, B., Beck, W., Jung, G., Cansier, A., and Winkelmann, G., Isolation and identification of hydroxamate siderophores of ericoid mycorrhizal fungi, *Biol. Metals,* 5, 51, 1992.

34. Cress, W.A., Johnson, G.V., and Barton, L.L., The role of endomycorrhizal fungi in iron uptake by *Hilaria jamesii, J. Plant Nutr.,* 9, 547, 1986.

35. Denny, H.J. and Wilkins, D.A., Zinc tolerance in *Betula* spp. IV. The mechanism of ectomycorrhizal amelioration of zinc toxicity, *New Phytol.,* 106, 545, 1987.

36. Wilkins, D.A., The influence of sheathing (ecto-)mycorrhizas of trees on the uptake and toxicity of metals, *Agric. Ecosyst. Environ.,* 35, 245, 1991.

37. Dighton, J., Clint, G.M., and Poskitt, J.M., Uptake and accumulation of ^{137}Cs by upland grassland soil fungi: a potential pool of Cs immobilization, *Mycol. Res.,* 95, 1052, 1991.

38. Dehn, B. and Schüepp, H., Influence of VA mycorrhizae on the uptake and distribution of heavy metals in plants, *Agric. Ecosyst. Environ.,* 29, 79, 1989.

39. Linderman, R.G., Microbial interactions with the rhizosphere microflora: the mycorrhizosphere effect, *Phytopathology,* 78, 366, 1988.

40. Linderman, R.G., Vesicular-arbuscular mycorrhizae and soil microbial interactions, in *Mycorrhizae in Sustainable Agriculture,* Bethlenfalvay, G.J. and Linderman, R.G., Eds., ASA Spec. Publ. 54, Soil Science Society of America, Madison, WI, 1992, chap. 3.

41. Schwab, S.M., Menge, J.A., and Leonard, R.T., Quantitative and qualitative effects of phosphorus on extracts and exudates of sudangrass roots in relation to vesicular-arbuscular mycorrhiza formation, *Plant Physiol.,* 73, 761, 1983.

42. Young, J.L. and Jarrell, W.M., Mycorrhizal systems and "other" elements, especially micronutrients, in *Proc. 6th North Am. Conf. Mycorrhizae,* Molina, R., Ed., College of Forestry, Oregon State University, Corvallis, 1985, 170.

43. Menge, J.A., Jarrell, W.M., Labanauskas, C.K., Ojala, J.C., Huszar, C., Johnson, E.L.V., and Sibert, D., Predicting mycorrhizal dependency of Troyer citrange on *Glomus fasciculatus* in California citrus soils and nursery mixes, *Soil Sci. Soc. Am. J.,* 46, 762, 1982.

44. Berry, C.R., Survival and growth of pine hybrid seedlings with *Pisolithus* ectomycorrhizae on coal spoils in Alabama and Tennessee, *J. Environ. Qual.,* 11, 709, 1982.

45. Kothari, S.K., Marschner, H., and Römheld, V., Effect of a vesicular-arbuscular mycorrhizal fungus and rhizosphere micro-organisms on manganese reduction in the rhizosphere and manganese concentrations in maize (*Zea mays* L.), *New Phytol.,* 117, 649, 1991.

46. Arines, J., Porto, M.E., and Vilariño, A., Effect of manganese on vesicular-arbuscular mycorrhizal development in red clover plants and on soil Mn-oxidizing bacteria, *Mycorrhiza,* 1, 127, 1992.

47. Bethlenfalvay, G.J. and Franson, R.L., Manganese toxicity alleviated by mycorrhizae in soybean, *J. Plant Nutr.,* 12, 953, 1989.

48. Crowley, D.E., Reid, C.P.P., and Szaniszlo, P.J., Microbial siderophores as iron sources for plants, in *Iron Transport in Microbes, Plants and Animals,* Winkelmann, G., Van der Helm, D., and Neilands, J.B., Eds., VCH Publishers, Weinheim, Germany, 1987, chap. 20.

49. Crowley, D.E., Römheld, V., Marschner, H., and Szaniszlo, P.J., Root-microbial effects on plant iron uptake from siderophores and phytosiderophores, *Plant Soil,* 142, 1, 1992.

50. Mitchell, R.J., Garrett, H.E., Cox, G.S., and Atalay, A., Boron and ectomycorrhizal influences on mineral nutrition of container-grown *Pinus ehinata* Mill., *J. Plant Nutr.,* 13, 1555, 1990.

51. Rhodes, L.H. and Gerdemann, J.W., Translocation of calcium and phosphate by external hyphae of vesicular-arbuscular mycorrhizae, *Soil Sci.,* 126, 125, 1978.

52. Bradley, R., Burt, A.J., and Read, D.J., The biology of mycorrhiza in the Ericaceae. VIII. The role of mycorrhizal infection in heavy metal resistance, *New Phytol.,* 91, 197, 1982.

53. Colpaert, J.V. and Van Assche, J.A., The effects of cadmium on ectomycorrhizal *Pinus sylvestris* L., *New Phytol.,* 123, 325, 1993.

54. Turnau, K., Kottke, I., and Oberwinkler, F., Element localization in mycorrhizal roots of *Pteridium aquilinum* (L.) Kuhn collected from experimental plots treated with cadmium dust, *New Phytol.,* 123, 313, 1993.

55. Meisch, H.-U., Beckmann, I., and Schmitt, J.A., A new cadmium-binding phosphoglycoprotein, cadmium-mycophosphatin, from the mushroom, *Agaricus macrosporus, Biochem. Biophys. Acta,* 745, 259, 1983.

56. Morselt, A.F.W., Smits, W.T.M., and Limonard, T., Histochemical demonstration of heavy metal tolerance in ectomycorrhizal fungi, *Plant Soil,* 96, 417, 1986.

57. Cumming, J.R. and Weinstein, L.H., Nitrogen source effects on Al toxicity in nonmycorrhizal and mycorrhizal pitch pine *(Pinus rigida)* seedlings. I. Growth and nutrition, *Can. J. Bot.,* 68, 2644, 1990.

58. Cumming, J.R. and Weinstein, L.H., Aluminum-mycorrhizal interactions in the physiology of pitch pine seedlings, *Plant Soil,* 125, 7, 1990.

59. Lambert, D.H., Cole, H., Jr., and Baker, D.E., The role of boron in plant response to mycorrhizal infection, *Plant Soil,* 57, 431, 1980.

60. Mitchell, R.J., Garrett, H.E., Cox, G.S., Atalay, A., and Dixon, R.K., Boron fertilization, ectomycorrhizal colonization, and growth of *Pinus echinata* seedlings, *Can. J. For. Res.,* 17, 1153, 1987.

61. Gildon, A. and Tinker, P.B., Intercations of vesicular-arbuscular mycorrhizal infection and heavy metals in plants. I. The effects of heavy metals on the development of vesicular-arbuscular mycorrhizas, *New Phytol.,* 95, 247, 1983.

62. Denny, H.J. and Wilkins, D.A., Zinc tolerance in *Betula* spp. III. Variation in response to zinc among ectomycorrhizal associates, *New Phytol.,* 106, 535, 1987.

63. Sanders, I.R. and Fitter, A.H., The ecology and functioning of vesicular-arbuscular mycorrhizas in co-existing grassland species. II. Nutrient uptake and growth of vesicular-arbuscular mycorrhizal plants in a semi-natural grassland, *New Phytol.,* 120, 525, 1992.

64. Vogt, K.A., Publicover, D.A., and Vogt, D.J., A critique of the role of ectomycorrhizas in forest ecology, *Agric. Ecosyst. Environ.,* 35, 171, 1991.

65. Fitter, A.H., Costs and benefits of mycorrhizas: implications for functioning under natural conditions, *Experientia,* 47, 350, 1991.

66. Jakobsen, I. and Rosendahl, L., Carbon flow into soil and external hyphae from roots of mycorrhizal cucumber plants, *New Phytol.,* 115, 77, 1990.

67. Colpaert, J.V., Van Assche, J.A., and Luijtens, K., The growth of the extramatrical mycelium of ectomycorrhizal fungi and the growth response of *Pinus sylvestris* L., *New Phytol.,* 120, 127, 1992.

68. Johnson, N.C., Copeland, P.J., Crookston, R.K., and Pfleger, F.L., Mycorrhizae: possible explanation for yield decline with continuous corn and soybean, *Agron. J.,* 84, 387, 1992.

69. Peng, S., Eissenstat, D.M., Graham, J.H., Williams, K., and Hodge, N.C., Growth depression in mycorrhizal citrus at high-phosphorus supply, *Plant Physiol.,* 101, 1063, 1993.

CHAPTER 9

Contribution of VA Mycorrhiza to Zinc Uptake in Plants

A.K. Sharma, P.C. Srivastava, and B.N. Johri

INTRODUCTION

Vesicular-arbuscular (VA) mycorrhizae are a group of fungi having endotrophic associations with the roots of most higher plants except those belonging to the Order Centrospermae and in the families Brassicaceae, Fumariaceae, Cyperaceae, and Commelinaceae.[1] The physiological significance of VA mycorrhizal association for the higher plants is increased nutrient uptake, water absorption, drought tolerance, and root disease resistance.[2-4] An increased uptake of nutrients by mycorrhizal plants growing in deficient soils has been reported for phosphates,[5-7] and zinc and copper.[8-15] At high phosphorus doses, the VA mycorrhizal association also ensures sufficient uptake of zinc and copper by plants[9,11] and thereby prevents phosphorus-induced deficiency of these micronutrients.

Zinc is one of the cations whose deficiency is fairly widespread throughout the world. In India alone, about 47% of the soils are known to be deficient in this micronutrient.[16] Several soil factors such as alkaline pH, high $CaCO_3$, poor drainage conditions, high sesquioxide (Fe and Al oxides) content (especially in ferralitic soils), and heavy doses of P fertilizers are known to induce zinc deficiency in crops and therefore result in poor yields. In such countries normal yield in deficient soils might be achieved with the use of appropriate mycorrhizal fungi. In this chapter we present evidence to show VA mycorrhizal-associated improvement in zinc uptake in higher plants.

Basically, the passage of a nutrient from soil to growing plants involves three steps. These include (1) movement of nutrient from soil to surface of growing roots, i.e., transport; (2) movement of nutrient from the surfaces to the inside of the roots, i.e., absorption; and (3) transport of absorbed nutrient from the root to shoot, i.e., translocation. Enhancement of either or both of the first

0-87371-942-5/94/$0.00+$.50

two steps causes higher uptake by plants growing under limited nutrient supply. The third step, translocation, is dependent on plant metabolism and plays an important role in supplying the nutrient to the shoots. The translocation step may only be limiting under the influence of other nutrients with an antagonistic effect on the nutrient in question. In subsequent sections, each of these steps will be discussed with reference to zinc supply in VA mycorrhizal plants.

ZINC TRANSPORT TO VA MYCORRHIZAL ROOTS

As envisioned by Barber[17] for plant nutrients, zinc can be transported to the root surface by three basic processes, namely, mass flow, root interception, and diffusion.

Mass Flow

The plant absorbs water through the root and creates a suction, under which water flows towards it. Zinc in the soil solution moves along with this flow of water to the plant roots. Zinc accumulation at the root surface in this manner is said to be mobilized by mass flow or convection. It is calculated as the product of concentration of zinc in the soil solution and the amount of water absorbed by the plant. Since the concentration of water-soluble zinc is normally very low in soil (14 to 92 μg kg^{-1} soil)[18] the contribution of mass flow to zinc supply in plants is expected to be low.[19] However, mycorrhizal plants have been shown to absorb more water than nonmycorrhizal plants,[20] which suggests that mycorrhizal plants may derive more zinc through mass flow. To estimate the relative significance of mass flow in zinc transport to mycorrhizal plants, Sharma and Srivastava[14] grew mycorrhizal *(Glomus macrocarpum)* and nonmycorrhizal greengram (*Vigna sativa* wilezek) in zinc-deficient clay loam (diethylene triamine pentaacetic acid [DTPA, pH 7.3] extractable zinc = 0.93 mg kg^{-1}) and zinc-sufficient sandy loam (DTPA extractable zinc = 1.46 mg kg^{-1}) fertilized with varying doses of zinc sulfate (0, 2.5, 5.0, and 10.0 mg Zn kg^{-1} soil). The equilibrium concentration at field capacity moisture content in soils receiving 0, 2.5, 5.0, and 10.0 mg Zn kg^{-1} soil, respectively, was 30, 32, 38, and 50 μg dm^{-3} Zn in clay loam and 100, 122, 160, and 201 μg dm^{-3} Zn in sandy loam soil. They demonstrated that VA mycorrhizal plants absorbed a greater volume of water from soil than nonmycorrhizal plants (Figure 1). The increase in the volume of water absorbed by mycorrhizal plants varied from 0.193 to 0.346 dm^{-3} in clay loam and from 0.160 to 0.282 dm^{-3} in sandy loam soil. The calculated transport of zinc to greengram plants through mass flow was relatively higher in sandy loam than clay loam, especially at higher doses. At the highest zinc dose (10 mg Zn kg^{-1} soil), the total zinc uptake by nonmycorrhizal greengram was 0.46 mg per pot for clay loam and 0.12 mg per pot for sandy loam; the uptake by mycorrhizal plants was 0.76 mg per pot for clay loam and 0.55 mg per pot for sandy loam. Though mycorrhizal greengram absorb a greater volume of water from soil, their higher zinc demand still can not be met by mass flow alone. As shown in Figure 2, at 10 mg Zn kg^{-1} soil, the percent contribution of mass flow to the total zinc uptake in mycorrhizal greengram remained at only 7.5% in clay loam and 25.6% in sandy loam; on the other hand, for nonmycorrhizal plants the contribution of mass flow to the total zinc uptake was 9.9% in clay loam and 73.5% in sandy loam.

Root Interception

In soil, plant roots grow and occupy space by displacing soil solution and also by pushing the soil particles aside. They thus come into close contact with zinc ions present in soil solution and with other cation exchange positions. The contributions of root interception to zinc supply in plants can be computed by multiplying the root volume with the zinc concentration in soil solution at a given soil moisture content. Since mycorrhizal infection often increases root volume under low status of available phosphorus,[11] mycorrhizal plants can acquire more zinc than nonmycorrhizal plants. As shown in Figure 1, in sandy loam soil testing low in available phosphorus (Olsen's P = 4.1 mg kg^{-1}) mycorrhizal greengram plants have 37.4 to 70.2% higher root volume than nonmycorrhizal controls and are likely to derive more zinc through root interception.[14] In clay loam soil which had medium available phosphorus status (Olsen's P = 12.0 mg kg^{-1} soil), mycorrhizal association did not result in a significant increase in root volume. The overall relative contribution of root interception to zinc supply to plants remains fairly low, i.e., less than 1.0% (Figure 2). Oliver and Barber[19] have also shown that root interception made a relatively small contribution to zinc supply in soybean, as compared to mass flow and diffusion.

Diffusion

When mass flow and root interception do not mobilize the required amount of zinc to plants, a depletion zone is created around the plant root. Wilkinson et al.[21] obtained autoradiographic evidence of the creation of zinc depletion zones around barley roots. Due to this depletion, zinc diffuses along a concentration gradient. This type of zinc transport to plant roots is called diffusion. The extent of apparent diffusion is estimated by subtracting the sum of the zinc transported through mass flow and root interception from the total zinc uptake. In the case of greengram, it was shown that in none of the treatments had the joint contribution of mass flow and root interception reached 100% of zinc supply to plants (Figure 2). In these cases diffusion played a critical role to zinc supply. For mycorrhizal plants, the increase in zinc uptake through mass flow and root interception, due to improved water absorption and root volume, respectively, could not meet more than 26% of the total zinc supply even under

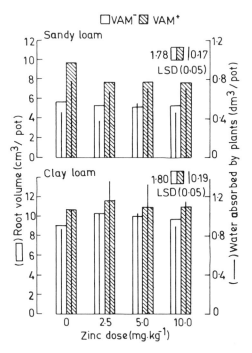

Figure 1. Effect of vesicular-arbuscular mycorrhizal inoculation and levels of zinc application on root volume and water absorption by greengram plants. Sandy loam and clay loam soils were zinc sufficient and deficient, respectively. Legends: VAM–, nonmycorrhizal; VAM+, mycorrhizal; LSD least significant differences at $p = 0.05$. (Redrawn from Sharma, A.K. and Srivastava, P.C., *Biol. Fertil. Soils*, 11, 52, 1991, with the permission of Springer-Verlag, New York.)

10 mg Zn per kilogram of soil treatment in sandy loam soil (Figure 2). The percent contribution by apparent diffusion to the total zinc uptake was consistently greater for mycorrhizal over nonmycorrhizal plants.

Thus, VA mycorrhizal plants receive a major portion of their zinc supply through diffusion rather than via mass flow or root interception. How VA mycorrhizal plants absorb the major portion of zinc by diffusion, however, needs to be considered. The VA mycorrhizal association brings about some architectural changes in the root system of host plants. Hetrick et al.[22] demonstrated that a vesicular-arbuscular fungus, *Glomus etunicatum*, associated with big blue stem *(Andropogon gerardii)*, increased the specific root length but reduced the relative amount of branching in the root system. Apparently mycorrhizal plants develop a more exploratory root pattern with extramatricular hyphae, which may permit extraction of nutrients from a larger volume of soil. Conversely, nonmycorrhizal plants maintain a more highly branched pattern of root growth and greater overlapping of nutrient depletion zones may occur around branched roots, which may become critical in direct extraction of nutrients by nonmycorrhizal plants from the soil. Beside this, Rhodes and

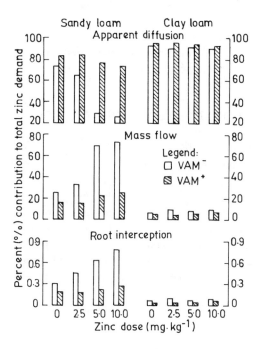

Figure 2.   Effect of vesicular-arbuscular mycorrhizal inoculation and level of zinc application on
the percent contribution by apparent diffusion, mass flow, and root interception processes to the total zinc uptake by greengram plants in two soils. Legends: VAM–,
nonmycorrhizal; VAM+, mycorrhizal. (Redrawn from Sharma, A.K. and Srivastava,
P.C., *Biol. Fertil. Soils*, 11, 52, 1991, with the permission of Springer-Verlag, New
York.)

Gerdemann[23] also reported that VA mycorrhizal roots extend extramatricular
hyphae to a distance of up to 8 cm to promote easier absorption of nutrients
from soil beyond the zone of nutrient depletion. Through extramatricular
hyphae, nutrients can be easily diffused from a distance to the root without
traversing a more tortuous path of low diffusivity through the soil. To estimate
the hyphal contribution in the acquisition of nutrients by mycorrhizal plants
from the soil, some workers[12,13,15] have employed the technique of spatial
separation between the root uptake and hyphal uptake zones. The higher
acquisition of phosphorus, zinc, and copper by mycorrhizal maize (*Zea mays*
L.),[12] and of phosphorus and copper by mycorrhizal white clover (*Trifolium
repens* L.),[13] has been ascribed to VA mycorrhizal hyphae. Kothari et al.[24]
reported that the VA mycorrhizal hyphae could transport as much as 48% of
total zinc uptake of mycorrhizal maize in a calcareous soil. Similarly, the
calculated delivery of copper by the VA mycorrhizal hyphae may range from
53 to 62% of the total copper uptake in mycorrhizal white clover.[13]

Apart from the direct involvement of VA mycorrhizal hyphae in zinc
transport to roots, an indirect effect in zinc transport can also be suspected

through a modification of the chemical environment of the rhizosphere through altered root exudation of the host plant[25] and the pH decrease. However, the relative importance of such indirect effects in zinc transport by mycorrhizal roots has to be confirmed through future experiments.

ZINC ABSORPTION BY VA MYCORRHIZAL ROOTS

After mobilization of zinc to the surface of VA mycorrhizal roots, the second step in its movement comprises the entry into the cell. Generally, the passage of zinc through limiting membranes is supposed to be mediated by a carrier(s) (transport proteins). These are believed to possess sites for zinc binding. The carrier-mediated uptake follows concentration-dependent Michaelis-Menten kinetics analogous to that of the enzymes. A typical relationship between the nutrient uptake rate and its concentration in the medium is shown in Figure 3. The mathematical relationship between V and S, analogous to Michaelis-Menten kinetics of enzymes, is as follows:

$$V = V_{max} S / (K_m + S)$$

where
V = nutrient uptake rate,
V_{max} = maximal uptake rate,
K_m = nutrient concentration at which half of the maximal rate ($V_{max}/2$) is attained, and
S = nutrient concentration.

The numerical values of V_{max} and K_m can be obtained by fitting the data on V and S in Lineweaver-Burke, Hofstee, or Woolf plots.

Sharma et al.[26] investigated the concentration dependent kinetics of zinc absorption in VA *(Glomus macrocarpum)* mycorrhizal and nonmycorrhizal root segments (0.5 to 2.5 cm behind the tip of the primary and secondary roots) of 32-day-old corn *(Zea mays* L.) in a wide concentration range (75 μmol to 1.07 mol carrier-free ^{65}Zn m^{-3} in 2 mol m^{-3} CaCl$_2$, pH 6.0) for 1.5 h at 25°C. After the uptake period, the root segments enclosed in nylon bags were rinsed in 2 mol m^{-3} CaCl$_2$ solution and subjected to desorption in an identical but unlabelled solution for 45 min. Both the rinsing and desorption experiments were carried out in ice-jacketed solutions. Sharma et al.[26] observed that uptake rates of mycorrhizal corn roots were 12.1- to 9.3-fold greater than nonmycorrhizal roots at very low zinc concentrations (below 1×10^{-6} kmol m^{-3}) in the nutrient solution (Figure 4). In the range 1×10^{-6} to 1×10^{-5} kmol m^{-3} (0.06 to 0.6 mg dm^{-3}) of zinc, mycorrhizal roots maintained more than double the zinc uptake rate than nonmycorrhizal roots. Within the concentration range studied (75 μmol to 1.07 mol m^{-3}) zinc uptake by corn roots was mediated through five distinct concentration dependent phases (Figure 5).

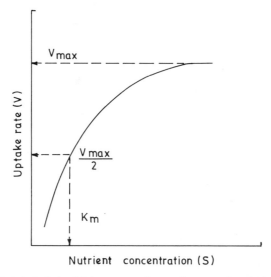

Figure 3.   A typical relationship between nutrient uptake rate and nutrient concentration.

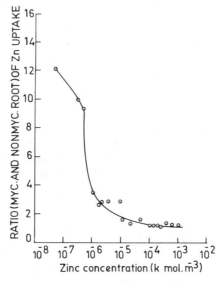

Figure 4.   Ratio of zinc uptake rate by VA mycorrhizal and nonmycorrhizal corn *(Zea mays)* roots. Zinc concentration, 7.5×10^{-8} to 1.07×10^{-3} kmol m^{-3}. (Redrawn from Sharma, A.K., Srivastava, P.C., Rathore, V.S., and Johri, B.N., *Biol. Fertil. Soils,* 13, 206, 1992, with the permission of Springer-Verlag, New York.)

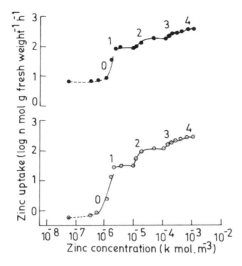

Figure 5. Isotherms (log V vs. log S for zinc uptake by VA mycorrhizal (.) and nonmycorrhizal (o) corn *(Zea mays)* roots. Numerals over the pattern of point depict zinc uptake phases. Zinc concentration range: 75 μmol to 1.07 mol m⁻³. (Redrawn from Sharma, A.K., Srivastava, P.C., Rathore, V.S., and Johri, B.N., *Biol. Fertil. Soils,* 13, 206, 1992, with the permission of Springer-Verlag, New York.)

In corn, below 1.5 mmol m^{-3} the uptake rate appeared as the lower part of the sigmoidal curve. Considering 0.05 g m^{-3} (0.76 mmol m^{-3}) as the physiologically optimum concentration of zinc, this part of the uptake isotherm represents the uptake rate of mycorrhizal roots under zinc deficiency situations. Between 1.5 to 4 mmol m^{-3} Zn the uptake rate increased linearly with the increase in the zinc concentration (phase 0). Zinc uptake rate in corn roots in the concentration range 4 mmol to 1.07 mol m^{-3} was represented by four phases: (1) 4 to 10 mmol m^{-3}; (2) 16 to 131 mmol m^{-3}; (3) 0.253 to 0.483 mol m^{-3}; and (4) 0.637 to 1.07 mol m^{-3} (Figure 5). Phases 3 and 4 showed early saturation. The transition between phase 0 and 1 was not clear as the former merged smoothly into the latter. The investigations done in our laboratory on zinc uptake kinetics in VA mycorrhizal and nonmycorrhizal (*Sorghum vulgare* Pers.) and French bean (*Phaseolus vulgaris* L.) have confirmed the multiphasic pattern for the concentration-dependent uptake kinetics of zinc.[27]

As shown in Figure 6, the Linweaver-Burke plots from the above data on maize root showed that phases 1 to 4 separately followed Michealis-Menten kinetics. A comparison of the computed values of V_{max} and K_m for the mycorrhizal and nonmycorrhizal corn roots indicated that in the relatively low concentration range (4 to 10 mmol m^{-3}) greater uptake by mycorrhizal roots could be attributed to larger numbers of uptake/carrier sites, i.e., higher value of V_{max} for inoculated roots (Table 1). In the higher concentration range (16 mmol to 1.07 mol m^{-3}) the greater specificity of the carrier (lower K_m values) was perhaps more important for the increased zinc uptake by mycorrhizal corn roots.

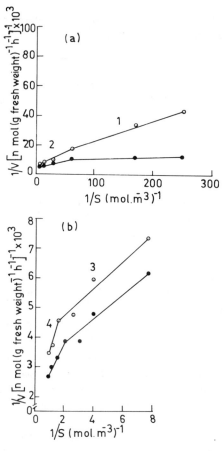

Figure 6. Lineweaver-Burke plots for zinc uptake by VA mycorrhizal (.) and nonmycorrhizal corn *(Zea mays)* roots: (a) zinc concentration of 4 to 131 mmol m⁻³; (b) zinc concentration of 0.254 to 1.07 mol m⁻³. (Redrawn from Sharma, A.K., Srivastava, P.C., Rathore, V.S., and Johri, B.N., *Biol. Fertil. Soils,* 13, 206, 1992, with the permission of Springer-Verlag, New York.)

Bowen et al.[28] reported that higher uptake of zinc by endomycorrhizal *Araucaria cunninghamii* was due to the greater absorbing power of the root surface. These researchers also observed very high zinc uptake in young, expanding tissue near the apex of rapidly growing roots. However, these workers doubted the generalization that mycorrhizal roots always exhibit higher zinc absorption rates than nonmycorrhizal roots. In the study carried out by Sharma et al.[26] identical parts of roots were used which had higher zinc absorption in mycorrhizal corn roots. Using the weighted mean square of the deviate as a measure of goodness of fit and also a run test for computing the probability of an uneven distribution of observation points, they further confirmed that the discontinuous isotherms of zinc uptake kinetics obtained for

Table 1.   Concentration Range and Kinetic Constants for Different Phases of Zinc Uptake by Mycorrhizal and Nonmycorrhizal Corn (Zea mays) Roots

Phase	Concentration range (mmol m^{-3})	V_{max} (g^{-1} FW h^{-1}) VAM+	V_{max} VAM−	K_m (mmol Zn m^{-3}) VAM+	K_m VAM−
0	1.5–4.0	—	—	—	—
I	4–10.0	83+/−6	30+/−2		
II	16–131	213+/−12	185+/−10	20+/−2	34+/−1
III	254–483	280+/−17	276+/−15	130+/−3	140+/−3
IV	637–1070	474+/−28	555+/−22	380+/−24	950+/−30

From Sharma, A.K., Srivastava, P.C., Rathore, V.S., and Johri, B.N., *Biol. Fertil. Soils,* 13, 206, 1992, with permission of Springer-Verlag, New York.

mycorrhizal and nonmycorrhizal roots could only be accounted for by a multiphasic model[29,30] consisting of many distinct phases, each following Michaelis-Menten kinetics, rather than a single + diffusion (a single saturable phase with a linear diffusion-controlled component) or dual + diffusion (two saturable phases with a linear diffusion-controlled component) models.[31-33]

Experiments done in our laboratory also show that zinc uptake in both mycorrhizal and nonmycorrhizal corn (*Zea mays* L.) root segments at 0.76 mmol m^{-3} Zn appears to be metabolic; in the presence of 10^{-2} mol m^{-3} 2,4-dinitrophenol there was 48.6 and 55.5% suppression in the zinc uptake rate, respectively.[27] Bowen et al.[28] also demonstrated that metabolically mediated absorption of zinc was 2.6 times greater for endomycorrhizal roots than uninfected roots of *Araucaria cunninghamii*. In the context of zinc absorption by VA mycorrhizal roots, certain points are not yet clear. For example, is the increased zinc uptake in VA mycorrhizal roots a result of extramatrical fungal hyphae or some physiological/ultrastructural modification of the host root cells? Cooper and Tinker[34] examined the uptake and translocation of zinc in VA mycorrhizal white clover *(Trifolium repens)* plants and concluded that the VA mycorrhizal hyphae were involved in transport of zinc to mycorrhizal plants, and the amount of transport was governed by the demand of the host plant rather than by the amount of the absorbing mycelium. However, the factors that regulate the transfer of zinc absorbed by mycorrhizal hyphae to the host cells, and which particular mechanism is involved in such transfer, is not known with certainty. Since mycorrhizal hyphae are known to translocate phosphorus mainly as polyphosphates,[35] zinc may bind with polyphosphates like other cations, viz., calcium, iron, and manganese,[36] and can be translocated with phosphorus into the plant roots. If this is so, the amount of zinc translocated may be limited by the amount of polyphosphate in the mycorrhizal hyphae or by the amount of zinc available to hyphae.[24] The transfer rate of zinc from fungal hyphae to the host may also be regulated by the growth rate of the host plant and the shoot demand for zinc. Further physiological investigations on the translocation of zinc in VA mycorrhizal plants may provide additional information on the contribution of VA mycorrhizae to zinc nutrition of plants.

Thus, VA mycorrhizal association augments zinc nutrition of plants under deficient situations by increasing its transport and subsequent uptake by roots. The VA mycorrhizal symbiosis may also offer better scope for managing zinc nutrition in plants and higher yields from zinc-deficient areas of poor productivity.

REFERENCES

1. Gerdemann, J.W., Vesicular-arbuscular mycorrhizae, in *The Development and Function of Roots,* Torrey, J.G. and Clarkson, D.T., Eds., Academic Press, London, 1975.
2. Gianinazzi-Pearson, V. and Gianinazzi, S., The physiology of vesicular-arbuscular mycorrhizal roots, *Plant Soil,* 71, 197, 1983.
3. Hayman, D.S., The physiology of vesicular-arbuscular endomycorrhizal symbiosis, *Can. J. Bot.,* 61, 944, 1983.
4. Jefferies, P., Use of mycorrhizae in agriculture, *Crit. Rev. Biotechnol.,* 5, 319, 1987.
5. Bowen, G.D., Bevege, D.I., and Mosse, B., Phosphate physiology of vesicular arbuscular mycorrhizas, in *Endomycorrhizae,* Sanders, F.E., Mosse, B., and Tinker, P.B., Eds., Academic Press, New York, 1975.
6. Sanders, F.E., Mosse, B., and Tinker, P.B., *Endomycorrhizas,* Academic Press, New York, 1975.
7. Abbott, L.K. and Robson, A.D., Growth stimulation of subterranean clover with vesicular-arbuscular mycorrhizas, *Aust. J. Agric. Res.,* 28, 69, 1977.
8. Gilmore, A.E., The influence of endotrophic mycorrhizae on the growth of pea seedlings, *J. Am. Soc. Hortic. Sci.,* 96, 35, 1971.
9. Lambert, D.H., Baker, D.E., and Cole, H., The role of mycorrhizae in the interactions of phosphorus with zinc, copper and other elements, *Soil Sci. Soc. Am. J.,* 43, 976, 1979.
10. Swaminathan, K. and Verma, K.C., Responses of three crops species to vesicular-arbuscular mycorrhizal infection on zinc deficient Indian soils, *New Phytol.,* 82, 481, 1979.
11. Gnekow, M.A. and Marschner, H., Role of VA mycorrhiza in growth and mineral nutrition of apple *(Molus pumila* var. *domestica)* rootstock cuttings, *Plant Soil,* 119, 285, 1989.
12. Kothari, S.K., Marschner, H., and Römheld, B., Direct and indirect effect of VA mycorrhizal fungi and rhizosphere microorganisms on acquisition of mineral nutrients by maize *(Zea mays* L.) in a calcareous soil, *New Phytol.,* 116, 637, 1990.
13. Li, X.L., Marschner, H., and George, E., Acquisition of phosphorus and copper by VA mycorrhizal hyphae and root to shoot transport in white clover, *Plant Soil,* 136, 49, 1991.

14. Sharma, A.K. and Srivastava, P.C., Effect of VAM inoculation on dry matter yield, total zinc uptake of moongbean (*Vigna radiata* L.) and zinc supply processes in soils, *Biol. Fertil. Soils,* 11, 52, 1991.

15. Sreenivasa, M.N., Selection of an efficient vesicular-arbuscular mycorrhizal fungus for chilli (*Capsicum annuum* L.), *Sci. Hortic.,* 50, 53, 1992.

16. Anon., XIIth Annu. Rep. All India Coordinated Scheme Micronutrients of Soils (1979–80), Indian Council of Agricultural Research, New Delhi, 1980.

17. Barber, S.A., A diffusion and mass flow concept of soil nutrient availability, *Soil Sci.,* 93, 39, 1962.

18. Srivastava, P.C., Kumar, S., Rathore, V.S., and Gangwar, M.S., Chemical fractions of labile zinc pool in relation to zinc uptake by rice in mollisols, *J. Nucl. Agric. Biol.,* 17, 39, 1988.

19. Oliver, S. and Barber, S.A., Mechanism for movement of Mn, Fe, B, Zn, Cu, Al, and Sr from one soil to the surface of soybean roots, *Proc. Soil Sci. Soc. Am.,* 30, 368, 1966.

20. Safir, G.R., Broyer, J.S., and Gerdemann, J.W., Nutrient status and mycorrhizal enhancement of water transport in soybean, *Plant Physiol.,* 49, 700, 1972.

21. Wilkinson, H.F., Loneragan, J.F., and Quirk, J.P., The movement of zinc to plant roots, *Proc. Soil Sci. Soc. Am.,* 32, 831, 1968.

22. Hetrick, B.A.D., Leslie, J.F., Wilson, G.T., and Kitt, D.G., Physical and topological assessment of effects of a vesicular-arbuscular mycorrhizal fungus on root architecture of big glue stem, *New Phytol.,* 110, 85, 1988.

23. Rhodes, L.H. and Gerdemann, J.W., Phosphate uptake zones of mycorrhizal and nonmycorrhizal onion, *New Phytol.,* 75, 551, 1975.

24. Kothari, S.K., Marschner, H., and Römheld, V., Contribution of VA mycorrhizal hyphae in acquisition of phosphorus and zinc by maize grown in a calcareous soil, *Plant Soil,* 131, 177, 1991.

25. Mosse, B., Advances in the study of vesicular-arbuscular mycorrhizae, *Annu. Rev. Phytopathol.,* 11, 171, 1973.

26. Sharma, A.K., Srivastava, P.C., Rathore, V.S., and Johri, B.N., Kinetics of zinc uptake by mycorrhizal (VAM) and nonmycorrhizal corn (*Zea mays* L.) roots, *Biol. Fertil. Soils,* 13, 206, 1992.

27. Sharma, A.K., Srivastava, P.C., Rathore, V.S., and Johri, B.N., unpublished data, 1992.

28. Bowen, G.D., Skinner, M.F., and Becete, D.I., Zinc uptake by mycorrhizal and uninfected root of *Pinus radiata* and *Araucaria cunninghamii, Soil Biol. Biochem.,* 6, 141, 1974.

29. Nissen, P., Uptake of sulphate by roots and leaf slices of barley, mediated by single multiphasic mechanisms, *Physiol. Plant.,* 24, 315, 1971.

30. Nissen, P. and Nissen, O., Validity of multiphasic concept of ion absorption on plants, *Physiol. Plant.,* 57, 47, 1983.

31. Epstein, E., Rains, D.W., and Elzam, D.E., Resolution of dual mechanism of potassium absorption by barley roots, *Proc. Natl. Acad. Sci. U.S.A.,* 49, 684, 1963.

32. Kochain, L.V. and Lucas, W.J., Potassium transport in corn roots. I. Resolution of kinetics into a saturable and a linear component, *Plant Physiol.,* 70, 1723, 1982.

33. Borstlap, A.C., The use of model-fitting in interpretation of 'dual uptake' isotherms, *Plant Cell Environ.*, 6, 407, 1983.

34. Cooper, K.M. and Tinker, P.B., Transfer and translocation of nutrients in vesicular-arbuscular mycorrhizas. II. Uptake and translocation of phosphorus, zinc, and sulphur, *New Phytol.*, 81, 43, 1978.

35. Cox, G., Sanders, F.E., Tinker, P.B., and Wild, J.A., Ultrastructural evidence relating to host-endophyte transfer in a vesicular-arbuscular mycorrhiza, in *Endomycorrhizas*, Sanders, F.E., Mosse, B., and Tinker, P.B., Eds., Academic Press, New York, 1975.

36. White, J.A. and Brown, M.F., Ultrastructural and X-ray analysis of phosphorus granules in a vesicular-arbuscular mycorrhizal fungus, *Can. J. Bot.*, 57, 2812, 1979.

CHAPTER 10

Control of Mn Status in Plants and Rhizosphere: Genetic Aspects of Host and Pathogen Effects in the Wheat Take-All Interaction

Zdenko Rengel, Judith F. Pedler, and Robin D. Graham

INTRODUCTION

The soilborne ascomycete *Gaeumannomyces graminis* (Sacc.) v. Arx and Oliver var. *tritici* Walker *(Ggt)* is the etiologic agent of take-all, a root-rotting disease that causes large economic losses in wheat and barley production worldwide.[1] In Australia alone, take-all costs cereal growers about AUS$200 million in lost yield annually.[2] The problem may increase in the future because of an increased reliance on minimum-tillage practices to reduce soil degradation. Reduced tillage may increase the incidence and severity of the take-all disease of cereals with a corresponding loss in yield.[3]

In the field, take-all is observed as patches of severely stunted plants that senesce prematurely and show characteristic "white-heads" containing either shriveled grain or no grain at all.[4] Roots of diseased plants have brown to black necrotic lesions caused by accumulation of fungal hyphae in the stele. The blocked stele prevents upward transport of water and nutrients in the xylem elements. In addition, invasion and disruption of the phloem deprives apical root meristems of assimilate supply so they cease to elongate and die.[5] The devastating effect of take-all on the host plant is due to loss of its functional root system since the *Ggt* fungus produces no toxin.[6]

MEASURES TO CONTROL TAKE-ALL

A classical approach in breeding for resistance has not been fruitful in the case of take-all since only limited variability in susceptibility of wheat genotypes

0-87371-942-5/94/$0.00+$.50

to take-all has been demonstrated. Resistance has therefore been sought in related species. Rye[7] and some wild cereal relatives[8] have take-all resistance genes that could be transferred to wheat. Laboratory tests for resistance to take-all have been developed,[9] but the results of these tests will eventually have to be repeated in the field. However, differences among wheat genotypes in the frequency of white-heads due to take-all are not consistent in different environments.[10]

Control of take-all by applying fungicides[11] or biological control agents such as fluorescent pseudomonads[12] to seed prior to seeding has lacked consistency in suppressing disease in the field. Since successive wheat cropping increases the incidence of take-all,[13,14] some degree of disease control in the field can be achieved by crop rotation,[15] while managing the wheat stubble may[16] or may not[3] be effective in reducing the disease.

The lack of effective sources of cultivar resistance or chemical control of take-all shifted emphasis to fertilization management as a way of reducing crop damage. Well-nourished plants suffer less damage than plants of poor nutrient status. Such a result may simply be due to well-nourished plants producing more roots and consequently "escaping" from the fungal attack, rather than roots of well-nourished plants being more resistant to disease (infection of roots was not assessed in most of the earlier studies, for example see Reference 17). However, more recent research that has measured root infection showed that fertilization of wheat with NH_4–N, Mg, P, K, and various micronutrients reduced the incidence and severity of take-all in both controlled environments and in the field[13,18-28] (for earlier literature see Reference 29).

Fertilization with NH_4–N not only improves the nutrition of a host plant to render it more resistant to *Ggt*, but also changes populations of the soil microflora, such as fluorescent *Pseudomonas* spp. antagonistic to take-all.[24,25] While a mechanistic explanation for the role of NH_4–N in reducing take-all is lacking, it appears that an increase in fluorescent pseudomonads antagonistic to take-all is related both to the unknown products of root necrosis[30] and to effects of NH_4–N fertilization per se.[24,25]

Interactions between micronutrients and the take-all fungus are also very complex. Generally, nutrient-deficient plants are highly susceptible to take-all, and alleviating the deficiency then increases resistance to infection by this fungus.[22] However, supplying micronutrients in amounts greater than those needed for optimum plant growth does not cause a further reduction in infection of wheat roots[18,19] until, in the extreme, additions are large enough to deleteriously affect the saprophytic growth of the fungus.[31]

Of all the nutrient elements mentioned for their influence on take-all, we chose Mn for further study in our laboratory because soil conditions conducive to Mn deficiency are the same as those conducive to the take-all disease.[32,33] No other element fits this criterion as closely.

MANGANESE AND TAKE-ALL

Mn-deficient wheat plants are more susceptible to take-all;[27,33,34] increased availability of Mn in soil and its uptake by plants reduces severity of the disease.[31] Graham and Rovira[34] suggested three possible mechanisms by which Mn^{2+} increases plant resistance to take-all:

1. Mn^{2+} may be directly toxic to the fungus itself.
2. Mn^{2+} may act through the physiology of the plant by increasing the photosynthetic rate with a corresponding increase in the carbon supply to roots and increased exudation of soluble organic compounds into the soil, therefore affecting rhizosphere microflora.
3. Mn^{2+} may increase synthesis of ligneous defense products in roots.

Toxicity of Mn to the *Ggt* Fungus

The high tolerance of *Ggt* to Mn indicates that the mechanism of a Mn-related decrease in take-all does not reside in direct toxicity of Mn to the fungus.[31,32] The same conclusion is supported by the reduction in take-all achieved with increasing seed Mn content,[27] which eliminates Mn deficiency in the plant[27,35,36] but which has little impact on soil Mn.

Mn Effects on Rhizosphere Microflora

The hypothesis that Mn increases resistance to take-all through increasing the quantity or changing the composition of root exudates does not appear to have been tested; however, some indirect evidence suggests that this hypothesis may be of only secondary importance. While Graham and Rovira[34] suggested that root exudates may influence soil microflora in a way that would result in increased competition with the *Ggt* fungus, and therefore reduce its potential to infect cereal roots, Wilhelm[26] found that sterile excised root pieces precultured with Mn were more resistant to *Ggt* penetration than sterile root pieces precultured without Mn. Clearly, in the latter example, the Mn-related increase in resistance was observed without the presence of other microorganisms. However, rhizosphere microflora may play a secondary role in the cereal/take-all interaction, especially in conjunction with Mn availability or fertilization.[37,38]

Mn Effects on Root Phenolics and Lignin

The third hypothesis of Graham and Rovira[34] suggested an interaction between Mn and the production of phenolics and lignitubers in roots as a mechanism of Mn-influenced increase in resistance to take-all. Soluble phenolics

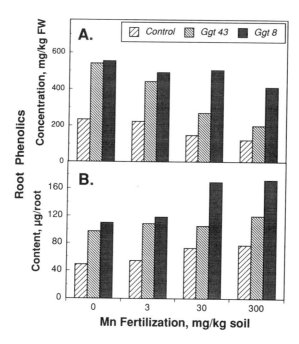

Figure 1.   Effects of Mn fertilization and *Ggt* inoculum on concentration (A) and total content
(B) of soluble phenolics in wheat roots grown for 28 days in Wangary sand in a growth
chamber.[46] Data were averaged over all four genotypes tested (Bayonet, Millewa,
Aroona, and C8MM) because the variable genotype was only marginally significant
($p < 0.055$) for phenolics concentration. Tukey's $HSD_{0.05}$ values for phenolics concentration were (in mg/kg root fresh weight): Mn, 66; inoculum, 52; Mn × inoculum, 144;
and for total phenolics content (in mg/root): Mn, 18; inoculum, 14; genotype, 18; Mn
× inoculum, 41. (From Rengel, Z., Graham, R.D., and Pedler, J.F., *Plant Soil,* 151,
255, 1993. With permission from Kluwer Academic, Dordrecht.)

(derived from the intermediates of the shikimic acid pathway) and lignin
(formed by condensation of 4-hydroxylated cinnamyl alcohols) have an important role in plant defense against various pathogens.[39-41] Phenylalanine ammonia-lyase (an enzyme involved in metabolism of phenolics) and peroxidase (an
enzyme involved in production of H_2O_2 that is necessary for polymerization of
monomeric units into lignin chains) are stimulated by or dependent on Mn^{2+}
ions.[42,43] A deficiency of Mn may therefore prevent plants from effectively
synthesizing phenolics and lignin, thereby rendering them more susceptible to
fungal infection.[40,44]

Concentration and Content of Phenolics and Lignin

Mn deficiency reduced the concentration of soluble phenolics and lignin in
noninfected wheat roots grown in nutrient solution.[45] However, a somewhat
different effect was observed in a recent study with soil-grown wheat roots
infected by the *Ggt* fungus.[46] In this study, plants deficient in Mn had greater

Table 1.   Root Lesions and Lignin Content in Root Tissue of Four Wheat Genotypes Grown in Sandy Wangary Soil in a Growth Chamber for 28 Days[46]

Variable	Total length of *Ggt* lesions (mm)	Lignin content[a] (Abs_{280}/root system)
Mn, mg/kg soil		
0	38	0.14
3	28	0.12
30	23	0.25
300	22	0.28
Genotype[b]		
Bayonet	30	0.14
Millewa	27	0.16
Aroona	26	0.22
C8MM	23	0.27
Tukey's $HSD_{0.05}$		
Mn	6	0.07
Genotype	6	0.07

Note: Mn fertilization and the genotype effect caused a similar trend in total length of lesions when roots were infected with a highly virulent isolate (*Ggt* 8) or a weakly virulent isolate (*Ggt* 43); data presented here were averages for both *Ggt* isolates.

[a] For lignin content, data were pooled over the two *Ggt* isolates (8 and 43) and no *Ggt* treatment because the inoculum variable was nonsignificant ($p < 0.402$).

[b] Mn-inefficient: Bayonet and Millewa; Mn-efficient: Aroona and C8MM; see also Table 3).

concentrations (but accumulated smaller total amounts) of phenolics (Figure 1) and lignin (Table 1) in their roots than plants grown in Mn-sufficient conditions.[46]

The relationship between Mn and the production of phenolics and lignin in soil-grown wheat roots infected by the take-all fungus is complicated by the fact that the accumulation of phenolics in roots is greatly elicited by take-all infection (a phenomenon which is common in plant-pathogen interactions[40,41,47]), even under Mn-deficient conditions. In the study with four wheat genotypes grown in sandy Wangary soil fertilized with 0, 3, 30, or 300 mg Mn per kilogram of soil and infected with various *Ggt* strains,[46] the concentration of soluble phenolics in roots was proportional ($r = 0.85$) to the total length of root stelar lesions as an index of take-all infection (Figure 2).

Increased accumulation of cell-wall lignin is often a cause of increased resistance to pathogen penetration.[48,49] In the case of take-all infection, host cells deposit lignin on the innermost surface of their cell walls around the site of hyphal penetration[6] to form closed tubular structures. These structures (lignitubers) slow down hyphal penetration rather than completely prevent it.[50]

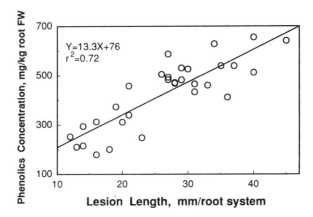

Figure 2. The relationship between the total length of root-stellar lesions and root phenolics concentration. Data from the Wangary-soil experiment with four wheat genotypes (see Figure 1), two *Ggt* isolates (*Ggt* 8 and *Ggt* 43), and four Mn treatments (0, 3, 30, and 300 mg Mn per kilogram of soil) are shown.[46] Every point represents the average of four replicates. (From Rengel, Z., Graham, R.D., and Pedler, J.F., *Plant Soil,* 151, 255, 1993. With permission from Kluwer Academic, Dordrecht.)

Shorter root lesions developed on roots that accumulated more lignin (Figure 3).[46] Lignin concentration dropped (probably as a consequence of "dilution" due to greater production of dry matter, data not shown) while total lignin content in roots increased due to Mn fertilization (Table 1). Mn-efficient genotypes (C8MM and Aroona) accumulated more lignin in roots than Mn-inefficient genotypes (Bayonet and Millewa). There was a *Ggt* isolate-dependent negative relationship (r = –0.85 for *Ggt* 8 and r = –0.84 for *Ggt* 43) between total lesion length and root lignin content (Figure 3). Slopes of the two lines representing the relationship were the same, while the highly virulent isolate, *Ggt 8,* caused longer root lesions at a particular lignin content (greater intercept, Figure 3) than the weakly virulent isolate, *Ggt* 43.

Rate of Accumulation of Phenolics and Lignin

In another experiment,[51] wheat seedlings of the Mn-inefficient genotype Bayonet and the Mn-efficient genotype C8MM were grown in nutrient solution without added Mn for 18 days, and then transferred to a Mn-deficient sandy soil (Wangary) fertilized with Mn at 0 or 30 mg/kg. At the time of transfer from nutrient solution to soil, plants were severely deficient in Mn as revealed by visual symptoms as well as by measuring the induction kinetics of chlorophyll a fluorescence on the adaxial surface of attached youngest emerged blades according to Simpson and Robinson[52] (data not shown). Upon transfer of plants from the 0 mg Mn nutrient solution to soil, concentrations (data not shown) as well as accumulation rates of phenolics and lignin dropped (Figure 4), possibly because of a different growth pattern that might have been due to reduced

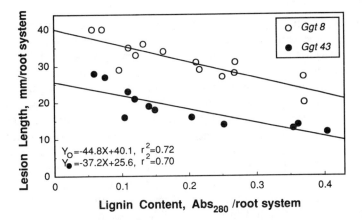

Figure 3.   The relationship between root lignin content and total length of root-stellar lesions caused by *Ggt* isolates 8 (highly virulent) and 43 (weakly virulent).[46] Data on all four wheat genotypes (see Figure 1) and four Mn treatments (see Figure 2) are combined. Every data point represents the average of four replicates. (From Rengel, Z., Graham, R.D., and Pedler, J.F., *Plant Soil,* 151, 255, 1993. With permission from Kluwer Academic, Dordrecht.)

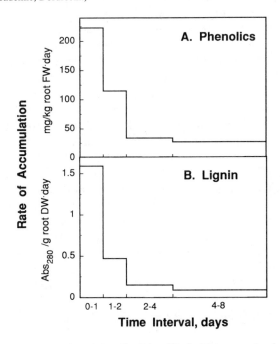

Figure 4.   Rate of accumulation of phenolics (A) and lignin (B) in roots of two wheat genotypes (Bayonet and C8MM) in consecutive time intervals during 8 days growth in rhizotrons in sandy Wangary soil fertilized with either 0 or 30 mg Mn per kilogram of soil.[51] Plants were pregrown for 18 days in nutrient solution without added Mn and then transferred to soil. Data were averaged over all variables that were nonsignificant. Tukey's $HSD_{0.05}$ values were 160 and 0.61 for the rate of accumulation of phenolics and lignin, respectively.

supply of essential nutrients from soil compared to the supply from the solution. The same result was observed regardless of Mn fertilization, indicating that Mn was not the factor limiting accumulation of phenolics and lignin.

The positive relationship between the rates of accumulation of phenolics and lignin (Figure 5) is in accordance with the synthesis of lignin through polymerization of phenolic monomers.[39] Interestingly, the slope of this relationship was steeper for C8MM than for Bayonet, indicating more efficient mechanisms for conversion of phenolics into lignin in the Mn-efficient and take-all resistant genotype C8MM. A first inspection suggests that this result is unlikely to be related to root Mn since no difference was found between the two genotypes in Mn concentration in their root tissues (data not shown). However, the roots of the two genotypes may have quite different amounts of inactive (precipitated) Mn in the root cell walls, as described later; alternatively, a differential efficiency in the conversion of phenolics into lignin may exist, i.e., caused by differences in Mn-peroxidase activity between the two genotypes. Further research to elucidate this point is warranted.

It appears that the limited supply of residual Mn in soil not amended with Mn was sufficient for unimpeded biosynthesis of phenolic compounds as precursors for lignin formation. In the 0 mg Mn treatment, the concentration of soluble phenolics significantly increased due to infection with the take-all fungus (234 vs. 554 mg/kg root fresh weight), and lignin concentration was higher in the 0 mg than in the 300 mg/kg Mn soil treatment (8.7 vs. 5.4 L/cm·g).[46] A low Mn requirement for phenolic biosynthesis and lignin formation is therefore suggested. In the other experiment,[51] relatively high concentrations of phenolics and lignin in roots after 18 d of growth in the 0 mg Mn solution (311 mg phenolics per kilogram of root fresh weight and 3.1 L/cm·g for lignin) and low, highly deficient Mn concentrations in roots (3.1 mg/kg Mn dry weight) and shoots (4.8 mg/kg Mn dry weight) (the critical concentration of Mn in wheat leaf tissue is 11 to 16 mg/kg[53]) at the same time also support the indication of a low internal Mn requirement for phenolics and lignin biosynthesis in wheat roots. Mn could possibly have been replaced as a cofactor in enzymes involved in the biosynthesis of phenolics and lignin by other divalent cations (mostly Mg^{2+}, see Reference 44) under these circumstances. However, the possible short-term effects of Mn on the biosynthesis of phenolics and lignin and, consequently, on resistance to take-all, need to be looked at more closely.

MANGANESE OXIDATION BY THE *Ggt* FUNGUS

Although macronutrient requirements of *Ggt* have been assessed,[54] and enzyme activity evaluated,[55,56] these have done little to link the etiology and habit of the fungus with the widely varying virulence of isolates, and the apparently random nature of many outbreaks of the take-all disease. Mn may provide such a link with regard to both host and pathogen.

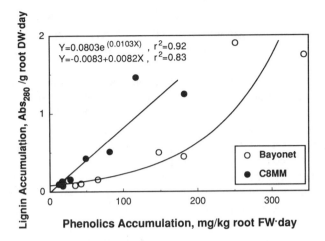

Y=0.0803e$^{(0.0103X)}$, r^2=0.92
Y=-0.0083+0.0082X , r^2=0.83

○ Bayonet
● C8MM

Lignin Accumulation, Abs$_{280}$ /g root DW·day

Phenolics Accumulation, mg/kg root FW·day

Figure 5. The relationship between the rates of accumulation of phenolics and lignin in root tissue of two wheat genotypes during 8 d of growth (19- to 26-d-old plants) in sandy Wangary soil fertilized with Mn at 0 or 30 mg/kg soil.[51] Every point represents the average of four replicates.

The *G. graminis* fungi, in particular *Ggt,* are Mn oxidizers;[26,57,58] oxidation of Mn^{2+} in both agar and soil has been observed. Fungal isolates vary in oxidizing power as well as in the pattern and intensity of oxidation. The oxide, MnO$_2$, may appear as dark nodes on the hyphae, as concentric dark rings as the fungus grows on agar plates, or as a diffuse precipitate throughout the agar medium. The oxide may form in the medium beyond the growth of the hyphae. In soil, nodules of MnO$_2$ were identified at *Ggt* infection sites on wheat roots.[26] Furthermore, using a *G. graminis* var. *graminis* isolate, the Mn oxide precipitates have been shown to be associated specifically with hyphopodia and specialized hyphae, but not with runner hyphae.[58]

Relationship Between Mn-Oxidizing Capacity and Virulence of *Ggt* Isolates

There is evidence of a positive correlation between the Mn-oxidizing capacity and the virulence of *Ggt* isolates.[57] Isolates of *Ggt* were screened for virulence on wheat in pots containing Mn-deficient soil.[59] An agar disc (taken from the edge of a vigorously growing fungal colony and completely covered with *Ggt*) of the same size as the pot diameter was placed about 30 mm below the seed. In this system, roots could not escape contact with the inoculum, thus providing a test of virulence since infection of the roots is only escaped by the weakness of the isolate. After 3 to 4 weeks, plant tops were not visibly affected by the inoculum; virulence of the fungal isolates was measured as the total length of black root lesions. In addition, *Ggt* isolates were ranked for

Mn-oxidizing capacity after growing them for 14 days at 20°C on plates of weak potato-dextrose agar supplemented with $MnSO_4$ to give 50 mg/kg Mn.

Initially, 14 isolates of *Ggt* from South Australian soils were tested; a positive relationship ($r^2 = 0.59$) was obtained between the ranking of isolates for Mn oxidation and the total length of lesions caused by different *Ggt* isolates on wheat roots.[57] Subsequently, using subcultures of *Ggt* isolates from the initial experiment and an additional 11 isolates from New South Wales soils, this positive relationship was confirmed (Table 2). Interestingly, some of the isolates which had been subcultured showed an attenuation, not only of virulence (widely recognized to occur through subculturing), but also of Mn-oxidizing capacity. The clearest comparison between the subcultured *Ggt* 8 and the nonsubcultured *Ggt* 8 isolates showed a distinct reduction in both virulence and oxidation scores due to subculturing (Table 2). Variation in virulence of *Ggt* isolates has not previously been clearly correlated to any function of the fungus, including enzyme production,[56] melanization or the paleness of hyphae, or growth rate.[60]

Temperature has a strong effect on both virulence and Mn-oxidizing capacity of *Ggt* isolates from Indiana soils (U.S.).[61] Oxidation capacity and virulence were positively related to temperature for some isolates (an enhancement in both parameters with an increase in temperature from 15 to 25°C), while this relationship was negative for other isolates. The temperature-dependence of the virulence and Mn oxidation capacity of these *Ggt* isolates may have been due to seasonal climate differences in North America and a greater range of environmental niches available for the isolates to exploit. Temperature conditions have previously been shown to cause differential infection of cereal roots by *Ggt* and *Phialophora graminicola*,[62] with *Ggt* causing more infection at lower temperatures. The differences in Mn-oxidation capacity of *Ggt* isolates could account for some of the temperature effects on their virulence.

Ggt isolates are also known to vary in virulence with respect to the host from which they were isolated, with more virulent strains often being isolated from plants more resistant to take-all, such as rye and ryegrass.[63] Incidentally, these more resistant plants are usually also more tolerant to Mn-deficient conditions.[64]

Mechanism of Mn Oxidation By the *Ggt* Fungus

Oxidation of Mn is widely known in soil fungi and bacteria;[65] several mechanisms for induction of Mn oxidative capabilities have been proposed, including defense against the potential toxic effects of Mn^{2+} and utilization of energy released upon oxidation.

A variety of fungi require Mn as an essential nutrient, with optimum growth achieved at around 1 mg/kg.[66] In the case of *Ggt*, growth was minimal in soil where the available Mn (determined after DTPA extraction) was 0.008 mg/kg dry soil, with all other nutrients present in nonlimiting amounts.[83] Addition of

Table 2. Comparison of Mn-Oxidizing Capacity with Virulence of Selected Ggt Isolates

Ggt isolate[a]	First experiment Mn-oxidation rank[b]	First experiment Lesion length[c] (mm/plant)	Second experiment Mn-oxidation rank[b]	Second experiment Lesion length[c] (mm/plant)
Control[d]	0	0	0	0
82	2	4	1	0
233	7	8	14	27
43	11	17	5	2
8 (Subcultured)	13	17	11	4
500	14	21	20	39
8 (Not subcultured)	—	—	23	41

[a] Results are drawn from an initial experiment using 14 isolates[57] and a second experiment of 26 isolates[83] including the initial 14 (which had undergone several subculturings in the interim) and an isolate of Ggt 8 which had not been subcultured (autoclaved rye seed inoculated with the Ggt 8 and kept dry at $-2°C$ for the duration of subculturing of other isolates).

[b] Visual ranking of the oxidation capacity ranges from no Mn oxide formed (0) to the greatest formation (14 in the initial and 26 in the second experiment).

[c] The total lesion length (measured by the total length of root lesions and averaged over ten replicates) was significantly related to the Mn-oxidizing capacity ($r^2 = 0.59$ for the 1st and $r^2 = 0.53$ for the 2nd experiment).

[d] The control for both the Mn-oxidizing capacity and the virulence test was uninoculated potato-dextrose agar.

4.5 mg Mn per kilogram of dry soil doubled the growth of hyphae (data not shown).

Nothing is known about the relationship between soil Mn levels and Mn oxidation by *Ggt* isolates. In contrast, oxidation of Mn by virulent isolates of *Ggt* grown on agar was observed where more than 40 mg/kg Mn was present,[83] while the levels at which addition of Mn to agar begins to stunt *Ggt* growth are 70 to 100 mg/kg,[26] and growth is not completely halted until Mn levels reach 400 mg/kg agar. Therefore, the existence of a Mn-oxidizing capability, at least by agar-grown *Ggt,* is unlikely to be solely as a defense against toxic levels of Mn^{2+}.

Mn Oxidation and Pathogenic Action of *Ggt*

Oxidation of Mn by *Ggt* may be an important part of its pathogenic action, perhaps through affecting the Mn status of the plant and developing a localized Mn deficiency in the rhizosphere. Depletion of plant-available Mn in the rhizosphere by oxidizing microorganisms can cause Mn deficiency in plants,[67,68] and weakened, Mn-deficient plants are more susceptible to *Ggt*[26-28,69] (see also Table 1). If the *Ggt* fungus causes the depletion of Mn in the rhizosphere, its pathogenic action may be much more subtle than previously thought. Combatting such an action would require plant roots to be less susceptible to Mn deficiency, a possibility which is discussed later in this chapter.

DIFFERENCES AMONG CROP GENOTYPES IN TOLERANCE TO MN DEFICIENCY

Soils known to cause Mn deficiency on susceptible crops are usually impoverished siliceous and calcareous sandy soils of neutral or higher pH, which favor chemical and microbial oxidation and immobilization of soluble (i.e., available) Mn^{2+}. Even these soils generally contain large reserves of total Mn relative to that removed in harvest, so that Mn deficiency of susceptible crops is due to insufficient availability of soil Mn to plants, rather than an absolute shortage of soil Mn. Discussion on soil biogeochemistry of Mn is beyond the scope of this chapter, and the reader is referred to other sources.[70-73]

On Mn-deficient soils, Mn fertilizers have remarkably low residual value (they are quickly converted to the plant-unavailable forms) making them unsuitable for correction of deficiency. An alternative strategy for correcting problems associated with Mn-deficient soils is to breed crop cultivars for greater tolerance to this soil condition. Genotypic differences in tolerance to Mn-deficient soils have been recognized since the deficiency itself was first identified in the 1920s; it has been reported widely that such differences, at least partly, are under genetic control and could therefore be exploited in plant breeding.[64,74-78] A Mn-efficient genotype, in an agronomic sense, is one which is able to grow and yield well without added Mn fertilizer in a soil which is limiting in available Mn for another, standard, genotype. Mn-efficient genotypes are usually able to extract more Mn from these 'deficient' soils.[64,74-75]

Breeding for Tolerance to Mn Deficiency in South Australia

Manganese deficiency is the most intractable trace element problem in crops grown on most South Australian soils. A number of Mn-inefficient cultivars, released unwittingly by cereal breeders in this state in recent years (Galleon and Skiff barley, Coorong triticale, and Condor, Bayonet, and Millewa wheat), has stimulated interest in breeding for efficiency. Table 3 presents a selection of results from recent field variety trials on Mn-deficient soils using paired subplots of + or – Mn treatments for each entry. Efficiency is calculated as 100(–Mn/+Mn) for yield or Mn uptake. A wide range of efficiencies is indicated at each site, and provides evidence that considerable improvement in Mn efficiency is possible in modern cultivars with the germplasm already available. Further improvement would be possible through the transfer of Mn efficiency from rye, which is outstandingly efficient by comparison with wheat. From our studies on rye addition lines, it appears that efficiency is carried on the 2R chromosome.[64]

Genetics of Mn Efficiency

Until recently, little was known of the genetics of Mn efficiency in higher plants. Recent experiments with barley suggest the involvement of simple mechanisms of Mn efficiency controlled by a single, major, dominant gene.[79,80] In the study of 72 barley genotypes from a world collection, the pedigree relationships within the most efficient group and within the inefficient group were consistent with single, major-gene inheritance.[84] In the study in which progeny from the cross between barley genotypes Weeah (Mn-efficient) and WI-2585 (Mn-inefficient) were visually assessed for the leaf-chlorosis score,[79] the distribution of F_2 individuals followed a 3:1 ratio characteristic of a single dominant gene for Mn efficiency without overdominance (Figure 6). In further work with F_3 individuals, this conclusion was further substantiated, and a narrow-sense heritability of 71% for Mn efficiency was found.[85]

RESISTANCE OF MN-EFFICIENT WHEAT GENOTYPES TO TAKE-ALL

Genetic improvement in Mn efficiency may improve Mn nutrition of a host plant and therefore increase its resistance to take-all. Such a hypothesis follows logically from the two observations already discussed, namely, that in deficient soils the addition of Mn decreases disease[27,46] and that Mn uptake from

Table 3. Relative Manganese Efficiency of Wheat Genotypes Grown at Two Locations in South Australia[a]

Genotype	Location of experiment	Vegetative yield			Mn concentration		Grain yield		
		-Mn	+Mn (t/ha)	E-Mn	-Mn	+Mn (mg/kg)	-Mn	+Mn (t/ha)	E-Mn
C8MM	Karkoo						1.73	1.85	95
Condor	Karkoo						0.76	1.40	54
Bayonet	Karkoo						0.48	1.00	48
Aroona	Marion Bay	0.90	1.45	62	9.5	15	0.83	2.25	37
Halberd × Schomburgk	Marion Bay	0.77	1.19	65	8.7	15	0.80	1.89	42
Machete	Marion Bay	0.57	1.16	50	8.1	13	0.62	1.57	39
Millewa	Marion Bay	0.29	0.89	42	7.6	12	0.11	1.13	10
LSD (Genotype × Mn)		0.37			2.9		0.42		

[a] Three wheats were grown at Karkoo in 1984–85 and four at Marion Bay (MB) in 1989 (both locations in South Australia). The data show vegetative and grain yields, the efficiencies derived from both harvests, and the Mn concentrations in the young leaves at the first harvest.[83,86] Efficiency (E-Mn) is calculated as 100(−Mn/+Mn). LSD for the MB data only.

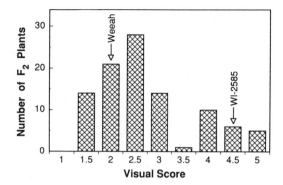

Figure 6. Frequency distribution for a leaf-chlorosis score of 100 F_2 individuals from the Mn-efficient by Mn-inefficient barley cross, Weeah × WI-2585, grown in Mn-deficient soil.[79] (From McCarthy, K.W. et al., in *Int. Symp. Manganese in Soils and Plants: Contributed Papers,* Webb, M.J. et al., Eds., Manganese Symposium Inc., Adelaide, 1988, 121. With permission from Kluwer Academic, Dordrecht.)

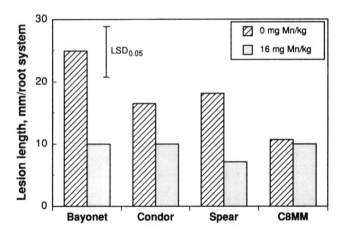

Figure 7. Resistance of wheat genotypes to take-all in Mn-deficient Wangary soil receiving no Mn fertilizer is correlated positively with their Mn efficiency.[28] Such an effect is dissipated by Mn fertilization. (From Wilhelm, N.S., Graham, R.D., and Rovira, A.D., *Plant Soil,* 123, 267, 1990. With permission from Kluwer Academic, Dordrecht.

environments having low Mn availability is enhanced in efficient cultivars[26,28,46] (see also Table 1). This hypothesis has been tested and confirmed in our laboratory. Wilhelm et al.[28] showed that Mn-efficient genotypes of wheat were more resistant to take-all than inefficient ones when grown in Mn-deficient soil (but not different under conditions of Mn adequacy, see also Table 1 and Figure 7). Recently, these results have been confirmed with a wider range of genotypes.[83]

It appears, therefore, that the genetic diversity already known to exist for tolerance to Mn deficiency is sufficient to improve the resistance of wheat plants to diseases that are aggravated by Mn deficiency. Such a statement may be applicable to other crop plants as well.[22,32] In general agreement with this line of reasoning is the observation stemming from the gene-linkage studies[81] that susceptibility to powdery mildew in cucumber was genetically linked to, or pleiotropic with, Mn deficiency-induced chlorosis.

CONCLUDING REMARKS

In the absence of effective sources of cultivar resistance or chemical means of controlling take-all, fertilization management is proving to be important for reducing crop damage due to this disease. Improved Mn status of a host plant increases resistance to *Ggt* infection. Plant Mn status is dependent not only on the level of Mn fertilization, but also on the genotype of both the host (Mn-efficient cultivars extract more Mn from soils with limited availability of that nutrient) and the *Ggt* fungus (isolates that are more powerful oxidizers of Mn are also more virulent by virtue of reducing plant availability of Mn in the rhizosphere).

Consistent mechanistic explanations of the interactions between Mn fertilization, Mn-efficiency of cereal genotypes, and Mn-oxidation capacity and virulence of *Ggt* isolates are still lacking. Research to address these interactions should, among other topics, concentrate on the short-term biosynthesis of phenolics and lignin in the infection court, quantification of Mn-oxidation capacity of *Ggt* isolates in the rhizosphere, the relationship between genetics of Mn efficiency and take-all resistance of cereal genotypes, and influence of Mn fertilization and Mn-efficient genotypes on populations of rhizosphere microflora which can influence virulence of the *Ggt* fungus.

ACKNOWLEDGMENTS

Continuous financial support from the Australian Research Council and Grains Research Development Corporation is gratefully acknowledged.

REFERENCES

1. Garrett, S.D., Introduction, in *Biology and Control of Take-all,* Asher, M.J.C. and Shipton, P.J., Eds., Academic Press, London, 1981, 1.
2. C.S.I.R.O. Division of Soils, *Annual Report,* C.S.I.R.O., Canberra, Australia, 1991.
3. de Boer, R.F., Steed, G.R., and Macauley, B.J., Effects of stubble and sowing treatments on take-all of wheat in north-eastern Victoria, *Aust. J. Exp. Agric.,* 32, 641, 1992.

4. McAlpine, D., Take-all and white-heads in wheat (*Ophiobolus graminis* Sacc.), *J. Dep. Agric. Victoria Aust.*, 2, 420, 1904.

5. Clarkson, D.T., Drew, M.C., Ferguson, I.B., and Sanderson, J., The effect of the take-all fungus, *Gaeumannomyces graminis*, on the transport of ions by wheat roots, *Physiol. Plant Pathol.*, 6, 75, 1975.

6. Weste, G., The process of root infection by *Ophiobolus graminis*, *Trans. Br. Mycol. Soc.*, 59, 133, 1972.

7. Wallwork, H., Screening for resistance to take-all in wheat, triticale and wheat-triticale hybrid lines, *Euphytica*, 40, 103, 1989.

8. Conner, R.L., MacDonald, M.D., and Whelan, E.D., Evaluation of take-all resistance in wheat-alien amphiploid and chromosome substitution lines, *Genome*, 30, 597, 1988.

9. Solel, Z., Ben-Ze'ev, I.S., and Dori, S., Features of resistance to take-all disease in cereal species evaluated by laboratory assays, *J. Phytopathol.*, 130, 219, 1990.

10. Penrose, L.D.J., Disease in wheat genotypes infected with *Gaeumannomyces graminis* var. *tritici*, *Ann. Appl. Biol.*, 118, 513, 1991.

11. Smiley, R.W., Wilkins, D.E., and Klepper, E.L., Impact of fungicide seed treatments on Rhizoctonia root rot, take-all, eyespot, and growth of winter wheat, *Plant Dis.*, 74, 782, 1990.

12. Weller, D.M., Biological control of soilborne plant pathogens in the rhizosphere with bacteria, *Annu. Rev. Phytopathol.*, 26, 379, 1988.

13. Lucas, P. and Sarniguet, A., Soil receptivity to take-all. Influence of some cultural practices and soil chemical characteristics, *Symbiosis*, 9, 51, 1990.

14. Werker, A.R., Gilligan, C.A., and Hornby, D., Analysis of disease-progress curves for take-all in consecutive crops of winter wheat, *Plant Pathol.*, 40, 8, 1991.

15. Rothrock, C.S. and Cunfer, B.M., Influence of small grain rotations on take-all in subsequent wheat crop, *Plant Dis.*, 75, 1050, 1991.

16. Murray, G.M., Heenan, D.P., and Taylor, A.C., The effect of rainfall and crop management on take-all and eyespot of wheat in the field, *Aust. J. Exp. Agric.*, 31, 645, 1991.

17. Trolldenier, G., Effect of varied NPK nutrition and inoculum density on yield losses of wheat caused by take-all, in *Ecology and Management of Soil-borne Plant Pathogens*, Parker, C.A., Rovira, A.D., Moore, K.J., Wong, P.T.W., and Kollmorgen, J.F., Eds., American Phytopathological Society, St. Paul, MN, 1985, 218.

18. Brennan, R.F., Effect of copper application on take-all severity and grain yield of wheat in field experiments near Esperance, Western Australia, *Aust. J. Exp. Agric.*, 31, 255, 1991.

19. Brennan, R.F., The role of manganese and nitrogen nutrition in the susceptibility of wheat plants to take-all in Western Australia, *Fert. Res.*, 31, 35, 1992.

20. Brennan, R.F., Effect of superphosphate and nitrogen on yield and take-all of wheat, *Fert. Res.*, 31, 43, 1992.

21. Brennan, R.F., The effect of zinc fertilizer on take-all and the grain yield of wheat grown on zinc-deficient soils of the Esperance region, Western Australia, *Fert. Res.*, 31, 215, 1992.

22. Graham, R.D. and Webb, M.J., Micronutrients and disease resistance and toler-
    ance in plants, in *Micronutrients in Agriculture*, 2nd ed., Mortvedt, J.J., Ed., Soil
    Science Society of America, Madison, WI, 1991, 329.
23. Leggett, M.E., Sivasithamparam, K., and McFarlane, M.J., Effect of nitrogen
    supply on rhizosphere interactions and take-all disease of wheat, *Can. J.
    Microbiol.*, 37, 42, 1991.
24. Sarniguet, A., Lucas, P., and Lucas, M., Relationships between take-all, soil
    conduciveness to the disease, populations of fluorescent pseudomonads and
    nitrogen fertilizers, *Plant Soil*, 145, 17, 1992.
25. Sarniguet, A., Lucas, P., Lucas, M., and Samson, R., Soil conduciveness to take-
    all of wheat. Influence of the nitrogen fertilizers on the structure of populations
    of fluorescent pseudomonads, *Plant Soil*, 145, 29, 1992.
26. Wilhelm, N.S., Investigations into *Gaeumannomyces graminis* var. *tritici* Infec-
    tion of Manganese-Deficient Wheat, Ph.D. thesis, University of Adelaide, Aus-
    tralia, 1991.
27. Wilhelm, N.S., Graham, R.D., and Rovira, A.D., Application of different sources
    of manganese sulphate decreases take-all *(Gaeumannomyces graminis* var. *tritici)*
    of wheat grown in a manganese deficient soil, *Aust. J. Agric. Res.*, 38, 1, 1988.
28. Wilhelm, N.S., Graham, R.D., and Rovira, A.D., Control of Mn status and
    infection rate by genotype of both host and pathogen in the wheat take-all
    interaction, *Plant Soil*, 123, 267, 1990.
29. Huber, D.M., The role of nutrition in the take-all disease of wheat and other small
    grains, in *Soil-borne Plant Pathogens: Management of Diseases with Macro and
    Micro Elements*, Engelhard, A.W. Ed., American Phytopathological Society, St.
    Paul, MN, 1989, 46.
30. Rovira, A.D. and Wildermuth, G.B., The nature and mechanism of suppression,
    in *Biology and Control of Take-all*, Asher, M.J.C. and Shipton, P.J., Eds.,
    Academic Press, London, 1981, 385.
31. Marschner, P., Ascher, J.S., and Graham, R.D., Effect of manganese-reducing
    rhizosphere bacteria on the growth of *Gaeumannomyces graminis* var. *tritici* and
    on manganese uptake by wheat (*Triticum aestivum* L.), *Biol. Fertil. Soils*, 12, 33,
    1991.
32. Graham, R.D., Effects of nutrient stress on susceptibility of plants to disease with
    particular reference to the trace elements, *Adv. Bot. Res.*, 10, 221, 1983.
33. Huber, D.M. and Wilhelm, N.S., The role of manganese in resistance to plant
    diseases, in *Manganese in Soils and Plants*, Graham, R.D., Hannam, R.J., and
    Uren, N.C., Eds., Kluwer Academic, Dordrecht, 1988, 155.
34. Graham, R.D. and Rovira, A.D., A role for manganese in the resistance of wheat
    plants to take-all, *Plant Soil*, 78, 441, 1984.
35. Longnecker, N.E., Marcar, N.E., and Graham, R.D., Increased manganese con-
    tent of barley seeds can increase grain yield in manganese-deficient conditions,
    Aust. J. Agric. Res., 42, 1065, 1991.
36. Marcar, N.E. and Graham, R.D., Effect of seed manganese content on the growth
    of wheat *(Triticum aestivum)* under manganese deficiency, *Plant Soil*, 96, 165,
    1986.
37. Huber, D.M., Wagner, J.E., Nashaar, H.E.L., and Moore, L.W., Interactions of
    a peat carrier and potential biological control agents, *Phytopathology*, 76, 1104,
    1986.

38. Huber, D.M., Woodhead, S.H., and Mburu, D.N., Rhizosphere interactions with *Bacillus subtilis* as a biological control agent, in *Proc. Biological Control Symposium,* 4th Int. Congr. Plant Pathology, Melbourne, Australia, 1983.

39. Friend, J., Plant phenolics, lignification and plant disease, *Prog. Phytochem.,* 7, 197, 1981.

40. Matern, U. and Kneusel, R.E., Phenolic compounds in plant disease resistance, *Phytoparasitica,* 16, 153, 1988.

41. Niemann, G.J., van der Kerk, A., and Niessen, W.M.A., Free and cell wall-bound phenolics and other constituents from healthy and fungus-infected carnation (*Dianthus caryophyllus* L.) stems, *Physiol. Mol. Plant Pathol.,* 38, 417, 1991.

42. Engelsma, G., A possible role of divalent manganese in the photoinduction of phenylalanine ammonia-lyase, *Plant Physiol.,* 50, 599, 1972.

43. Gross, G.G., Janse, C., and Elstner, E.F., Involvement of malate, monophenols, and the superoxide radical in hydrogen peroxide formation by isolated cell walls from horseradish (*Armoracia lapathifolia* Gilib), *Planta,* 136, 271, 1977.

44. Burnell, J.N., The biochemistry of manganese in plants, in *Manganese in Soils and Plants,* Graham, R.D., Hannam, R.J., and Uren, N.C., Eds., Kluwer Academic, Dordrecht, 1988, 125.

45. Brown, P.H., Graham, R.D., and Nicholas, D.J.D., The effects of manganese and nitrate supply on the levels of phenolics and lignin in young wheat plants, *Plant Soil,* 81, 437, 1984.

46. Rengel, Z., Graham, R.D., and Pedler, J.F., Manganese nutrition and accumulation of phenolics and lignin as related to differential tolerance of wheat genotypes to the take-all fungus, *Plant Soil,* 151, 255, 1993.

47. Grand, C., Sarni, F., and Lamb, C.J., Rapid induction by fungal elicitor of the synthesis of cinnamyl-alcohol dehydrogenase, a specific enzyme of lignin synthesis, *Eur. J. Biochem.,* 169, 73, 1987.

48. Asada, Y. and Matsumoto, I., Induction of disease resistance in plants by a lignification-inducing factor, in *Molecular Determinants of Plant Diseases,* Nishimura, S., Vance, C.P., and Doke, N., Eds., Springer-Verlag, Heidelberg, 1987, 223.

49. Bruce, R.J. and West, C.A., Elicitation of lignin biosynthesis and isoperoxidase activity by pectic fragments in suspension cultures of castor bean, *Plant Physiol.,* 91, 889, 1989.

50. Skou, P.J., Morphology and cytology of infection process, in *Biology and Control of Take-all,* Asher, M.J.C. and Shipton, P.J., Eds., Academic Press, London, 1981, 175.

51. Rengel, Z., Graham, R.D., and Pedler, J.F., Time-course of Mn effects on biosynthesis of phenolics and lignin in wheat roots as a defence barrier to the take-all fungus, submitted.

52. Simpson, D.J. and Robinson, S.P., Freeze-fracture ultrastructure of thylakoid membranes in chloroplasts from manganese-deficient plants, *Plant Physiol.,* 74, 735, 1984.

53. Graham, R.D., Davies, W.J., and Ascher, J.S., The critical concentration of manganese in field-grown wheat, *Aust. J. Agric. Res.,* 36, 145, 1985.

54. Sivasithamparam, K. and Parker, C.A., Physiology and nutrition in culture, in *Biology and Control of Take-all,* Asher, M.J.C. and Shipton, P.J., Eds., Academic Press, London, 1981, 125.

55. Weste, G., Extracellular enzyme production by various isolates of *Ophiobolus graminis* and *O. graminis* var *avenae*. I. Enzymes produced in culture, *Phytopathology*, 67, 189, 1970.

56. Weste, G., Extracellular enzyme production by various isolates of *Ophiobolus graminis* and *O. graminis* var *avenae*. II. Enzymes produced within host tissue, *Phytopathology*, 67, 327, 1970.

57. Buchhorn, S.C., Evidence of the Relationship Between Manganese and Take-All of Wheat, Honours thesis, University of Adelaide, Australia, 1988.

58. Arnott, H., Rosemann, T., Graham, R.D., and Huber, D.M., An experimental study of manganese mineralisation in the take-all fungus, *Gaeumannomyces graminis, Mycol. Soc. Am. Newsl.*, 42, 3, 1991.

59. MacDonald, H.J. and Rovira, A.D., Division of Soils, Tech. Memorandum 12/85, C.S.I.R.O., Canberra, 1988.

60. Asher, M.J.C., Pathogenic variation, in *Biology and Control of Take-all,* Asher, M.J.C. and Shipton, P.J., Eds., Academic Press, London, 1981, 199.

61. Rosemann, T.S., Graham, R.D., Arnott, H.J., and Huber, D.M., The interaction of temperature with virulence and manganese oxidizing potential in the epidemiology of *Gaeumanommyces graminis, Phytopathology*, 81, 1215, 1991.

62. Smiley, R.W., Fowler, M.C., and Reynolds, K.L., Temperature effects on take-all of cereals caused by *Phialophora graminicola* and *Gaeumannomyces graminis, Phytopathology*, 76, 923, 1986.

63. Dewan, M.M. and Sivasithamparam, K., Differences in the pathogenicity of isolates of the take-all fungus from roots of wheat and ryegrass, *Soil Biol. Biochem.*, 22, 119, 1990.

64. Graham, R.D., Genotypic differences in tolerance to manganese deficiency, in *Manganese in Soils and Plants,* Graham, R.D., Hannam, R.J., and Uren, N.C., Eds., Kluwer Academic, Dordrecht, 1988, 261.

65. Timonin, M.I., Illman, W.I., and Hartererink, T., Oxidation of manganous salts by soil fungi, *Can. J. Microbiol.*, 18, 793, 1972.

66. Thind, K.S. and Midan, M., Effect of various trace elements on the growth and sporulation of four fungi, *Proc. Indian Natl. Sci Acad. Part B*, 43, 115, 1977.

67. Barber, D.A. and Lee, R.B., The effect of micro-organisms on the absorption of manganese by plants, *New Phytol.*, 73, 97, 1983.

68. Bromfield, S.M., The properties of a biologically formed Mn oxide, its availability to oats and its solution by root washings, *Plant Soil*, 9, 325, 1958.

69. Huber, D.M., Immobilization of Mn predisposes wheat to take-all, *Phytopathology*, 77, 1715, 1987.

70. Bartlett, R.J., Manganese redox reactions and organic reactions in soils, in *Manganese in Soils and Plants,* Graham, R.D., Hannam, R.J., and Uren, N.C., Eds., Kluwer Academic, Dordrecht, 1988, 59.

71. Norwell, W.A., Inorganic reactions of manganese in soils, in *Manganese in Soils and Plants,* Graham, R.D., Hannam, R.J., and Uren, N.C., Eds., Kluwer Academic, Dordrecht, 1988, 37.

72. Uren, N.C., Rhizosphere reactions of aluminium and manganese, *J. Plant Nutr.*, 12, 173, 1989.

73. Uren, N.C. and Reisenauer, H.M., The role of root exudates in nutrient acquisition, *Adv. Plant Nutr.*, 3, 79, 1988.

74. Bansal, R.L., Nayyar, V.K., and Takkar, P.N., Field screening of wheat cultivars for manganese efficiency, *Field Crops Res.,* 29, 107, 1991.

75. Kaur, N.P., Takkar, P.N., Arora, C.L., and Nayyar, V.K., Relative Mn efficiency of wheat cultivar-HD 2009, *Indian J. Plant Physiol.,* 32, 306, 1989.

76. Marcar, N.E. and Graham, R.D., Genotypic variation for manganese efficiency in wheat, *J. Plant Nutr.,* 10, 2049, 1987.

77. Nyborg, M., Sensitivity to manganese deficiency of different cultivars of wheat, oat and barley, *Can. J. Plant Sci.,* 50, 198, 1970.

78. Vose, P.B. and Griffiths, D.J., Manganese and magnesium in the grey speck syndrome of oats, *Nature,* 191, 299, 1961.

79. McCarthy, K.W., Longnecker, N.E., Sparrow, D.H.B., and Graham, R.D., Inheritance of manganese efficiency in barley (*Hordeum vulgare* L.), in *Int. Symp. Manganese in Soils and Plants: Contributed Papers,* Webb, M.J., Nable, R.O., Graham, R.D., and Hannam, R.J., Eds., Manganese Symposium Inc., Adelaide, 1988, 121.

80. Longnecker, N.E., Graham, R.D., McCarthy, K.W., Sparrow, D.H.B., and Egan, J.P., Screening for manganese efficiency in barley (*Hordeum vulgare* L.), in *Genetic Aspects of Plant Mineral Nutrition,* El Bassam, N. et al., Eds., Kluwer Academic, Dordrecht, 1990, 273.

81. Robinson, R.W., Linkage relations of genes for tolerance to powdery mildew in cucumber, *Curcubit Genet. Coop.,* 1, 11, 1978.

82. Graham, R.D., Breeding wheats for tolerance to micronutrient deficient soils: present status and priorities, in Proc. Int. Conf. Wheat for Nontraditional Warm Climate Areas, Iguassu Falls, Brazil, 1991.

83. Pedler, J.F., Graham, R.D., Webb, M.J., and Rengel, Z.R., unpublished data.

84. Graham, R.D., Longnecker, N.E., and Sparrow, D.H.B., unpublished data.

85. McCarthy, K.W., Longnecker, N.E., Graham, R.D., and Sparrow, D.H.B., unpublished data.

86. Graham, R.D. and Ascher, J.S., unpublished data.

CHAPTER 11

The Interactions of Manganese and Iron

H.M. Reisenauer

INTRODUCTION

Early interest in Fe and Mn stemmed from reports at the turn of the century describing their roles and interrelationships in correcting chlorotic disorders of the important agricultural crops of the time. Then, in 1942, Somers and Shive[1] reported their studies of Fe-Mn relationships and noted that "Symptoms of Fe toxicity correspond to those of Mn deficiency, and symptoms of Mn toxicity correspond to those of Fe deficiency." Current interpretation would assign Somers and Shive's symptoms of Fe toxicity to Mn deficiency at high levels of Fe, and of the Mn toxicity to an "excess Mn"-induced Fe deficiency that is correctable by foliar Fe treatment,[2,3] or to Mn toxicity per se.[4] Their publication incited much additional experimentation, however, to date, our understanding of these important relationships in plants and of the complex soil and plant chemistry of Mn and Fe is yet incomplete.

Nutrient interactions have large effects on the yield potential of agricultural crops and on the health and well-being of animals, and are secondary only to supply in determining the acquisition and utilization of nutrients. They result from mutual or reciprocal influences of an increase in the supply of one nutrient or growth factor upon the response of an organism to a change in the supply of another factor. Thus, response to two interacting factors supplied together can be much greater, or less, than the sum of the responses to individual additions.

Olsen[5] has divided nutrient interactions mechanistically into three types:

1. Those in which an increased supply of one nutrient may either enhance or reduce a crop's requirement for, and response to, the addition of another nutrient.

0-87371-942-5/94/$0.00+$.50
© 1994 by CRC Press, Inc.

2. Those resulting from antagonistic or synergistic effects of one element or factor that influence the absorption or utilization of the other.
3. Those in which the absorption or utilization of two or more elements are affected differently as a result of reactions within the soil or the plant, or other changes in the growth system.

Type 1 interactions are most evident among the effects of macronutrients on crop growth and are of lesser concern to trace element nutrition. Antagonistic type 2 effects of Mn and Fe are observed in most liquid culture experiments that include variables of Mn and Fe. They were clearly illustrated by the data of Somers and Shive[1] (Figure 1), and others,[6-11] and undoubtedly are a principal, but not always readily recognized mechanism, in many experiments. Type 3 interactions of Mn and Fe are observed in systems in which additional factors are involved. These additional factors can include the effects of pH, temperature, competing ions, root growth rates, chelates, root exudates, etc. The recognition and understanding of the nature of interactions is critical to the diagnosis and correction of nutritional disorders, to the attainment of maximum yields, and to the development of meaningful models describing plant-soil systems.

PLANT ACQUISITION OF MN AND FE

Levels of supplied Mn and Fe, and their accumulation in plants in relation to requirements, determine the nature and intensity of their participation in interactions. Absorption of an ion or molecule into a plant root involves movement of the particle to the root surface, followed by transit through the cell wall "free space"* to the outer cell membrane (the plasmalemma), and finally transport into the living system, the cytoplasm. For most nutrients the latter step is against diffusion gradients and requires the input of metabolic energy. Plant accumulation and utilization of Fe and Mn differs widely among species,[12-15] and is affected to a large degree by environmental factors.[16-18] The rate of accumulation of a nutrient is determined by the rate of its arrival at the root surface (concentration in the bathing solution), ion adsorption, and precipitation, chelation, reduction, and other reactions within the free space and at the surface of the plasmalemma, and finally by the kinetics of ion transport throughout the cytoplasm. In plants with mycorrhizal associations, nutrient acquisition by roots is supplemented by absorption and transport within the fungal mycelium, where the ion is transferred directly to the cytoplasm.[19] In other cases, destruction of the plasmalemma by toxic levels of some ions, or physical deformation,[20,21] allows direct entry of soil solution into xylem vessels and the plant's nutrient transport system.

* Those parts of root tissues in which ions exchange relatively rapidly with the surrounding solution.

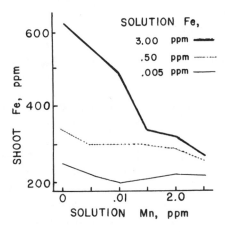

Figure 1. Total Fe in soybean shoots as affected by concentration of Fe and Mn in the culture solution. (Redrawn from Somers, I.I. and Shive, J.W., *Plant Physiol.*, 17, 582, 1942, with permission of American Society of Plant Physiologists, Rockville, MD.)

The roles of the free space of roots and its associated solid phases in supplying nutrients to plants are not well understood. The cell walls of roots and associated mucigel are known to possess cation exchange properties[22-24] and to contain precipitates of Fe and Mn and other elements.[17,25-27] These deposits, root plaques, were first observed on the roots of plants growing in poorly aerated soils, and have been attributed to the oxidizing power of the roots of rice[28] and to the processes that have evolved for the alleviation of toxicities from excessive levels of Fe and Mn to other crops.[29-33] The mechanism of root plaque formation is not known,[34] nor is the chemistry of the extracellular oxidation and precipitation of Mn by microorganisms.[35,36]

Ryan and Hariq[37] and Bienfait and associates[38] have proposed that plaques form from the release and precipitation of Fe from Fe-chelate complexes absorbed by the charged sites of root surfaces. The precipitation of MnO_2 in the free space is known to increase with pH,[39] and with the Mn^{2+} concentration of the associated solution, and to be restricted to the immediate vicinity of the root.[32] Its mechanism is assumed to differ from that in usual aqueous systems since precipitation in the latter does not occur below pH 8.6.[32]

Root plaques are now generally recognized as nutrient sources to plants.[26,38,40-42] The existence of these root-surface deposits along with the comparatively high and variable concentrations of both elements in root tissues, as reported by Vlamis and Williams,[7] Levan and Rhia,[32] St-Cyr and Crowder,[39] and Munns et al.,[43] and the earlier evidence for nutritional roles of solid phase compounds presented by Hewitt,[2] Leeper,[44] Chapman,[45] Jenny,[22] and Uren[46,47] suggest the need for further study. The possible contribution of root plaques to nutrient interactions is apparent. Mechanisms for the utilization of these Fe and Mn compounds have been reviewed recently.[48-50]

Plant Mn Requirements

Plant requirements for Mn are met at shoot tissue levels of 20 to 40 ppm and toxic reactions result when tissues have accumulated from 200 to 5300 ppm Mn.[15,51-53] Reuse of the element within the plant is limited; thus plants require a continuous supply if all requirements are to be met.[16,43] Symptoms of Mn deficiency that appear first in the new growth at early stages may be difficult to differentiate visually from those of Fe deficiency.[3,54] Uptake of the element from soils is determined most by the concentration of Mn^{2+} in the soil solution, the level of readily exchangeable Mn, and the amount and reactivity of insoluble Mn in the soil.[49,55-59] The concentration of Mn^{2+} in the soil solution can vary by orders of magnitude within very short time spans,[55,60] and plant accumulations of toxic amounts are not uncommon. Insoluble forms in soils utilized by plants include the readily reducible MnO_2, which occurs both as a soil constituent and precipitant at root surfaces, as well as $Mn-PO_4$[61] and Mn-Ca-CO_3 salts.[62] Plant uptake from soils is markedly influenced by pH (Figure 2), and increases with soil temperature.[18,63-66] Plant Mn content is reduced by added Fe,[53,67,68] and by competing divalent cations including, Ca,[69] Mg,[69-71] and to a lesser degree by Zn and Cu.[72,73] Experiments evaluating the effects of other cations on Mn uptake from nutrient solutions have frequently given results conflicting with those from soils.[74] These differences result from secondary reactions of added treatments with soil constituents which, as demonstrated by the data of Petrie and Jackson,[75] can be appreciable. Those nutrient cations that exert affects sufficient to provide significant interactions include H^+, Fe, and Mg.

Plant Fe Requirements

Plant uptake of Fe, and the prediction and correction of Fe deficiency of crops growing in the world's extensive areas of calcareous soils,[76-78] is of continuing concern to plant nutritionists. In addition, Fe toxicity is the second most severe yield limiting factor in paddy rice.[79] The solubility of Fe^{3+} in other than strongly acid or flooded soils is much too low to supply crop needs, and either chelation or reduction to Fe^{2+} is necessary if crop requirements are to be met.[80,81] In addition, transport of the element to the root and its utilization within the plant are depressed by high levels of heavy metals (mainly Cu, Zn, and Mn) and PO_4^{3-} and HCO_3^-.[2,73,82-85] Excessively high levels of Fe^{2+} occur in agricultural soils only under strongly reducing conditions.[81]

Plant species, and even cultivars within a species, differ widely in their ability to mobilize the sparingly soluble Fe compounds of soils.[12,53,86] Iron-efficient plants have adapted both morphological and chemical mechanisms for mobilization and uptake of Fe from the rhizosphere. These, as summarized by Marschner,[20] Brown,[87] Chaney,[88] Romheld and Marschner,[89] Longnecker and Welch,[90] and De Vos et al.,[91] include acidification of the rhizosphere soil, formation of rhizodermal

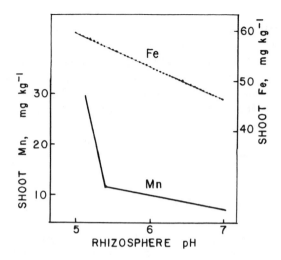

Figure 2. The effects of rhizosphere pH on Fe and Mn in the shoots of dwarf French beans grown in a brown earth soil. (Adapted from Sarkar, A.N. and Wyn Jones, R.G., *Plant Soil*, 66, 361, 1982. With permission of Kluwer Academic Publishers, Dordrecht.)

transfer cells that are sites of release of both H^+ and chelates, enhancement of Fe^{3+} reduction at the plasma membrane, and enhanced release of phytosiderophores. Differences that exist between species have been classified into two strategies: Strategy I is for dicots and nongraminaceous monocots and Strategy II is for the grasses. Strategy I is characterized by: (1) release of H^+ ions and reducing compounds from roots, and (2) increase in reduction of Fe^{3+} to Fe^{2+} at the root plasmalemma, followed by absorption of the reduced form.[92,93] At this point, competition between divalent cations as with Mn^{2+} may occur. Strategy II is characterized by the release of phytosiderophores from roots during Fe-deficiency stress. The phytosiderophores mobilize Fe^{3+}, making it available to plants.[94] It may be absorbed as the Fe^{3+}-phytosiderophore,[84,89] or the complex may be adsorbed within the apoplast and reduced to Fe^{2+}, as with Strategy I. The primary role of Mn in plant redox reactions,[95] along with the suggested involvement of the superoxide radical in the reduction of Fe^{3+} at the plasmalemma[96] and other plant stress responses,[97-100] imply further interdependence of Mn and Fe in plant responses.

Root characteristics and functions that directly impact plant accumulation of Mn and Fe are confined to their apical and subapical zones; thus root growth rates are important as they determine the volume of soil contacted.[93,101] These are also the portions of the roots most active in nutrient absorption and release of exudates and excretions.[102] Roots in soil extend their length between 1 and 20 mm/day by elongating the cells behind the root cap and meristem. Nutrient uptake by nonstressed roots in solution culture occurs along the whole length of the root,[103] but in soil the bulk of the uptake of the less mobile elements,[104] such as Fe and Mn in low Mn soils, would appear to be from the root apex (root tip to the oldest root hairs). Other functions of roots in soils of direct concern to this study include the secretion or excretion of H^+,[105] and HCO_3^-,[85,106] reductants, and of a wide range of organic compounds.[102] Plants have been observed to change the pH of the soil in their immediate vicinity by as much as two pH units.[107] They commit as much as 40 to 50% of the C fixed in photosynthesis to their roots, and about half of this C finds its way into the soil as root products. These effects, combined with those resulting from nutrient absorption, physical deformations, oxygen consumption, and associated root actions produce marked modifications of the rhizosphere soil.

INTERACTIONS OF MN AND FE

The interactions of Mn and Fe result from differences in their reactions within the soil, in their absorption and utilization by plants, and the influences of environmental factors and other ions on these reactions. In experiments including adequate levels of both nutrients, increased concentration of either element has most frequently decreased plant uptake or utilization of the other.[11,53,67] These interactive effects, however, can be complicated by rhizosphere acidification resulting from Fe or P deficiencies or N source effects; by changes in soil physical properties including temperature, moisture levels, and aeration; by additions of synthetic chelates; by the competitive effects of other nutrient cations; by the effects in increased levels of available silicate; and by microorganisms associated with the plant roots. Accordingly, observed interactions of Mn and Fe differ widely between soils and nutrient solutions, and cannot be interpreted as a simple antagonism but rather as the end result of the combined effects of the several properties of the system on the uptake and functioning of the two elements.

Rhizosphere Acidification

The chemical and biological characteristics of the cylinder of soil immediately surrounding a plant root are subject to appreciable modification by plant processes. Those characteristics altered most include pH, levels of available nutrients, microbial population, moisture content, aeration, and physical properties. Soil pH changes of two or more units are common.[19,81,107,108] They are a direct result of excess cation-over-anion intake or absorption of NH_4^+.[109,110] The effects of rhizosphere pH on uptake of Mn and Fe into the shoots of dwarf French beans grown in a brown earth soil are shown in Fig 2.[111] Tissue contents of both elements increased with acidification, but the rate at which Mn was absorbed increased much more rapidly below pH 5.5 than did that of Fe, resulting in a rapid change in the tissue Fe/Mn ratio. This relationship shown is essentially the same as that reported by Godo and Reisenauer,[112] Sillanpaa,[113] and Rollwagen and Zasoski.[114] In an aqueous culture system free of precipitated oxides, uptake of both elements would be expected to decrease with increasing acidity. This is expected from the competition between H^+ and Mn^{2+}, as demonstrated in the experiments of Robson and Loneragan[31] and others.[115,116] In the solution culture experiments of Jolley et al.,[117] Fe stress acidification in the presence of precipitated Fe-oxide maintained plant Fe levels, while Mn decreased as a result of cation competition and probable depletion of solution Mn. The effects of acidification of soil systems on plant Mn uptake are dominated by effects on soil Mn availability (enhanced reduction of MnO_2 by root exudates[112]) and may result in a severalfold increase in the Mn content of plant tops.[19]

Soil Physical Properties

The soil physical properties most strongly influencing Fe and Mn availability include temperature, moisture content, and compaction.[57] Increased soil temperature favors absorption of both elements; however, the effects on the Mn accumulation are relatively much greater than those on Fe.[66] The most significant effect of temperature on plant Mn relationships is its effect on tolerance to excesses of the element. Thus Rufty and co-workers[64] observed leaf tissue levels of 700 to 1200 ppm associated with appearance of visual toxicity symptoms on plants grown at day/night temperatures of 22/18°C, and 5000 to 8000 ppm at 30/26°C. Low soil temperatures are commonly associated with deficiencies of Mn.[66] In the experiments of Parker,[118] the temperature decrease caused by a surface mulch of corn residue reduced the Mn content of the following crop by an average of 25%. Comparable temperature effects on Fe uptake are much less dramatic.[18] Soil moisture content through it effects on aeration and HCO_3^- level,[85,106] and compacting through its effects on exudation,[50] root morphology, and growth rates,[119] influence absorption of both elements. Aeration and compaction influence Fe solubility and plant uptake proportionately more than Mn. The toxic levels of Mn that build up rapidly after flooding soils that contain plentiful supplies of soluble organic matter,[60] or in sterilized soils,[120] are temporary,[58,121] but initially highly significant.

Effects of Chelates

Metal chelating compounds are important to plant nutrition because they increase the solubility and buffer the concentration of metal ions in soil solutions, and affect many important chemical and biological processes. They arise in soils as products of microbial metabolism (siderophores), as chelating plant root exudates (phytosiderophores), or as synthetic compounds added as components of fertilizers.[78] Chelated compounds of Fe are stable over a wide range of conditions.[122] In contrast, there are no known stable complexes of Mn^{2+} in aqueous systems at neutral or acidic pH.[37,123,124] Bartlett[125,126] has proposed that soluble organic acid complexes of Mn^{3+}, a product of reverse dismutation, may serve as highly reactive and mobile Mn sources in the rhizosphere. Their role in soil processes has not been established.

Much is known of the reactions and ionic equilibria of synthetic chelates in aqueous systems,[80] but there is comparatively little information on their adsorption by soil constituents,[127] their biodegradation,[37] their oxidation by soil constituents as MnO_2, or their absorption and mobility in plants.[20] The interactions of Mn and Fe, observed in both nutrient solution and soil systems, are markedly influenced by the great differences in the stabilities and adsorption reactions of chelate complexes.[37,122,123]

Addition of chelates to nutrient solutions has been a common practice since the early 1950s (see review by Hewitt[2]). Their effects are known to differ between crop species,[128,129] but in general, at chelate concentrations not exceeding 50 μM (or approximately stoichiometric to Fe) growth and shoot content of Fe have been increased, root Fe and leaf Mn have been decreased, and root Mn increased.[68,128,130-133] At higher concentrations (twice stoichiometric and beyond) chelates compete for Fe, Zn, and Cu, reducing their concentration in both root and leaf tissues.[128,134] Added chelates have been observed to both worsen Mn deficiency,[66,129,135,136] and to reduce Mn toxicity.[137-140]

Biologically produced chelates are important to the mineral nutrition of both plants and microorganisms in soils. They are known to be present in soils,[141] to form highly stable complexes with Fe,[20,94] and to mobilize Fe, Zn, Cu, and Mn in soils.[142,143] The proposed mechanisms of chelate effects are largely speculative. The frequently noted decreases in Mn uptake with increased Fe supply (see review of Moraghan et al.[140]) has been explained by Baxter and Osman[68] to result from soil acidification associated with Fe deficiency stress response. Laurie et al.[132] have presented a model describing differences between chelates related to kinetic reactivity. Since complexes with divalent cations reach equilibrium more rapidly than those with trivalent ions, the reactions or absorption reactions for Fe^{3+} would lag behind those of the divalent cations, thus favoring the latter. The reactions determining the direction and nature of Fe-Mn interactions involve the complex chemistry of both the rhizosphere and of nutrient absorption processes within the plant.

Effects of Competing Nutrient Elements

The interactions of Mn and Fe are subject to measurable modification by changes in the level of those elements that interact directly, or via another element, with either Mn or Fe. In solution systems, significant effects of most of the nutrient elements have been reported.[9,31,54,74,115,144-146] In contrast, data from soil systems have included significant effects for H^+ (Figure 2), Mg,[70] and the heavy metal-induced Fe deficiencies.[2,73,147] Differences between soil and solution systems can be attributed to buffering of nutrient levels in soils by precipitation and adsorption reactions, and resulting reductions in the intensity of treatment variables. Added Mg, as with enhanced supply of silicate and increased ambient temperature and light intensity, reduces the toxic reactions of Mn.[70,148] Magnesium supply and light intensity are involved in the regulation of superoxide and peroxide scavenging enzymes in chloroplasts.[71] The magnitude of the Mg protective effect is substantial, and according to Le Bot and associates,[70] can be predicted from the tissue Mg:Mn ratio. Previous proposals for use of tissue ratios have met with varying degrees of success.[1,6,135,149,150]

Effects of Added Silicate

Silicon, although not formally established as an essential plant nutrient,[151] is necessary for the normal growth of many species, and affects the mineral nutrition and disease resistance of many others.[152] Additions of Si to nutrient solutions increase the tolerance of cereals for high levels of Mn,[153] and decreases the absorption of both Mn and Fe.[154] It is generally accepted that the mechanisms of the two effects differ. The mechanism for increased Mn tolerance acts by preventing the element from collecting into localized areas of leaves which then become necrotic; whereas decreases in the amounts of absorbed Fe and Mn result from increases in the oxidation and precipitation of Mn and Fe^{3+} oxides at root surfaces.[154-156] Wallace[157] has suggested that uptake of Si by gramineae is a significant factor in affecting rhizosphere pH (1% Si in a 5000 kg ha^{-1} crop is equivalent to 357 kg ha^{-1} of lime) and in influencing the absorption of other plant nutrients.

Mycorrhizal Effects

Mycorrhiza is a symbiotic association between a fungus and a root cell in which the fungus receives sustenance from the plant, and in turn, increases the absorption of water and most, but not all, nutrient elements. Precise evaluation of the role of the association in nutrient absorption is complicated by the near universal occurrence of the fungi in soils, and by the very large effects of mycorrhiza on crop uptake of P and growth in low P soils.[158-160] Thus, satisfactory crop growth in the absence of fungi requires soil sterilization and/or large additions of P fertilizer, both of which produce important differences in soil properties.[161,162]

Although species and cultivars of plants differ in their responses to mycorrhizal infection, it is commonly assumed that vesicular-arbuscular mycorrhiza (VAM) fungi will increase uptake of those nutrients that are present in the soil solution at low concentrations and that move to plant roots primarily by diffusion.[159] Increased uptake of P, Zn, and Cu have been commonly demonstrated;[162,163] effects on Fe have been positive but small,[164] and those on Mn strongly negative.[165-167] Bethlenfalvay and Franson[166] have suggested that the significantly higher levels of Mn in non-VAM plants were related to higher levels of root exudation by non-VAM plants and the roles of these exudates in solubilizing soil MnO_2. Their contribution to interactions between Fe and Mn can be appreciable.

SUMMARY

Although the interaction of Mn and Fe may appear to be a mutual antagonism, quantitatively the relationship is much more complex and its many aspects not well understood. The interaction is markedly influenced by plant, soil, and environmental factors including rhizosphere pH, soil physical characteristics, levels of synthetic and natural chelates, level of available Si, the actions of rhizosphere microorganisms, and the levels of other nutrients. There is great need for additional information on many aspects of the Mn-Fe relationships of plants — particularly on the nature, formation, and roles of root plaque deposits; on plant and rhizosphere redox chemistry; on the nature and reactions of metal chelate complexes at the root surfaces; and on the contributions of rhizosphere microorganisms to Mn and Fe solubility of these elements in soils and their uptake by plants.

REFERENCES

1. Somers, I.I. and Shive, J.W., The iron-manganese relation in plant metabolism, *Plant Physiol.,* 17, 582, 1942.
2. Hewitt, E.J., Sand and Water Culture Methods Used in the Study of Plant Nutrition, Tech. Commun. No. 22 (Rev. 2nd ed.), Commonwealth Agricultural Bureaux, Farnham Royal, England, 1966.
3. Hannam, R.J. and Ohki, K., Detection of manganese deficiency and toxicity in plants, in *Manganese in Soils and Plants,* Graham, R.D., Hannam, R.J., and Uren, N.C., Eds., Kluwer Academic, Dordrecht, 1988, chap. 16.
4. Kohno, Y., Foy, C.D., Fleming, A.L., and Krizak, D.T., Effect of Mn concentration on the growth and distribution of Mn and Fe in two bush bean cultivars grown in solution culture, *J. Plant Nutr.,* 7, 1, 1984.
5. Olsen, S.R., Micronutrient interactions, in *Micronutrients in Agriculture,* Mortvedt, J.J., Giordano, P.M., and Lindsay, W.L., Eds., Soil Science Society of America, Madison, WI, 1972.
6. DeKock, P.C. and Inkson, R.H.E., Manganese content of mustard leaves in relation to iron and major nutrient supply, *Plant Soil,* 17, 183, 1962.
7. Vlamis, J. and Williams, D.E., Iron and manganese relations in rice and barley, *Plant Soil,* 20, 221, 1964.
8. Gupta, U.C., Effects of manganese and lime on yield and on the concentrations of manganese, molybdenum, boron, copper, and iron in the boot stage tissue of barley, *Soil Sci.,* 114, 131, 1972.
9. Ohki, K., Mn and B effects on micronutrients and P in cotton, *Agron. J.,* 67, 204, 1975.
10. Patil, J.D. and Patil, N.D., Effect of calcium carbonate and organic matter on the growth and concentration of iron and manganese in sorghum *(Sorghum bicolor), Plant Soil,* 60, 295, 1981.
11. Zaharieva, T., Kasabov, D., and Romheld, V., Responses of peanut to iron-manganese interaction in calcareous soil, *J. Plant Nutr.,* 11, 6, 1988.

12. Millikan, C.R., Plant varieties and species in relation to the occurrence of deficiencies and excesses of certain nutrient elements, *J. Austr. Inst. Agric. Sci.,* 27, 220, 1961.

13. Vose, P.B., Varietal differences in plant nutrition, *Herb. Abstr.,* 33, 1, 1963.

14. Brown, J.C. and Jones, W.E., Fitting plants nutritionally to soils. III. Sorghum, *Agron. J.,* 69, 110, 1977.

15. Edwards, D.G. and Asher, C.J., Tolerance of crop and pasture species to manganese toxicity, in *Plant Nutrition 1982,* Proc. Ninth Int. Plant Nutrition Colloquium, Harwick University, England, Scaife, A., Ed., Commonwealth Agricultural Bureaux, Slough, England, 1982, 145.

16. Clarkson, D.T., The uptake and translocation of manganese by plant roots, in *Manganese in Soils and Plants,* Graham, R.D., Hannan, R.J., and Uren, N.C., Eds., Kluwer Academic, Dordrecht, 1988, chap. 7.

17. Kochian, L.V., Mechanisms of micronutrient uptake and translocation in plants, in *Micronutrients in Agriculture,* 2nd ed., Mortvedt, J.J., Cox, F.R., Shuman, L.M., and Welch, R.M., Soil Science Society of America, Madison, WI, 1991, 229.

18. Moraghan, J.T. and Mascagni, H.J., Jr., Environmental and soil factors affecting micronutrient deficiencies and toxicities, in *Micronutrients in Agriculture,* 2nd ed., Mortvedt, J.J., Cox, F.R., Shuman, L.M., and Welch, R.M., Eds., Soil Science Society of America, Madison, WI, 1991, chap. 11.

19. Marschner, H., *Mineral Nutrition in Higher Plants,* Academic Press, London, 1986.

20. Marschner, H., Symposium summary and future research areas, in *Iron Nutrition and Interactions in Plants,* Chen, Y. and Hadar, Y., Eds., Kluwer Academic, Dordrecht, 1991, 365.

21. Bell, P.F., Chaney, R.L., and Angle, J.S., Free metal activity and total metal concentration as indices of micronutrient availability to barley (Hordeum vulgare (L.) 'Klages'), in *Iron Nutrition and Interactions in Plants,* Chen, Y. and Hadar, Y., Eds., Kluwer Academic, Dordrecht, 1991, 69.

22. Jenny, H., Pathways of ions from soil into root according to diffusion models, *Plant Soil,* 25, 265, 1966.

23. Lauchli, A., Apoplastic transport in tissues, in *Encyclopedia of Plant Physiology,* Vol. 2B, Luttage, U. and Pitman, M.G., Eds., Springer-Verlag, Berlin, 1976, 3.

24. Haynes, R.J., Ion exchange properties of roots and ionic interactions within the root apoplasm. Their role in ion accumulation by plants, *Bot. Rev.,* 46, 75, 1980.

25. Clarkson, D.T. and Sanderson, J., Sites of adsorption and translocation of iron in barley roots, *Plant Physiol.,* 61, 546, 1978.

26. Linehan, D.J., Micronutrient cation sorption by roots and uptake by plants, *J. Exp. Bot.,* 35, 1571, 1984.

27. Taylor, G.J., Crowder, A.A., and Rodden, R., Formation and morphology of an iron plaque on the roots of *Typha latifolia* L. grown in solution culture, *Am. J. Bot.,* 71, 666, 1984.

28. Doi, Y., Studies on the oxidizing power of roots of crop plants. I. The difference with species of crop plants and wild grasses, *Proc. Crop Sci. Soc. Jpn.,* 2, 12, 1952.

29. Armstrong, W. and Boatman, D.J., Some field observations relating to growth of bog plants to conditions of soil aeration, *J. Ecol.,* 55, 101, 1967.

30. Bartlett, R.J., Iron oxidation proximate to plant roots, *Soil Sci.,* 92, 372, 1961.
31. Robson, A.D. and Loneragan, J.F., Sensitivity of annual *Medicago* species to manganese toxicity as affected by calcium and pH, *Aust. J. Agric. Res.,* 21, 223, 1970.
32. Levan, M.A. and Riha, S.J., The precipitation of black oxide coatings on flooded conifer roots of low internal porosity, *Plant Soil,* 95, 33, 1986.
33. Horiguchi, T., Mechanism of manganese toxicity and tolerance of plants. II. Deposition of oxidized manganese in plant tissues, *Soil Sci. Plant Nutr.,* 33, 595, 1987.
34. Kuo, S. and Mikkelsen, D.S., Effect of P and Mn on growth response and uptake of Fe, Mn, and P by sorghum, *Plant Soil,* 62, 15, 1981.
35. Bromfield, S.M., The oxidation of manganous ions under acid conditions by an acidophilous actinomycete from acid soil, *Aust. J. Soil Res.,* 16, 91, 1978.
36. Ghiorse, W.C., The biology of manganese transforming microorganisms in soil, in *Manganese in Soils and Plants,* Graham, R.D., Hannan, R.J., and Uren, N.C., Eds., Kluwer Academic, Dordrecht, 1988, chap. 5.
37. Ryan, J. and Hariq, S.N., Transformation of incubated micronutrient chelates in calcareous soils, *Soil Sci. Soc. Am. J.,* 47, 806, 1983.
38. Bienfait, H.F., van den Briel, W., and Mesland-Mul, N.T., Free space iron pools in roots. Generation and mobilization, *Plant Physiol.,* 78, 596, 1985.
39. St-Cyr, L. and Crowder, A.A., Manganese and copper in the root plaque of *Phragmites australis* (Cav.) Trin. ex Studel, *Soil Sci.,* 149, 191, 1990.
40. Hershey, D.R. and Paul, J.L., Iron nutrition of the broadleaf evergreen shrub, *Euonymus japonica* Thunb., *J. Plant Nutr.,* 7, 641, 1984.
41. Longnecker, N. and Welch, R.M., Accumulation of apoplastic iron in plant roots. A factor in the resistance of soybeans to iron-deficiency induced chlorosis?, *Plant Physiol.,* 92, 17, 1990.
42. Zhang, F., Romheld, V., and Marschner, H., Role of the root apoplasm for iron acquisition by wheat plants, *Plant Physiol.,* 97, 1302, 1991.
43. Munns, D.N., Johnson, C.M., and Jacobson, L., Uptake and distribution of manganese in oat plants. II. A kinetic model, *Plant Soil,* 19, 193, 1963.
44. Leeper, G.W., Relationship of soils to manganese deficiency of plants, *Nature,* 134, 972, 1934.
45. Chapman, H.D., Absorption of iron from finely ground magnetite by citrus seedlings, *Soil. Sci.,* 48, 309, 1939.
46. Uren, N.C., Chemical reduction of an insoluble higher oxide of manganese by plant roots, *J. Plant Nutr.,* 4, 65, 1981.
47. Uren, N.C., Chemical reduction at the root surface, *J. Plant Nutr.,* 5, 4, 1982.
48. Marschner, H., Romheld, V., and Kissel, M., Different strategies in higher plants in mobilization and uptake of iron, *J. Plant Nutr.,* 9, 695, 1986.
49. Uren, N.C., Rhizosphere reactions of aluminum and manganese, *J. Plant Nutr.,* 12, 173, 1989.
50. Mozafar, A., Contact with ballotini (glass spheres) stimulates exudation of iron reducing and iron chelating substances from barley roots, in *Iron Nutrition and Interactions in Plants,* Chen, Y. and Hadar, Y., Eds., Kluwer Academic, Dordrecht, 1991, 135.
51. Mulder, E.G. and Gerretsen, F.C., Soil manganese in relation to plant growth, *Adv. Agron.,* 4, 221, 1952.

52. Labanaushas, C.K., *Manganese, Diagnostic Criteria for Plants and Soils,* Chapman, H.D., Ed., Division of Agricultural Sciences, University of California, Berkeley, 1966, 264.

53. Foy, C.D., Chaney, R.L., and White, M.C., The physiology of metal toxicity in plants, *Annu. Rev. Plant Physiol.,* 29, 511, 1978.

54. Argawala, S.C., Chatterjee, C., Gupta, S., and Nautiyal, N., Iron-manganese interaction and its relation to boron levels in tomato plants, *Plant Soil,* 55, 377, 1980.

55. Leeper, G.W., Factors affecting availability of inorganic nutrients in soils with special reference to micronutrient metals, *Annu. Rev. Plant Physiol.,* 3, 1, 1952.

56. Geering, H.R., Hodgson, J.F., and Sdano, C., Micronutrient cation complexes in soil solution, *Soil Sci. Soc. Am. J.,* 33, 81, 1969.

57. Cheng, B.T. and Quellette, G.J., Manganese availability in soil, *Soils Fert.,* 34, 589, 1971.

58. Reisenauer, H.M., Determination of plant-available soil manganese, in *Manganese in Soils and Plants,* Graham, R.D., Hannan, R.J., and Uren, N.C., Eds., Kluwer Academic, Dordrecht, 1988, chap. 6.

59. Warden, B.T. and Reisenauer, H.M, Fractionation of soil manganese forms important to plant availability, *Soil Sci. Soc. Am. J.,* 55, 345, 1991.

60. Graven, E.H., Attoe, O.J., and Smith, D., Effect of liming and flooding on manganese toxicity in alfalfa, *Soil Sci. Soc. Am. Proc.,* 29, 702, 1965.

61. Boyle, F.W., Jr. and Lindsay, W.L., Manganese phosphate equilibrium relationships in soils, *Soil Sci. Soc. Am. J.,* 50, 588, 1986.

62. Jauregui, M.A. and Reisenauer, H.M., Calcium carbonate and manganese dioxide as regulators of available manganese and iron, *Soil Sci.,* 134, 105, 1982.

63. Heenan, D.P. and Carter, O.G., Influence of temperature on the expression of manganese toxicity by two soybean varieties, *Plant Soil,* 47, 219, 1977.

64. Rufty, T.W., Miner, G.S., and Raper, C.D., Jr., Temperature effects on growth and manganese tolerance in tobacco, *Agron. J.,* 71, 638, 1979.

65. Ghazali, N.J. and Cox, F.R., Effect of temperature on soybean growth and manganese accumulation, *Agron. J.,* 73, 363, 1981.

66. Moraghan, J.T., Manganese deficiency in soybeans as affected by FeEDDHA and low soil temperature, *Soil Sci. Soc. Am. J.,* 49, 1584, 1985.

67. Kirsch, R.K., Harward, M.E., and Petersen, R.G., Interrelationships among iron, manganese, and molybdenum in the growth and nutrition of tomatoes grown in culture solutions, *Plant Soil,* 12, 259, 1960.

68. Baxter, J.C. and Osman, M., Evidence for the existence of different uptake mechanisms in soybean and sorghum for iron and manganese, *J. Plant Nutr.,* 11, 51, 1988.

69. Lohnis, M.P., Effect of magnesium and calcium supply on the uptake of manganese by various crop plants, *Plant Soil,* 12, 339, 1960.

70. Le Bot, J., Gross, M.J., Caravalho, M.J.G.P.R., van Beusichem, M.L., and Kirkby, E.A., The significance of the magnesium to manganese ratio in plant tissues for growth and alleviation of manganese toxicity in tomato *(Lycopersicon esculentum)* plants, *Plant Soil,* 124, 205, 1990.

71. Goss, M.J. and Caravalho, M.J.G.P.R., Manganese toxicity. The significance of magnesium for the sensitivity of wheat plants, *Plant Soil,* 139, 91, 1992.

72. Safaya, N.M., Phosphorus-zinc interaction in relation to absorption rates of phosphorus, zinc, copper, manganese, and iron in corn, *Soil Sci. Soc. Am. J.*, 430, 719, 1976.

73. Amberger, A., Gutser, R., and Wunsch, A., Iron chlorosis induced by high copper and manganese supply, *J. Plant Nutr.*, 5, 715, 1982.

74. Vlamis, J. and Williams, D.E., Ion competition in manganese uptake by barley plants, *Plant Physiol.*, 37, 650, 1962.

75. Petrie, S.E. and Jackson, T.L., Effects of fertilization on soil solution pH and manganese concentration, *Soil Sci. Soc. Am. J.*, 48, 315, 1984.

76. Vose, P.B., Iron nutrition in plants. A world overview, *J. Plant Nutr.*, 5, 233, 1982.

77. Chen, Y. and Barak, P., Iron nutrition of plants in calcareous soils, *Adv. Agron.*, 35, 217, 1982.

78. Wallace, A., Rational approaches to control of iron deficiency other than plant breeding and choice of resistant cultivars, in *Iron Nutrition and Interactions in Plants*, Chen, Y. and Hadar, Y., Eds., Kluwer Academic, Dordrecht, 1991, 323.

79. Ponnamperuma, F.N., The chemistry of submerged soils, *Adv. Agron.*, 24, 29, 1972.

80. Lindsay, W.L., *Chemical Equilibria in Soils*, John Wiley & Sons, New York, 1979.

81. Mengel, K. and Kirkby, E.A., *Principles of Plant Nutrition*, 4th ed., International Potash Institute, Bern, Switzerland, 1987.

82. DeKock, P.C. and Hall, A., The phosphorus-iron relationship in genetical chlorosis, *Plant Physiol.*, 30, 293, 1955.

83. Kashirad, A. and Marschner, H., Iron nutrition of sunflower and corn plants in mono and mixed culture, *Plant Soil*, 41, 91, 1974.

84. Marschner, H., Romheld, V., Horst, W.J., and Martin, P., Root-induced changes in the rhizosphere. Importance for the mineral nutrition of plants, *Z. Pflanzenernaehr. Bodenkd.*, 49, 441, 1986.

85. Dofing, S.M., Penas, E.J., and Maranville, J.W., Effect of bicarbonate on iron reduction by soybean roots, *J. Plant Nutr.*, 12, 797, 1989.

86. Jolley, V.D. and Brown, J.C., Iron efficient and inefficient oats. I. Differences in phytosiderophore release, *J. Plant Nutr.*, 12, 423, 1989.

87. Brown, J.C., Mechanisms of iron uptake by plants, *Plant Cell Environ.*, 1, 294, 1978.

88. Chaney, R.L., Diagnostic practices to identify iron deficiency in higher plants, *J. Plant Nutr.*, 7, 47, 1984.

89. Romheld, V. and Marschner, H., Mobilization of iron in the rhizosphere of different plant species, in *Advances in Plant Nutrition*, Vol. 2, Tinker, B. and Lauchli, A., Eds., Praeger Publishers, New York, 1986, 155.

90. Longnecker, N. and Welch, R., The relationships among iron-stress response, iron-efficiency and iron uptake of plants, *J. Plant Nutr.*, 9, 3, 1986.

91. DeVos, C.R., Lubberding, H.J., and Bienfait, H.F., Rhizosphere acidification as a response to iron deficiency in bean plants, *Plant Physiol.*, 81, 842, 1986.

92. Chaney, R.L., Brown, J.C., and Tiffin, L.O., Obligatory reduction of ferric chelates in iron uptake by soybeans, *Plant Physiol.*, 50, 208, 1972.

93. Sijmons, P.C. and Bienfait, H.F., Development of Fe^{3+} reduction activity and H^+ extrusion during growth of iron-deficient bean plants in a rhizostat, *Biochem. Physiol. Pflanzen,* 181, 283, 1986.

94. Romheld, V., The role of phytosiderophores in acquisition of iron and other micronutrients in graminaceous species. An ecological approach, *Plant Soil,* 130, 127, 1991.

95. Salisbury, F.B. and Ross, C.W., *Plant Physiology,* 2nd ed., Wadsworth Publishing, Belmont, CA, 1978, chap. 9.

96. Cakmak, I., van de Wetering, D.A.M., Marschner, H., and Bienfait, H.F., Involvement of superoxide radical in extracellular ferric reduction by iron-deficient bean roots, *Plant Physiol.,* 85, 310, 1987.

97. Schoner, S. and Krause, G.H., Protective systems against active oxygen species in spinach. Response to cold acclimation in excess light, *Planta,* 180, 383, 1990.

98. Tsang, E.W.T., Bowler, C., Herouart, D., van Camp, W., Villarroel, R., Genetello, C., van Montagu, M., and Inze, D., Differential regulation of superoxide dismutases in plants exposed to environmental stress, *Plant Cell,* 3, 783, 1991.

99. Bowler, C., van Montagu, M., and Inze, D., Superoxide dismutase and stress tolerance, *Annu. Rev. Plant Physiol. Plant Mol. Biol.,* 43, 83, 1992.

100. Polle, A., Chakrabarti, K., Chakrabarti, S., Seifert, F., Schramel, P., and Rennenberg, H., Antioxidants and manganese deficiency in needles of Norway spruce (*Picea abies* L.), *Plant Physiol.,* 99, 1084, 1992.

101. Marschner, H., Mechanisms of manganese acquisition by roots from soils, in *Manganese in Soils and Plants,* Graham, R.D., Hannam, R.J., and Uren, N.C., Eds., Kluwer Academic, Dordrecht, 1988, chap. 13.

102. Uren, N.C. and Reisenauer, H.M., The role of root exudates in nutrient acquisition, in *Advances in Plant Nutrition,* Vol. 3, Praeger Publishers, New York, 1988, 79.

103. Clarkson, D.T. and Hanson, J.B, The mineral nutrition of higher plants, *Annu. Rev. Plant Physiol.,* 31, 239, 1980.

104. Bray, R.H., Confirmation of the nutrient mobility concept of soil-plant relationships, *Soil Sci.,* 95, 1963.

105. Spanwick, R.M., Electrogenic ion pumps, *Annu. Rev. Plant Physiol.,* 32, 267, 1981.

106. Brown, J.C., Lunt, O.R., Holmes, R.S., and Tiffin, L.O., The bicarbonate ion as an indirect cause of iron chlorosis, *Soil Sci.,* 88, 260, 1959.

107. Barber, S.A., *Soil Nutrient Bioavailability, a Mechanistic Approach,* John Wiley & Sons, New York, 1984, chap. 6.

108. Moorby, H., White, R.E., and Nye, P.H., The influence of phosphate nutrition on H^+ ion efflux from roots of young rape plants, *Plant Soil,* 105, 247, 1988.

109. Haynes, R.J. and Goh, K.M., Ammonium and nitrate nutrition of plants, *Biol. Rev.,* 53, 465, 1978.

110. Gahoonia, T.S., Claassen, N., and Jungk, A., Mobilization of phosphate in different soils by ryegrass supplied with ammonium or nitrate, *Plant Soil,* 140, 241, 1992.

111. Sakar, A.N. and Wyn Jones, R.G., Effect of rhizosphere pH on the availability and uptake of Fe, Mn, and Zn, *Plant Soil,* 66, 361, 1982.

112. Godo, G.H. and Reisenauer, H.M., Plant effects on soil manganese availability, *Soil Sci. Soc. Am.,* 44, 993, 1980.

113. Sillanpaa, M., Micronutrients and the nutrient status of soils: a global study, *FAO Soils Bull.*, 48, 3, 1982.

114. Rollwagen, B.A. and Zasoski, R.J., Nitrogen source effects on rhizosphere pH and nutrient accumulation by Pacific Northwest conifers, *Plant Soil*, 105, 79, 1988.

115. Jariel, D.M., Wallace, S.U., Jones, U.S., and Samonte, H.P., Growth and nutrient composition of maize genotype in acid nutrient solutions, *Agron. J.*, 83, 612, 1991.

116. Yan, F., Schubert, S., and Mengel, K., Effect of low root medium pH on net proton release, root respiration, and root growth of corn (*Zea Mays* L.) and broad bean (*Vicia faba* L.), *Plant Physiol.*, 99, 415, 1992.

117. Jolley, V.D., Brown, J.C., and Nugent, P.E., A genetically related response to iron deficiency stress in muskmellon, in *Iron Nutrition and Interactions in Plants*, Chen, Y. and Hadar, Y., Eds., Kluwer Academic, Dordrecht, 1991, 117.

118. Parker, D.T., Influence of mulching on the manganese content of corn plant tissue, *Agron. J.*, 54, 303, 1962.

119. Lindberg, S. and Pettersson, S., Effects of mechanical stress on uptake and distribution of nutrients in barley, *Plant Soil*, 83, 295, 1985.

120. Leeper, G.W., The forms and reactions of manganese in soils, *Soil Sci.*, 63, 79, 1947.

121. Gotoh, S. and Patrick, W.H., Jr., Transformation of manganese in a waterlogged soil as affected by redox potential and pH, *Soil Sci. Soc. Am. Proc.*, 36, 738, 1972.

122. Norvell, W.A., Reactions of metal chelates in soils and nutrient solutions, in *Micronutrients in Agriculture*, 2nd ed., Mortvedt, J.J., Cox, F.R., Shuman, L.M., and Welch, R.M., Eds., Soil Science Society of America, Madison, WI, 1972, chap. 7.

123. Yamaguchi, K.S. and Sawyer, D.T., The redox chemistry of manganese(III) and (IV) complexes, *Isr. J. Chem.*, 25, 164, 1985.

124. Richtie, G.S.P., The chemical behaviour of aluminum, hydrogen and manganese in acid soils, in *Soil Acidity and Plant Growth*, Robson, A.D., Ed., Academic Press, Sidney, 1989, chap. 1.

125. Bartlett, R.J., Soil redox behavior, in *Soil Physical Chemistry*, Sparks, D.L., Ed., CRC Press, Boca Raton, FL, 1986, 180.

126. Bartlett, R.J., Manganese redox reactions and organic interactions in soils, in *Manganese in Soils and Plants*, Kluwer Academic, Dordrecht, 1988, chap. 4.

127. Henrot, J. and Wieder, R.K., Processes of iron and manganese retention in laboratory peat microcosms subjected to acid mine drainage, *J. Environ. Qual.*, 19, 3212, 1990.

128. Guinn, G. and Joham, H.E., Effects of two chelating agents on absorption and translocation of Fe, Cu, Mn, and Zn by the cotton plant, *Soil Sci.*, 94, 220, 1962.

129. Murphy, L.S. and Walsh, L.M., Correction of micronutrient deficiencies with fertilizers, in *Micronutrients in Agriculture*, Mortvedt, J.J, Giordano, P.M., and Lindsay, W.L., Eds., Soil Science Society of America, Madison, WI, 1972, chap. 15.

130. Heenan, D.P. and Campbell, L.C., Manganese and iron interactions on their uptake and distribution in soybean (*Glycine max* (L.) Merr.), *Plant Soil*, 70, 317, 1983.

131. Korcak, R.F., Influence of micronutrient and phosphorus levels and chelator to iron ratio on growth, chlorosis, and nutrition of *Vaccinium ashei* Reade and *V. elliottii* Chapman, *J. Plant Nutr.*, 12, 1311, 1989.

132. Laurie, S.H., Tancock, N.P., McGrath, S.P., and Sanders, J.R., Influence of complexation on the uptake by plants of iron, manganese, copper and zinc. II. Effect of DPTA in a multi-metal and computer simulation study, *J. Exp. Bot.*, 42, 515, 1991.

133. Moraghan, J.T., Iron-manganese relationships in white lupine grown on a calciaquoll, *Soil Sci. Soc. Am. J.*, 56, 471, 1992.

134. Laurie, S.H., Tancock, N.P., McGrath, S.P., and Sanders, J.R., Influence of complexation on the uptake by plants of iron, manganese, copper, and zinc. I. Effect of EDTA in a multi-metal and computer simulation study, *J. Exp. Bot.*, 42, 509, 1991.

135. Wallace, A. and Alexander, G.V., Manganese in plants as influenced by manganese and iron chelates, *Commun. Soil Sci. Plant Anal.*, 4, 51, 1973.

136. Hue, N.V., A possible mechanism for manganese phytotoxicity in Hawaii soils amended with a low-manganese sewage sludge, *J. Environ. Qual.*, 17, 473, 1988.

137. Grasmanis, V.O. and Leeper, G.W., Toxic manganese in near-neutral soils, *Plant Soil*, 25, 41, 1966.

138. Moraghan, J.T. and Freeman, T.J., Influence of FeEDDHA on growth and manganese accumulation in flax, *Soil Sci. Soc. Am. J.*, 42, 455, 1978.

139. Moraghan, J.T., Manganese toxicity in flax growing on certain calcareous soils low in available iron, *Soil Sci. Soc. Am. J.*, 43, 1177, 1979.

140. Moraghan, J.T., Freeman, T.P., and Whited, D., Influence of FeEDDHA and soil temperature on the growth of two soybean varieties, *Plant Soil*, 95, 57, 1986.

141. Bossier, P. and Verstraete, W., Detection of siderophores in soil by a direct bioassay, *Soil Biol. Biochem.*, 18, 481, 1986.

142. Treeby, M., Marschner, H., and Romheld, V., Mobilization of iron and other micronutrient cations from a calcareous soil by plant-borne, microbial, and synthetic metal chelators, *Plant Soil*, 114, 217, 1989.

143. Zhang, F.S., Treeby, M., Romheld, V., and Marschner, H., Mobilization of iron by phytosiderophores as affected by other micronutrients, in *Iron Nutrition and Interactions in Plants*, Chen, Y. and Hadar, Y., Eds., Kluwer Academic, Dordrecht, 1991, 205.

144. Ishizuka, Y. and Ando, T., Interaction between manganese and zinc in growth of rice plants, *Soil Sci. Plant Nutr.*, 14, 201, 1968.

145. Heenan, D.P. and Campbell, L.C., Influence of potassium and manganese on growth and uptake of manganese by soybeans (*Glycine max* (L.) Merr. Cv. Bragg), *Plant Soil*, 61, 447, 1981.

146. Fleming, A.L., Enhanced Mn accumulation by snapbean cultivars under Fe stress, *J. Plant Nutr.*, 12, 715, 1989.

147. Warnock, R.E., Micronutrient uptake and mobility within corn plants (*Zea mays* L.) in relation to phosphorus-induced zinc deficiency, *Soil Sci. Soc. Am. Proc.*, 34, 765, 1970.

148. Cakmak, I. and Marschner, H., Magnesium deficiency and high light intensity enhance activities of superoxide dismutase, ascorbate peroxidase, and glutathione reductase in bean leaves, *Plant Physiol.*, 98, 1222, 1992.

149. Odurukwe, S.O. and Maynard, D.N., Mechanism of the differential response of Wf9 and Oh40B corn seedlings to iron nutrition, *Agron. J.*, 61, 694, 1969.

150. Alvarez-Tinaut, M.C., Leal, A., and Martinez, L.R., Iron-manganese interaction and its relation to boron levels in tomato plants, *Plant Soil*, 55, 377, 1980.

151. Epstein, E., *Mineral Nutrition of Plants: Principles and Perspectives*, John Wiley & Sons, New York, 1972.

152. Jones, L.H.P. and Handreck, K.A., Silica in soils, plants, and animals, *Adv. Agron.*, 19, 107, 1965.

153. Williams, D.E. and Vlamis, J., The effect of silicon on yield and manganese-54 uptake and distribution in the leaves of barley plants grown in culture solutions, *Plant Physiol.*, 32, 404, 1957.

154. Okuda, A. and Takahashi, E., The role of silicon, in *The Mineral Nutrition of the Rice Plant*, John Hopkins Press, Baltimore, MD, 1964, chap. 10.

155. Elawad, S.H., Street, J.J., and Gascho, G.J., Response of sugarcane to silicate source and rate. II. Leaf freckling and nutrient content, *Agron. J.*, 74, 484, 1982.

156. Verma, T.S. and Minhas, R.S., Effect of iron and manganese interaction on paddy yield and iron and manganese nutrition in silicon-treated and untreated soils, *Soil Sci.*, 147, 107, 1989.

157. Wallace, A., Participation of silicon in cation-anion balance as a possible mechanism for aluminum and iron tolerance in some gramineae, *J. Plant Nutr.*, 15, 1345, 1992.

158. Smith, S.S.E., Mycorrhizas of autotrophic higher plants, *Biol. Rev.*, 55, 475, 1980.

159. Tinker, P.B., Role of rhizosphere microorganisms in phosphorus uptake by plants, in *The Role of Phosphorus in Agriculture*, Khasawneh, F.E., Sample, E.C., and Kamprath, E.J., Eds., American Society of Agronomy, Madison, WI, 1980, chap. 22.

160. Howeler, R.H., Cadavid, L.F., and Burckhardt, E., Response of cassava to VA mycorrhizal inoculation and phosphorus application in greenhouse and field experiments, *Plant Soil*, 69, 327, 1982.

161. Abbott, L.K. and Robson, A.D., The effect of VA mycorrhizae on plant growth, in *VA Mycorrhiza*, Powell, C.L. and Bagyaraj, D.J., Eds., CRC Press, Boca Raton, FL, 1984, chap. 6.

162. Stribley, D.P., Mineral nutrition, in *Ecophysiology of VA Mycorrhizal Plants*, Safir, G.R., Ed., CRC Press, Boca Raton, FL, 1987, chap. 4.

163. Yost, R.S. and Fox, R.L., Influence of mycorrhizae on the mineral contents of cowpea and soybean grown in an oxisol, *Agron. J.*, 74, 475, 1982.

164. Kucey, R.M.N. and Janzen, H.H., Effects of VAM and reduced nutrient availability on growth and phosphorus and micronutrient uptake of wheat and field beans under greenhouse conditions, *Plant Soil*, 104, 71, 1987.

165. Biermann, B.J. and Lindermann, R.G., Increased geranium growth using pretransplant inoculation with a mycorrhizal fungus, *J. Am. Soc. Hortic. Sci.*, 108, 972, 1983.

166. Bethlenfalvay, G.J. and Franson, R.L., Manganese toxicity alleviated by mycorrhizae in soybeans, *J. Plant Nutr.*, 12, 953, 1989.

167. Lambert, D.H. and Weidensaul, T.C., Element uptake by mycorrhizal soybean from sewage-sludge-treated soil, *Soil Sci. Soc. Am. J.*, 55, 393, 1991.

CHAPTER 12

The Chemistry and Role of Metal Ion Chelation in Plant Uptake Processes

Stuart H. Laurie and John A. Manthey

INTRODUCTION

While 60 metallic elements are known to occur naturally, only a relatively small number are required by plants. Of these elements, Mn, Fe, Co, Cu, Zn, Mo, and Ni constitute the metal ion micronutrients; thus termed for the trace levels that are required for normal plant growth and reproduction.[1-3] These elements play critical roles in plant metabolism by being key components of enzyme active sites and by being involved in protein stabilization and activation (Table 1).[3-5] Proper plant and, ultimately, human nutrition requires an adequate supply of these trace metal micronutrients to crops.[6] The importance of micronutrients in agriculture is evident by the longstanding emphasis placed on research aimed at alleviating crop losses due to mineral deficiencies.[7] Micronutrient deficiencies in plants, with emphasis on iron and manganese, have recently been reviewed.[7-10]

The simplest forms of the required metal micronutrients in aqueous solution are the hydrated ions of Mn^{2+}, Fe^{3+}, Fe^{2+}, Co^{2+}, Cu^{2+}, Zn^{2+}, and Ni^{2+}, i.e., $M(H_2O)_6^{x+}$. In living systems, however, the amounts of these hydrated ions are generally exceedingly small. Yet, there is abundant evidence that their excess gives rise to toxicity and can be fatal. To avoid this, biological systems have developed tightly regulated controls of micronutrient uptake, translocation, and storage. Central to these controls is metal chelation, which regulates to a large degree the solubility and availability of metal micronutrients to plants. Thus, to understand the metabolic behavior of any trace metal we need to be able to determine the range and concentration of its various complexes (the speciation) as well as the trace metal's chemical and biological properties.

Table 1. Some Functional Biomacromolecules
          Containing Micronutrients Which
          Occur in Higher Plants

Micronutrient	Biomacromolecule
Fe	Cytochromes
	Cytochrome oxidase
	Ferredoxin
	Ascorbate oxidase
Cu	Cytochrome oxidase
	Plastocyanin
	Superoxide dismutase
Zn	Alcohol dehydrogenase
	Carbonic anhydrase
Mn	Superoxide dismutase

The objective of this chapter is to first briefly outline the chemical parameters that influence the solubilities of metal micronutrients in the rhizosphere, and thus, ultimately control the uptake of these elements by plants and microorganisms. The emphasis of this discussion is placed on metal ion chelation, and on the roles of micronutrient chelation in uptake by plants.

CHEMISTRY OF THE METAL ION MICRONUTRIENTS

Coordination Chemistry

The chemistry of the transition metal ion micronutrients is essentially that of coordination compounds, in which there is a central metal atom or ion covalently bonded to a number of other atoms, anions, or molecules. The number of atoms covalently bonded to the metal center defines the coordination number of the metal; this is mostly 4, 5, or 6. For a species (atomic or molecular) to function as a ligand it must possess at least one pair of valence electrons which it can donate to the metal ion, i.e., the ligands are Lewis bases. Ligands can donate either a single pair (monodentate) or more than one pair of electrons (polydentate).

The chelation properties of a particular metal are critically dependent on its oxidation state. Iron provides a good illustration of this in that Fe^{3+} has a strong preference for O-donor ligands such as phenolate, hydroxamate, and carboxylate. This is observed in the types of ligands in many synthetic Fe^{3+} chelating agents used in agriculture, and in microbially produced siderophores.[11] In contrast, Fe^{2+} preferentially chelates N and S ligands. Biological systems make subtle use of these differences in coordination in iron metabolism. An example of this is the fine tuning of cytochrome redox potentials by the integrated coordination spheres of the different heme active sites.

Thermodynamics of Coordination Compounds

The formation of coordination compounds in solution is a thermodynamically reversible process, such that the addition of each successive ligand can be described by a series of equilibrium reactions with corresponding equilibrium constants, viz.,

$$M(H_2O)_x + L \rightleftarrows M(H_2O)_{x-1}L + H_2O \tag{1}$$

$$M(H_2O)_{x-1}L + L \rightleftarrows M(H_2O)_{x-2}L_2 + H_2O \tag{2}$$

etc. (The charges have been omitted for clarity). The equilibrium constant for each step is given by

$$K_n = \left[M(H_2O)_{x-n}L_n\right] / \left[M(H_2O)_{x-n+1}L_{n-1}\right][L]^n \tag{3}$$

where $n = 1...6$, and the square brackets denote the equilibrium concentrations.

For thermodynamic and computational reasons, the equilibria are more usually described in terms of the formation reactions from the hydrated metal ions and ligands (omitting water molecules), i.e.,

$$M + nL \rightleftarrows ML_n \tag{4}$$

$$\beta_n = [ML_n] / [M][L]^n \tag{5}$$

β_n is the formation or stability constant. Compilations of K_n and β_n values are available in the literature.[12,13] In brief, the more positive the value of β_n the greater the extent of formation of the species ML_n. In general, especially for polydentate ligands, β_n values are very large. This is especially true for many synthetic Fe^{3+} chelates, e.g., those of HEDTA, EDTA, and DPTA, whose log stability constants range from 19 to 29, and the microbial ferric siderophores, whose log stability constants range above 30.[14]

For most ligands the β_n values follow the order $Mn^{2+} < Fe^{2+} < Co^{2+} < Ni^{2+} < Cu^{2+} > Zn^{2+}$, known as the Irving-Williams order. Hence, in this series Cu^{2+} ions form the strongest complexes. One of the most important considerations is that chelation is reversible and thus Le Chatelier's principle applies. Thus, regardless of the stability of the metal chelate, there is always some free metal ion and ligand in equilibrium with the chelated species. These may be minute amounts, well below analytical detection limits, but any reaction which alters their concentrations will upset the equilibrium. This is important in understanding micronutrient availability in both artificial systems and in the rhizosphere.

The equilibria resulting in maintaining constant levels of trace amounts of metal ions is the basis of the relatively newly applied metal chelate-buffered nutrient solutions.[15-19] Plant roots via uptake and adsorption act as sinks for the free metal ions, and it is only through these buffered systems that constant levels (typically far less than micromolar concentrations) of the metal ions can be maintained.

The impact of the reversibility of chelation can also be seen in soil systems, and thus extrapolated to soil solutions in the rhizosphere. There are many potential free metal ion sinks in soils which affect metal ion equilibria. For Fe^{3+}, precipitation at alkaline pH,[20] reduction to Fe^{2+},[21-23] adsorption of Fe-chelates to soil particles,[24,25] and absorption by the plant and rhizosphere microorganisms, all act as sinks which influence the equilibrium free metal ion concentration. Nearly all of these functions are influenced by pH, which is in dynamic flux at the root surface.

Kinetics of Coordination Compound Formation

A property often neglected in considering the influence of complexation, particularly in metabolic pathways, is kinetic reactivity. For divalent metal ions the attainment of equilibrium at ambient temperatures is usually achieved within the time of mixing the metal and ligand solutions. Such systems are referred to as kinetically labile, i.e., bond formation at the metal center is extremely rapid. Complexes of Fe^{3+} are usually labile in contrast to kinetically inert Cr^{3+} and Co^{3+} compounds. In these latter cases the bonding is slow, and attainment of equilibrium can take minutes, hours, or even days.

Also affecting the overall chelation rates are the relative concentrations of chelates and metal ions. These parameters controlling the rates of chelation strongly influence ligand exchange reactions in solution cultures and in the rhizosphere. Ligand exchange has important physiological implications in affecting metal ion availability in plant-microbe and microbe-microbe interactions. An example of this is the competition for Fe^{3+} through the production of specific microbial siderophores,[22,26-31] and in the case of the graminaceous plants, the phytosiderophores.[32]

Oxidation-Reduction Reactions

An important property of several transition metal micronutrients is the ability to exist in two or more oxidation states under physiological conditions. Indeed, the chemical behavior of these elements is very much dependent on the oxidation state; a small indication of this was previously noted in the differences between chelation of Fe^{2+} and Fe^{3+}. The ability of these metal micronutrients to exist in different oxidation states is extensively used in nature in the energy-coupled electron transfer processes of respiration and photosynthesis, and in the catalysis of many biologically important redox reactions.[5,33] The particular oxidation states of the metal micronutrients significantly control the biological activity and, ultimately, the availability of the element in living systems. An important consequence of redox state changes for Fe and Mn is the dramatic changes in solubility of these elements at root surfaces.[34] As reported elsewhere,[22,35-38] electron transfer processes generating either Fe^{2+} or Mn^{2+} species are critical to the absorption of these elements by most plants. The stabilities of individual metal chelates are profoundly affected by the oxidation states of the metal centers. This is illustrated in the large decrease in the stabilities of the Fe-siderophore chelates accompanying reduction of the Fe^{3+} metal center to Fe^{2+}. Electron transfer is used by many living systems to release and make available the Fe supplied to the organism as either soil-applied synthetic chelates or, for microorganisms, the ferrisiderophore compounds.[39]

METAL MICRONUTRIENTS IN THE SOIL ENVIRONMENT

The research into the mechanisms of metal ion micronutrient uptake by plants must ultimately be aimed at understanding these processes in the rhizosphere. The complexities of the chemical and biological factors that control ion uptake by soil-grown plants are obvious. The uptake of metal micronutrients is an integration of the root composition and biological activity, the microorganisms in the rhizosphere, and soil characteristics.[34,35,40] The transfer of metal ions from the soil particles to the plant can be described in simple form by

$$soil \rightarrow soil\ solution \rightarrow \frac{root\ surface}{(cell\ wall,\ mucilage)} \rightarrow \frac{root\ cell}{(apoplast, plasma\ membrane)}$$

Involved at each stage are the microorganisms in the rhizosphere. This scheme shows the importance of the soil solution as an interface between soil and roots. Detailed descriptions of root-soil interfaces are provided by Uren and Reisenauer,[35] and elsewhere.[41-43] As an interface, the soil solution is critical in the transfer of metal micronutrients to roots. Soil solutions can be characterized by methods described by Jeffery and Uren.[44] However, studies of the composition of soil solutions are hampered because isolation potentially disturbs the dynamic equilibria within the natural ecosystem. While the total (analytical) measurement of very low concentrations of metal ions in both dissolved (i.e., filterable) and in particulate forms can be carried out with accuracy, measurement of the exact chemical forms of each metal (their speciation) is not practically possible. At present, such information can only be estimated through computer modeling.

One of the major hurdles in understanding the soil solution is the dynamic state of the rhizosphere. Along with the microorganisms, plants exert enormous influences on the chemical composition of rhizospheres. These effects range from acidification and electron transfer, to the release of root exudates.[35,43] Root growth and differentiation of uptake active regions strongly modulate these effects on the rhizosphere. Particularly important are the root exudates, which provide the energy for the rhizosphere microorganisms;[35,42] hence, these are important factors in rhizosphere ecology.[45] The entire system outlined above, and reported in greater detail by Crowley and Gries,[32] is a function of plant growth and thus, integrated functions of time.

CHELATION AND MICRONUTRIENT SUPPLY TO PLANTS

Chelation has long been recognized as an important factor in metal ion uptake by plants. Micronutrient deficiencies are commonly associated with calcareous soils where precipitation renders many metal ions unavailable for plant absorption. Perhaps the most studied example of this is the solubility limitations of iron in carbonate soils. Yet even under nonalkaline conditions, plants can become severely iron deficient unless chelating agents (or other solubilizing agents, e.g., H^+ and reductants) are present in the soil solution. This latter observation explains the use of synthetic chelating agents in plant nutrition over the last 30 to 40 years. Many hundreds of papers have been published in this area. Yet, in spite of this intensive research on micronutrient uptake, there is still disagreement over the mechanisms of plant uptake of different micronutrients; in particular whether it is only the free metal ions that are absorbed at the root surface.[46] Linked to this are the still poorly understood roles that the rhizosphere microorganisms have in modifying the chelation and redox states of the various micronutrient metal ions.

The naturally occurring chelating agents in the rhizosphere are important influences in micronutrient cycling,[47] and may be broadly divided into two groups:[48] (1) biochemical, similar to those which occur in living organisms, and (2) complex polymers, generated by secondary reactions and which bear little resemblance to their analogues in living organisms. Included in the first group are the chelating agents found in plant root exudates (including the phytosiderophores of the graminaceous plants), substances released from decaying organic matter, and microbial chelating agents (mainly siderophores). The second group, which is actually derived from the first, is comprised of the humic substances (humates). The two groups are not necessarily mutually exclusive. Among the humates only the fulvic acids form metal complexes which are water soluble.

The chelating agents present in soil solutions, as reviewed by Stevenson,[47] can be assumed to increase metal ion solubility by virtue of their ability to form highly soluble metal chelates. Hodgson and co-workers measured the concentrations of Cu, Zn, and Mn and their degrees of complexation in several soil solutions.[49-51] Their results are summarized in Table 2. In general, various studies point to approximately 50% Zn, 90% Mn, and 99%

Table 2. Concentrations and Degree of Complexation of
          Cu, Zn, and Mn in Soil Solutions

Metal	Total concentration in soil solution/10^{-6} M	Degree of complexation (%)
Cu	< 0.01–0.6	89–99.8
Zn	< 0.03–3	28–99
Mn	< 0.02–68	84–99

Values taken from Hodgson and co-workers.[49-51]

Cu being in complexed forms. In contrast to Cu and Zn, most of the complexed Mn is in inorganic (oxide) forms rather than being organically bound.

Iron concentrations in soil solutions are normally in the range of 10^{-8} to 10^{-6} M, with a very high percentage as complexed species.[52] Crowley et al.[24] have shown that the majority of soluble iron in rhizospheres is likely chelated to microbial siderophores. They report that while the levels of siderophores in the bulk soil occur in the range 20 to 200 nM, the concentrations of these compounds in the rhizosphere, where microbial activity is high, are considerably greater, and may be dominant factors in dictating Fe supply in the rhizosphere. Similar considerations, as outlined by Crowley and Gries,[32] and Mori,[53] can be given to the phytosiderophores released by the graminaceous monocots during Fe deficiency.

Solution Micronutrient Uptake by Higher Plants

In the following section the mechanisms by which plants grown in nutrient solutions absorb metal ion micronutrients from chelated sources are examined. Preference is given to those studies using solution cultures for plant growth, because such solutions have the advantage of allowing for varying levels of control over nutrient supply.[54] It is anticipated that such studies yield information about metal ion absorption processes of roots in natural conditions, yet one is always aware of the shortcomings of hydroponic systems in accurately mimicking soil conditions. In spite of this, hydroponics remains a powerful tool in controlling metal ion speciation and supply to plants. This is especially true with the development and use of metal ion chelate-buffered solutions, and through the use of computer programming to accurately estimate metal nutrient concentrations and speciation.[15-19] The present discussion will be restricted to metal ion uptake by higher plants.

Iron

Iron deficiency in crops is a well-recognized agricultural problem where the crops are grown on calcareous soils. Not surprisingly, therefore, the uptake and translocation of iron by plants have been intensively studied. There are a number of reviews of this topic.[7,9,21,22,55-57]

There is abundant evidence that many dicotyledons reduce chelated Fe^{3+} to Fe^{2+} at the root surfaces, with subsequent transport of the Fe^{2+} species through the root plasmalemma.[21,22] As reported by Holden et al.,[58] and reviewed by Bienfait,[21] significant advances have been made in the characterization of the enzyme systems involved in these electron transfer processes catalyzed at the plasmalemma surface. Certain dicotyledons, e.g., tomato and soybean, have been found to release during iron deficiency reductants such as caffeic acid,[59] chlorogenic acid,[60] and D-galacturonic acid.[61] All of these compounds are capable of chelating Fe^{3+} prior to reduction. Little is known to what extent these compounds and other components of the cell surface, including the cell wall matrix[62,63] and cell wall-associated enzymes,[64] mediate iron solubility via pH, redox, and chelation processes at the root cell surface.

While extracellular Fe^{3+} reduction appears to be an essential step in iron uptake by dicotyledons, certain monocots acquire iron from ferric phytosiderophores. No appreciable levels of external reduction seems to occur prior to cellular uptake of the intact ferriphytosiderophores.[65] There is now some speculation whether dicots can similarly acquire iron as intact iron chelates. Reasons for this consideration are (1) the possible dominance of microbial siderophores in controlling iron speciation at root surfaces (as discussed above), and (2) the fact that these compounds, due to their extremely low reduction potentials,[66] are poor substrates for the Fe^{3+} chelate reduction at root surfaces. Without some mediating activity, these two considerations do not easily reconcile with obligatory Fe^{3+} reduction at the root surface of these important Fe chelates. Yet again, because of our poor understanding of the actual metal ion speciation at root surfaces, as modified by the roots and the rhizosphere microflora, little is known of the actual iron species acted upon by the root.[36]

To resolve these questions, research is needed where doubly labeled ^{55}Fe-^{14}C-chelates are supplied to the roots of dicotyledons grown under sterile conditions. This will permit determination of the ability of plants to acquire iron from ferrisiderophores without having to account for microbial activities which possibly alter the metal speciation.

Copper

The conclusion from a large number of studies on copper absorption by plant roots is that chelating substances enhance Cu solubility but inhibit Cu absorption by plants. The hydrated Cu ion, Cu_{aq}^{2+}, is considered the absorbed species.[67-69] Concentrations of Cu_{aq}^{2+} below 10^{-13} to 10^{-14} M result in copper deficiency. The nature of the Cu chelate is also important. Studies with barley showed that the larger the negative charge on the Cu chelate the greater the inhibition of Cu uptake.[70] These results can be correlated with the slower rates of dissociation of the more negatively charged chelates.

While the weight of evidence points to dissociation of Cu chelates as a prerequisite for Cu absorption,[71] studies with corn plants and radioactively labeled tetraethylenepentamine (a pentadentate chelate) indicated that the intact Cu chelate is taken up and translocated.[72] Secretion of low molecular weight organic compounds by roots which chelate Cu^{2+} has been observed by electron paramagnetic resonance spectroscopy studies of wheat plants.[71,73] Thus, care must be taken in contrasting the results of Cu absorption from culture solutions with those involving soil as the growth medium. This is true in light of the differences in the Cu chelating agents in the different systems, and the importance of chelation in influencing the uptake rates. Additionally, in soil, the more important aspect of Cu uptake by plants may be the supply of Cu to the soil solution. Supporting this is a report that the synthetic chelates, EDTA and DTPA, were found to enhance Cu concentrations in barley grown in soils.[74]

Zinc

Zinc, like copper, can form a wide range of complex species within the soil environment;[44] as with copper, chelation has the dual effect of increasing soluble zinc (and hence its availability) but decreasing plant uptake. A number of studies point to the aqueous ion $Zn(H_2O)_6^{2+}$ as the species that is preferentially absorbed by plant roots.[16,75] However, while chelates such as EDTA have been found to suppress zinc uptake in a number of different plants, there is a report of the chelate $[Zn(EDTA)]^{2-}$ being found in extracts from rye grass grown in an EDTA-containing nutrient solution.[76]

The critical minimum concentration of Zn^{2+} ions for healthy growth has been reported as either $10^{-10.6}$ or $10^{-7} M$. Halvorson and Lindsay[77] attempted to explain these differences in terms of a diffusion gradient within the root free space. Studies by Laurie et al.[78] (see below) are in close agreement with the lower value. A recent report of the use of metal chelate-buffered nutrient solutions showed that the dry weights of barley reached a maximum at Zn^{2+} activities at about $3 \times 10^{-11} M$.[16] In this study, zinc deficiencies were shown to have significant influences in the absorption and retention of a number of mineral nutrients; a phenomenon attributed to the role of zinc in maintaining the integrity of certain plasma membrane transport proteins.[15]

Manganese

The uptake of Mn by plants has been intensively investigated, and recently reviewed.[10] Significant advances have been made in establishing the critical levels of Mn needed in nutrient solutions for proper plant growth. Through the use of metal ion chelate buffer systems, Webb et al.[79] have established this level to be on the order of $10^{-8.3}$ M. Of the micronutrients, Mn forms the least stable chelates (see the Irving-Williams order). Not surprisingly, therefore, most of the Mn in soil solution is present as inorganic Mn oxides with minor organic species.[51]

Like iron, two possible oxidation states are important for manganese within the soil environment.[35,37] At above neutral pH in well aerated soils, Mn^{4+} is found as insoluble oxides. Reduction of Mn^{4+} to Mn^{2+} in the rhizosphere can occur through microbial activities, and is important in Mn supply to plants.[38,80,81] Similar to Zn^{2+}, it is assumed that the primary Mn^{2+} species absorbed by plants is the free metal.[82-84] Chelation of Mn^{2+} by fulvic acids has been reported to inhibit Mn uptake by corn and sunflower plants, although intact Mn-fulvic acid complexes appeared to be taken up by the plants.[85] On the other hand, chelation with EDTA has been reported to lead to increased Mn uptake by cotton and tea seedlings.[86] A recent experimental innovation has been the use of nuclear magnetic resonance spectroscopy to study Mn uptake in intact plants in a noninvasive manner.[87]

A Combined Multimetal Computer Simulation Approach

Most of the work cited in this chapter has concentrated on factors influencing the uptake of a single metal type and generally has not considered the influences from other metal ions present in the growth medium. Yet, mutual interactions between the micronutrients and other compounds are well documented.[88] Invariably, high concentrations of one metal ion can inhibit the uptake of some other metal; such inhibitions have been recorded for the pairs Fe-Mn, Cu-Mn, Cu-Zn, Ca-Zn, Ca-Mn, Ca-Fe, and Mg-Zn.

Most of the reports also do not consider the total speciation. For example, EDTA is a commonly added chelating agent to hydroponic growth solutions, and its use is to enhance Fe solubility by the formation of the chelated $[Fe(EDTA)]^-$ ion. However, the formation of the iron complex is very pH dependent and other metal ions can compete with Fe for the EDTA.

Laurie and co-workers[78] have recently combined modern analytical techniques and computer simulation to study simultaneously the uptake by barley plants of the macro- and micronutrient metal ions from the same medium. The absorptions of these metal ions were then correlated with the speciation and concentrations in the nutrient solutions.[78,89] These nutrient solutions were of well-defined compositions and contained a range of concentrations of either EDTA or DTPA. To model the speciation in the solutions a computer program, NUTRIENT, was written specifically for nutrient solutions. Its authenticity was established by comparison of its results with those from other programs and by comparison to experimental data.[90]

Table 3.   Mean Concentrations of Metal Ions in Barley Roots (μg/g of Dry Plant Material) as a Function of the EDTA/Fe Ratio in the Growth Solution After 15 Days of Growth[a]

EDTA/Fe ratio	Cu	Fe	Mn	Zn
1.00	7.1	3187.4	253.0	30.7
1.05	11.3	2184.3	360.2	34.7
1.10	35.6	1416.9	411.2	41.2
1.20	118.6	469.6	675.0	51.7
2.00	63.6	105.8	1135.4	17.5
LSD ($p < 0.01$)[b]	8.6	889.4	106.7	6.1

[a] Data taken from Reference 78.
[b] LSD = least significant difference values.

In this study, barley plants were harvested at regular intervals and the metal contents of the roots and shoots measured. Increasing the EDTA from an EDTA:Fe ratio of 1:1 to 2:1 resulted in a decrease in the concentrations of the simple hydrated ion forms of Cu^{2+}, Fe^{3+}, and Zn^{2+} by several orders of magnitude, and that of Mn^{2+} by a hundredfold. The variations in the metal uptake by the barley plants did not mimic the solution changes. This is shown by the metal ion content of the roots (Table 3). The data in Table 3 show that only Fe displayed the expected decrease in uptake as the solution complexation increased. The largest differences in root uptake were seen at the highest EDTA concentration, with all the plants becoming chlorotic as judged by their stunted growth and yellowing of leaves. However, the metal concentrations in the shoots were more consistent with zinc deficiency than with iron deficiency. Attribution of this chlorosis to zinc deficiency has been made previously.[91] Significantly, it was only in this treatment that the Zn_{aq}^{2+} concentration fell below the suggested critical level of $10^{-10.6}$ M for this ion (see previous section). The plants appeared to tolerate much lower levels of Cu and Fe than Zn.

An analogous study made with DTPA instead of EDTA showed significant differences in behavior consistent with uptake being dependent on the nature of the chelating agent.[89] The slower rates of uptake from the DTPA media as compared to the corresponding EDTA treatments pointed to the importance of kinetic (e.g., rate of dissociation of metal chelates) as well as thermodynamic factors in determining the metal ion uptake. From this, a simple model, shown in Scheme 1, was proposed to explain the mechanisms of metal uptake by roots. In this scheme, path **a** shows a free metal ion arriving at the plasma membrane. It is also probable that the complexed metal forms (indicated by ML) can reach the root surface, path **b**. There are then two possibilities: (1) dissociation at the plasma membrane immediately precedes metal ion absorption by the root, or (2) the metal complexes ML may

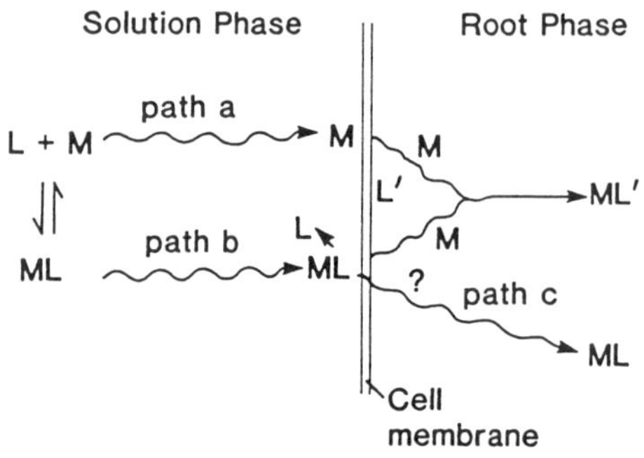

Scheme 1. See text for details. (From Laurie, S.H. et al., *J. Exp. Bot.*, 42, 515, 1991, by permission of Oxford University Press.)

be absorbed intact (path **c**). In path **b**, if the dissociation rate of ML at the plasma membrane is faster than the rate of uptake of M by the roots, then the uptake should be independent of the nature of the complexing agent. In this case, path **a** and path **b** would be indistinguishable. However, if the dissociation rate is slower than the uptake rate, the metal absorption would then depend on both kinetic and thermodynamic (stability constant) properties of the complex and, therefore, on the nature of the complexing ligand. The dependence on the nature of the ligands observed in our experiments suggest either path **b** or **c** could be involved. Under conditions of metal stress, priority could well be given to path **c** — as indicated in the reports mentioned above showing the possibility of uptake of intact chelated species by plant roots.

Once within the roots, the free metal ions must undergo further complexation to aid their transportation and to avoid precipitation of metal hydroxides. Within the roots there are a number of recognized complexing agents, as indicated by L' in the scheme.

REFERENCES

1. Eskew, D.L., Welch, R.M., and Cary, E.E., Nickel: an essential micronutrient for legumes and possibly all higher plants, *Science*, 222, 621, 1983.
2. Marschner, H., *Mineral Nutrition of Higher Plants*, Academic Press, New York, 1986.
3. Hewitt, E.J., A perspective of mineral nutrition: essential and functional metals in plants, in *Metals and Micronutrients: Uptake and Utilization by Plants*, Robb, D.A. and Pierpoint, W.S., Eds., Academic Press, New York, 1983, chap. 14.
4. Robb, D.A. and Pierpoint, W.S., Eds., *Metals and Micronutrients: Uptake and Utilization by Plants*, Academic Press, New York, 1983.
5. Römheld, V. and Marschner, H., Function of micronutrients in plants, in *Micronutrients in Agriculture*, 2nd ed., Mortvedt, J.J., Cox, F.R., Shuman, L.M., and Welch, R.M., Eds., Soil Science Society of America, Madison, WI, 1991, 297.

6. Van Campen, D.R., Trace elements in human nutrition, in *Micronutrients in Agriculture,* 2nd ed., Mortvedt, J.J., Cox, F.R., Shuman, L.M., and Welch, R.M., Eds., Soil Science Society of America, Madison, WI, 1991, 663.

7. Wallace, A. and Wallace, G.A., Some of the problems concerning iron nutrition of plants after four decades of synthetic chelating agents, *J. Plant Nutr.,* 15, 1487, 1992.

8. Mortvedt, J.J., Cox, F.R., Shuman, L.M., and Welch, R.M., Eds., *Micronutrients in Agriculture,* Soil Science Society of America, Madison, WI, 1991.

9. Chen, Y. and Hadar, Y., Eds., *Iron Nutrition and Interactions in Plants,* Kluwer Academic, Dordrecht, 1991.

10. Graham, R.D., Hannan, R.J., and Uren, N.C., Eds., *Manganese in Soils and Plants,* Kluwer Academic, Dordrecht, 1988.

11. Matzanke, B.F., Structures, coordination chemistry and function of microbial iron chelates, in *CRC Handbook of Microbial Iron Chelates,* Winkelmann, G., Ed., CRC Press, Boca Raton, FL, 1991, 15.

12. Sillen, L.G. and Martell, A.E., *Stability Constants,* Spec. Publ. No. 17, 1964; Suppl. Publ. No. 25, 1971, Chemical Society, London, 1971.

13. Mason, B. and Moore, G.B., *Principles of Geochemistry,* 4th ed., Wiley & Sons, New York, 1982.

14. Crumbliss, A.L., Aqueous solution equilibrium and kinetic studies of iron siderophore and model siderophore complexes, in *CRC Handbook of Microbial Iron Chelates,* Winkelmann, G., Ed., CRC Press, Boca Raton, FL, 1991, 117.

15. Norvell, W.A. and Welch, R.M., Growth and nutrient uptake by barley (*Hordeum vulgare* L. cv Herta): studies using an N-(2-hydroxyethyl)ethylenedinitrilotriacetic acid-buffered nutrient solution techniques. I. Zinc ion requirements, *Plant Physiol.,* 101, 619, 1993.

16. Norvell, W.A. and Welch, R.M., Growth and nutrient uptake by barley (*Hordeum vulgare* L.cv Herta): studies using an N-(2-hydroxyethyl)ethylenedinitrilotriacetic acid-buffered nutrient solution techniques. II. Role of zinc in the uptake and root leakage of mineral nutrients, *Plant Physiol.,* 101, 627, 1993.

17. Parker, D.R., Chaney, R.L., and Norvell, W.A., Chemical equilibrium models: application to plant nutrition research, in *Chemical Equilibrium and Reaction Models,* Loeppert, R.H., Ed., Soil Science Society of America, Madison, WI, in press.

18. Bell, P.F., Chaney, R.L., and Angle, J.S., Free metal activity and total metal concentrations as indices of micronutrient availability to barley (*Hordeum vulgare* L. 'Klages'), in *Iron Nutrition and Interactions in Plants,* Chen, Y. and Hadar, Y., Eds., Kluwer Academic, Boston, 1991, 69.

19. Chaney, R.L., Bell, P.F., and Coloumbe, B.A., Screening strategies for improved nutrient uptake and use by plants, *Hortic. Sci.,* 24, 565, 1989.

20. Schwertmann, U., Solubility and dissolution of iron oxides, *Plant Soil,* 130, 3, 1991.

21. Bienfait, F., Biochemical basis of iron efficiency reactions in plants, in *Iron Transport in Microbes, Plants and Animals,* Winkelmann, G., van der Helm, D., and Neilands, J.B., Eds., VCH Publishers, Weinheim, Germany, 1987, chap. 18.

22. Römheld, V., Existence of two different strategies for the acquisition of iron in higher plants, *Iron Transport in Microbes, Plants and Animals,* Winkelmann, G., van der Helm, D., and Neilands, J.B., Eds., VCH Publishers, Weinheim, Germany, 1987, chap. 19.

23. Wahid, P.A. and Kamalam, N.V., Reductive dissolution of crystaline and amorphous Fe(III) oxides by microorganisms in submerged soil, *Biol. Fertil. Soils,* 15, 144, 1993.

24. Crowley, D.E., Reid, C.P.P., and Szaniszlo, P.J., Microbial siderophores as iron sources for plants, *Iron Transport in Microbes, Plants and Animals,* Winkelmann, G., van der Helm, D., and Neilands, J.B., Eds., VCH Publishers, Weinheim, Germany, 1987, chap. 20.

25. Norvell, W.A., Reactions of metal chelates in soils and nutrient solutions, in *Micronutrients in Agriculture,* 2nd ed., Mortvedt, J.J., Cox, F.R., Shuman, L.M., and Welch, R.M., Eds., Soil Science Society of America, Madison, WI, 1991, 187.

26. Buyer, J.S., Kratzke, M.G., and Sikora, L.J., Microbial siderophores and rhizosphere ecology, in *The Biochemistry of Metal Micronutrients in the Rhizosphere,* Manthey, J.A., Crowley, D.E., and Luster, D.G., Eds., Lewis Publishers, Boca Raton, FL, ch. 6.

27. Höfte, M., Woestyne, M.V., and Verstraete, W., Role of siderophores in plant growth promotion and plant protection by fluorescent *Pseudomonads,* in *The Biochemistry of Metal Micronutrients in the Rhizosphere,* Manthey, J.A., Crowley, D.E., and Luster, D.G., Eds., Lewis Publishers, Boca Raton, FL, ch. 7.

28. Handelsman, J. and Parke, J.L., Mechanisms in biocontrol of soilborne plant pathogens, in *Plant Microbe Interact.,* 3, 27, 1984.

29. Bossier, P., Höfte, M., and Verstraete, W., Ecological significance of siderophores in soil, *Adv. Microbiol. Ecol.,* 10, 385, 1988.

30. Loper, J.E. and Buyer, J.S., Siderophores in microbial interactions on plant surfaces, *Mol. Plant-Microbe Interact.,* 4, 5, 1991.

31. Crowley, D.E., Römheld, V., Marschner, H., and Szaniszlo, P.J., Root-microbial effects on plant iron uptake from siderophores and phytosiderophores, *Plant Soil,* 142, 1, 1992.

32. Crowley, D.E. and Gries, D., Modeling of iron availability in the plant rhizosphere, in *The Biochemistry of Metal Micronutrients in the Rhizosphere,* Manthey, J.A., Crowley, D.E., and Luster, D.G., Eds., Lewis Publishers, Boca Raton, FL, ch. 14.

33. Moore, A.L. and Cottingham, I.R., Characterization of the higher plant respiratory chain, in *Metals and Micronutrients: Uptake and Utilization by Plants,* Robb, D.A. and Pierpoint, W.S., Eds., Academic Press, New York, 1983, 169.

34. Moraghan, J.T. and Mascagni, H.J., Environmental and soil factors affecting micronutrient deficiencies and toxicities, in *Micronutrients in Agriculture,* 2nd ed., Mortvedt, J.J., Cox, F.R., Shuman, L.M., and Welch, R.M., Eds., Soil Science Society of America, Madison, WI, 1991, 371.

35. Uren, N.C. and Reisenauer, H.M., The role of root exudates in nutrient acquisition, in *Advances in Plant Nutrition,* Vol. 3, Praeger Publishers, New York, 1988, 79.

36. Uren, N.C., Forms, reactions and availability of iron in soils, *J. Plant Nutr.,* 7, 165, 1984.

37. Uren, N.C., Chemical reduction of an insoluble higher oxide of manganese by plant roots, *J. Plant Nutr.*, 4, 65, 1981.

38. Kothari, S.K., Marschner, H., and Römheld, V., Effect of a vesicular-arbuscular mycorrhizal fungus and rhizosphere microorganisms on manganese reduction in the rhizosphere and manganese concentration in maize (*Zea mays* L.), *New Phytol.*, 117, 645, 1991.

39. Emery, T., Reductive mechanisms of iron assimilation, *Iron Transport in Microbes, Plants and Animals*, Winkelmann, G., van der Helm, D., and Neilands, J.B., Eds., VCH Publishers, Weinheim, Germany, 1987, 235.

40. Hemming, B.C., Microbial-iron interactions in the plant rhizosphere. An overview, *J. Plant Nutr.*, 9, 505, 1986.

41. Jenny, H., Pathways of ions from soil into root according to diffusion models, *Plant Soil*, 25, 265, 1966.

42. Rovira, A.D., Bowen, G.D., and Foster, R.C., The significance of rhizosphere microflora and mycorrhizas in plant nutrition, in *Inorganic Plant Nutrition*, Lauchli, A. and Bieleski, R.L., Eds., Springer-Verlag, New York, 1983, 61.

43. Marschner, H., Römheld, V., Horst, W.J., and Martin, P., Root induced changes in the rhizosphere. Importance for the mineral nutrition of plants, *Z. Pflanzernaehr. Bodenkd.*, 149, 441, 1986.

44. Jeffrey, J.J. and Uren, N.C., Copper and zinc species in the soil solution and effects of soil pH, *Aust. J. Soil Res.*, 21, 479, 1983.

45. Buyer, J.S. and Sikora, L.J., Rhizosphere interactions and siderophores, *Plant Soil*, 129, 101, 1990.

46. Wallace, A. and Wallace, G.A., in *9th Proc. Int. Plant Nutr. Colloq.*, Vol. 2, Scaife, A., Ed., C.A.B., Farnham Royal, U.K., 1982, 696.

47. Stevenson, F.J., Organic matter-micronutrient reactions in soil, in *Micronutrients in Agriculture*, 2nd ed., Mortvedt, J.J., Cox, F.R., Shuman, L.M., and Welch, R.M., Eds., Soil Science Society of America, Madison, WI, 1991, 145.

48. Stevenson, F.J. and Ardakani, M.S., Organic matter reactions involving micronutrients in soil, in *Micronutrients in Agriculture*, Mortvedt, J.J., Giordano, P.M., and Linday, W.L., Eds., Soil Science Society of America, Madison, WI, 1972, 79.

49. Hodgson, J.F., Geering, H.R., and Norvell, W.A., Micronutrient cation complexes in soil solution: partition between complexed and uncomplexed forms by solvent extraction, *Proc. Soil Sci. Soc. Am.*, 29, 665, 1965.

50. Hodgson, J.F., Linday, W.L., and Trierweiler, J.F., Micronutrient cation complexing in soil solution. II. Complexing zinc and copper in displaced solution from calcareous soils, *Proc. Soil Sci. Soc. Am.*, 30, 723, 1966.

51. Geering, H.R., Hodgson, J.F., and Sdano, C., Micronutrient cation complexes in soil solution. IV. The chemical state of manganese in soil solution, *Proc. Soil Sci. Soc. Am.*, 33, 81, 1969.

52. Loneragan, J.F., The availability and absorption of trace elements in soil-plant systems and their relation to movement and concentrations of trace elements in plants, in *Trace Elements in Soil-Plant-Animal Systems*, Nicholas, D.J.D. and Egan, A.R., Eds., Academic Press, London, 1975, 109.

53. Mori, S., Mechanisms of iron acquisition by graminaceous (strategy II) plants, in *The Biochemistry of Metal Micronutrients in the Rhizosphere*, Manthey, J.A., Crowley, D.E., and Luster, D.G., Eds., Lewis Publishers, Boca Raton, FL, ch. 15.

54. Epstein, E., *Mineral Nutrition of Plants: Principles and Perspectives,* Wiley International, New York, 1972, 36.

55. Vose, P.B., Iron nutrition in plants: a world overview, *J. Plant Nutr.,* 5, 233, 1982.

56. Clark, R.B., Iron deficiency in plants grown in the great plains of the U.S., *J. Plant Nutr.,* 5, 251, 1982.

57. Barton, L.L. and Hemming, B.C., Eds., *Iron Chelation in Plants and Soil Microorganisms,* Academic Press, New York, 1993.

58. Holden M.J., Luster, D.G., and Chaney, R.L., Enzymatic iron reduction at the root plasma membrane: partial purification of the NADH-Fe chelate reductase, in *The Biochemistry of Metal Micronutrients in the Rhizosphere,* Manthey, J.A., Crowley, D.E., and Luster, D.G., Eds., Lewis Publishers, Boca Raton, FL, ch. 18.

59. Olsen, R.A., Brown, J.C., Bennet, J.H., and Blumme, D., Reduction of Fe^{3+} as it relates to Fe chlorosis, *J. Plant Nutr.,* 5, 433, 1982.

60. Hether, N.H., Olsen, R.A., and Jackson, L.L., Chemical identification of iron reductants exuded by plant roots, *J. Plant Nutr.,* 7, 677, 1984.

61. Micera, G., Deiana, S., Dessi, A., Pusino, A., and Gessa, C., Oxidation of D-galacturonic acid by metal ions, *Inorg. Chim. Acta,* 120, 49, 1986.

62. Gessa, C. and Deiana, S., Fibrillar structure of Ca-polygalacturonate as a model for a soil-root interface. I. A hypothesis of the arrangement of the polymeric chains inside the fibrils, *Plant Soil,* 129, 211, 1990.

63. Gessa, C. and Deiana, S., Fibrillar structure of Ca-polygalacturonate as a model for a soil-root interface. II. A comparison with natural root mucilage, *Plant Soil,* 140, 1, 1992.

64. Manthey, J.A., Characterization of the nonenzymatic NADH-coupled Fe^{3+} reduction by lignin in citrus root cell walls, *Plant Physiol. Biochem.,* 30, 639, 1992.

65. Bar-Ness, E., Hadar, Y., Chen, Y., Römheld, V., and Marschner, H., Short-term effects of rhizosphere microorganisms on Fe uptake from microbial siderophores by maize and oat, *Plant Physiol.,* 100, 451, 1992.

66. Raymond, K.N., Muller, G., and Matzanke, B.F., Complexation of iron by siderophores. A review of their solution and structural chemistry and biological function, *Top. Curr. Chem.,* 123, 49, 1984.

67. Graham, R.D., Absorption of copper by plant roots, in *Copper in Soils and Plants,* Loneragan, J.F., Robson, A.D., and Graham, R.D., Academic Press, London, 1981, 141.

68. Dragun, J., Baker, D.E., and Risius, M.L., Growth and element accumulation by two single-cross corn hybrids as affected by copper in solution, *Agron. J.,* 68, 466, 1976.

69. Harrison, S.J., Lepp, N.W., and Phipps, D.A., Uptake patterns of free and complexed copper ions in excised roots of barley, *Z. Pflanzenphysiol.,* 113, 445, 1984.

70. Coombes, A.J., Phipps, D.A., and Lepp, N.W., Uptake patterns of free and complexed copper ions in excised roots of barley, *Z. Pflanzenphysiol.,* 82, 435, 1977.

71. Goodman, B.A. and Linehan, D.J., An electron paramagnetic resonance study of the uptake of Mn(II) and Cu(II) by wheat roots, in *The Soil-Root Interface,* Harley, J.L. and Scott-Russell, R., Academic Press, London, 1979, 67.

72. Smeulders, F. and van der Geign, S.C., *In situ* mobilization of heavy metals with tetraethylenepentamine in natural soils and its effect on toxicity and plant growth, *Plant Soil,* 70, 50, 1983.

73. Linehan, D.J., Micronutrient cation absorption by roots and uptake by plants, *J. Exp. Bot.,* 35, 1571, 1984.

74. Gonzales, O.C., Goecke, R.F., and Schalscha, E.B., Effect of synthetic chelators on micronutrient availability in two Chilean soils, *Agrochimica,* 16, 387, 1972.

75. Baker, D.A., Uptake of cations and their transport within the plant, in *Metals and Micronutrients: Uptake and Utilization by Plants,* Robb, D.A. and Pierpoint, W.S., Eds., Academic Press, New York, 1983, 3.

76. Bremmer, I. and Knight, A.H., The complexes of zinc, copper and manganese in ryegrass, *Br. J. Nutr.,* 24, 279, 1970.

77. Halvorson, A.D. and Lindsay, W.L., The critical Zn^{2+} concentration for corn and the nonabsorption of chelated zinc, *Soil Sci. Soc. Am. J.,* 41, 531, 1977.

78. Laurie, S.H., Tancock, N.P., McGrath, S.P., and Sanders, J.R., Influence of complexation on the uptake by plants of iron, manganese, copper and zinc. I. Effect of EDTA in a multi-metal and computer simulation study, *J. Exp. Bot.,* 42, 509, 1991.

79. Webb, M.J., Norvell, W.A., Welch, R.M., and Graham, R.D., Using a chelate-buffered nutrient solution to establish critical tissue levels and the solution activity of Mn^{2+} required by barley (*Hordeum vulgare* L. cv Herta), *Plant Soil,* 153, 195, 1993.

80. Marschner, P., Ascher, J.S., and Graham, R.D., Effect of manganese-reducing rhizosphere bacteria on the growth of *Gaeumannomyces graminis* var *tritici* and on manganese uptake by wheat (*Triticum aestivum* L.), *Biol. Fertil. Soils,* 12, 33, 1991.

81. Rengel, Z., Pedler, J.F., and Graham, R.D., Control of Mn status in plants and rhizosphere: genetic aspects of host and pathogen effects in the wheat take-all interaction, in *The Biochemistry of Metal Micronutrients in the Rhizosphere,* Manthey, J.A., Crowley, D.E., and Luster, D.G., Eds., Lewis Publishers, Boca Raton, FL, ch. 10.

82. Marschner, H., Mechanisms of manganese acquisition by roots from soils, in *Manganese in Soils and Plants,* Graham, R.D., Hannam, R.J., and Uren, N.C., Eds., Kluwer Academic, Dordrecht, 1988, 191.

83. Kochian, L.V., Mechanisms of micronutrient uptake and translocation in plants, in *Micronutrients in Agriculture,* 2nd ed., Mortvedt, J.J., Cox, F.R., Shuman, L.M., and Welch, R.M., Eds., Soil Science Society of America, Madison, WI, 1991, 229.

84. Clarkson, D.T., The uptake and translocation of manganese by plant roots, in *Manganese in Soils and Plants,* Graham, R.D., Hannam, R.J., and Uren, N.C., Eds., Kluwer Academic, Dordrecht, 1988, 101.

85. Karpukhin, A.R., Kauricher, I.S., Shestakov, E.I., and Shkrarin, B.I., Uptake of complex manganese-organic compounds by plants, *Izv. Timiryazevsk. S-kh. Acad.,* 3, 92, 1984; *Chem. Abstr.,* 101, 89750u, 1984.

86. Sardzhveladze, A.G. and Shuvalov, Y.N., Interaction of iron and manganese in the solution-tea plant system, *Subtrop. Kul't.,* 6, 34, 1983; *Chem. Abstr.,* 100, 190868t, 1984.

87. Ratkovic, S. and Vucinic, Z., The ^{1}H nmr relaxation method applied in studies of continual absorption of paramagnetic Mn^{2+} ions by roots of intact plants, *Plant Physiol. Biochem.,* 28, 617, 1990.

88. Robson, A.D. and Pitman, M.G., Inorganic plant nutrition, *Encyclopedia of Plant Physiology,* Vol. 15A, Lauchli, A. and Bideski, R.L., Eds., Springer-Verlag, New York, 1983, 147.

89. Laurie, S.H., Tancock, N.P., McGrath, S.P., and Sanders, J.R., Influence of complexation on the uptake by plants of iron, manganese, copper, and zinc. II. Effect of DTPA in a multi-metal and computer simulation study, *J. Exp. Bot.,* 42, 515, 1991.

90. McGrath, S.P., Sanders, J.R., Laurie, S.H., and Tancock, N.P., Experimental determinations and computer predictions of trace metal concentrations in dilute complex systems, *Analyst,* 111, 459, 1986.

91. Lindsay, W.L., Zinc in soils and plant nutrition, *Adv. Agron.,* 24, 147, 1972.

CHAPTER 13

Recent Aspects of Mn and Zn Absorption and Translocation in Cereals

Michael J. Webb

INTRODUCTION

The biochemical and biophysical aspects of Mn and Zn absorption and translocation have not been studied in detail for many years. This point has been made by authors of the two most recent reviews on Mn and Zn absorption and translocation.[1,2] Both of these reviews have looked at Mn and Zn absorption and translocation from a physiological point of view and have covered aspects of electrophysiology, absorption kinetics, and translocation of Mn and Zn at a cellular and organ level, and thus will not be repeated here.

The biochemical and physiological aspects of Mn and Zn absorption and translocation described above have, by necessity, been studied in isolation from the whole-plant system. Thus the main focus of this review will be longer-term Mn and Zn absorption and translocation by whole plants in systems that more closely resemble soil systems than those previously used for physiological studies. Much of this research has been driven by a desire to understand the consequences of uneven distribution of nutrients through soil profiles, especially in soils which are already marginal in their fertility for these particular micronutrients.

Mn and Zn have many functions within plants. The major role of Mn is in photosynthetic oxygen evolution by green plants, but it has other roles in enzyme activation. It is found as a tightly bound constituent of only a few proteins; the most important of these is the photosystem II complex where Mn has a redox function in its role in water-splitting and oxygen evolution.[3-5] This function has the highest requirement for Mn and is thus the first affected with the onset of Mn deficiency. Mn is also a component of Mn-superoxide dismutase.[6] Mn also has other roles in plants. It has been reported to activate some 36 enzymes[7] although it can often be replaced by other divalent cations, especially Mg^{2+}.[3,7] Mn is involved in the activation of at least four enzymes in the shikimic acid pathway, three of which are involved in the synthesis of lignin.[7] The involvement of Mn in synthesis of lignin and other secondary metabolites may have important roles in the resistance of plants to infection by pathogenic organisms.[8] Thus, an overall deficiency of Mn will have important consequences on photosynthesis and general health of plants, while localized deficiency may have important implications on disease resistance and tolerance.

Like Mn, Zn is found as a constituent of only a few plant enzymes[5] such as alcohol dehydrogenase, Cu-Zn-superoxide dismutase, carbonic anhydrase, and RNA polymerase, although it is associated with the activity of many more.[3] Zn has major roles in carbohydrate metabolism,[9] auxin metabolism,[10,11] protein synthesis,[12] and maintenance of root membrane integrity.[13,14] Recently, Zn has also been shown to have a role in resistance/tolerance to pathogenic organisms.[15-17]

The chemistry of Mn in soil and biological systems has been reviewed recently.[18-22] Briefly, Mn can have a variety of oxidation states, from II to VII, but the predominant oxidation states found in soils are Mn(II) and Mn(IV). Transformation between oxidation states II and IV occurs often through bacterial action.[23] Mn(IV) forms insoluble Mn oxides, which are unavailable to plants for absorption. Mn(II) oxides may also form which are highly insoluble and also unavailable to plants. It is generally believed that Mn is taken up only as the Mn^{2+} form.[1,5,24] The availability of Mn decreases with increasing pH as a result of decreasing Mn^{2+} concentrations.[25]

Zn, on the other hand, is only found in the one oxidation state, Zn(II). However, its availability, like that of Mn, also decreases with increasing pH. For Zn, this decrease in availability is a result of the formation of Zn hydroxides at pH levels above 7.7 and a 100-fold decrease in Zn^{2+} concentration with each unit increase in pH.[25] Like Mn, it is believed that Zn is taken up as the Zn^{2+} ion. However, the exact form of Zn taken up by graminaceous species capable of producing phytosiderophores is less certain, as the rate of Zn absorption is the same whether supplied as a salt, chelate, or phytosiderophore complex; an argument which, incidently, applies equally as well to Mn.[26]

ABSORPTION OF MN AND ZN FROM CHELATE-BUFFERED NUTRIENT SOLUTIONS

Mn Absorption

The kinetics of Mn absorption from solution culture has been studied previously.[27-30] In these studies, solution concentrations typically ranged from 1 μM to 1 mM, and although absorption rates demonstrate saturation kinetics,[1] the minimum concentration of Mn used in these experiments was usually in excess of that required for adequate growth (see Webb et al.[31]). Thus, while these studies may provide insights into the mechanisms of Mn absorption involved, they do not adequately address the absorption characteristics of roots in an environment that is representative of limiting Mn supply. Until recently, it has not been practical to study absorption from low concentrations in solution without giving serious consideration to depletion. Techniques such as programmed nutrient addition and flowing solution cultures have overcome some of these concerns, but this has not been without cost. Programmed nutrient addition techniques generally require some knowledge of absorption rates before an experiment is carried out, and flowing solution culture techniques are expensive in terms of both equipment and high quality nutrient salts required to maintain low levels of the micronutrients being studied. In spite of the associated difficulties, flowing solution culture, with four levels of Mn ranging from 0.009 to 6.19 μM, has been used to show that 0.2μM Mn was adequate for growth of Gabo wheat, Algerian and Mulga oats, Cyprus Barrel Medic, and Clare and Bacchus Marsh subterranean clover; and that 0.04μM Mn was adequate for Hunter River Lucerne and Avon oats.[53] These values were established after 21 days of growth and based on fresh matter production.

More recently, the technique of chelate buffering has become available.[32,33] This technique provides a convenient means of supplying low activity levels of micronutrient metals in a well-buffered environment, based on the assumption that it is only the free metal that is absorbed by plant roots.[1,2,24,32] Using this technique, Webb et al.[31] have determined that the critical Mn^{2+} activity in solution which can support growth of barley is about $10^{-8.3}M$. This value is supported by the results of Huang et al.[34] and other unpublished data[54] using similar techniques with barley when plants were grown at the same pH (6.00). This value from chelate-buffered studies is lower than that found in the flowing solution culture system cited above. Apart from the possibility of genotypic differences and solution pH influencing these critical levels of supply, the two techniques differ in how Mn is delivered to the root membrane surface. Although flowing solution culture employs rapidly moving solution to minimize depletion zones at the root surface by minimizing the unstirred layer, Mn must still diffuse through the apoplasm to the root membrane surface. Thus, the concentration of Mn^{2+} at the root membrane would be lower than that at the root surface. Presumably, this diffusion gradient would not exist in the chelate-buffered system as the apoplasm probably contains the chelate-Mn complex, which would be present in approximately 50-fold greater concentration than that of Mn^{2+}. Thus, the only limiting factor in maintaining a constant Mn^{2+} concentration would be the rate of dissociation of the chelate-Mn complex. This reaction is generally regarded as quite fast.[33]

The experiments of Webb et al.[31] and other unpublished data show that the Mn absorption rate (calculated using the formula of Williams[35]) is linear in the range from deficiency to adequacy (Figure 1a), suggesting that the capacity for Mn absorption is not a limiting factor. Clarkson[1] has also commented that plants have

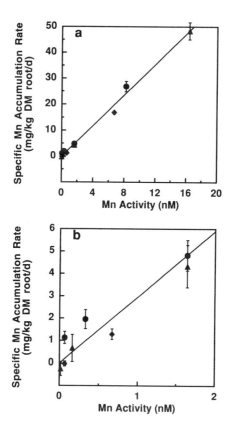

Figure 1.   Specific accumulation rate of Mn by barley grown at various levels of Mn^{2+} in a chelate-buffered nutrient solution. Accumulation rates calculated using the formula of Williams.[35] The results of three separate experiments are shown: (●) average of two genotypes ('Weeah' and WI2585),[54] (▲) one genotype ('Herta'),[31] (♦) average of three genotypes ('Weeah', 'Schooner', and 'Galleon').[56] All the data are shown on (a) but only the lower range on (b). On both figures the line of linear regression ($r^2 = 0.99$) for the whole set of data is shown. Standard errors of the mean are shown as vertical lines.

a high capacity for Mn absorption; a capacity which is much greater than their need. If the scale is expanded to resolve the lower range of Mn activities (Figure 1b), the absorption rate still complies to the same linear relationship. Thus, over a range of concentrations that are most likely to have effects on plant growth through effects on Mn nutritional status, the absorption rate of Mn appears to be a linear relationship with Mn supply. Although the Michaelis-Menten type models often employed in analysis of nutrient absorption would have predicted a linear component at low Mn concentrations, this had not previously been tested.

By examining the change in absorption rates with time, the above experiments show that Mn absorption by young roots is relatively slow and increases rapidly with age before reaching an apparent plateau (Figure 2). As this technique of estimating absorption rates (Williams[35]) depends on analysis of

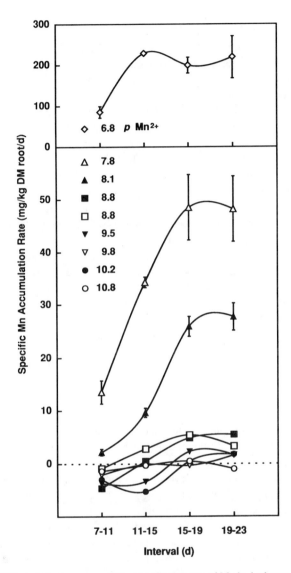

Figure 2. Effect of plant age on specific accumulation rate of Mn by barley grown at various levels of Mn^{2+} in a chelate-buffered nutrient solution. Accumulation rates were calculated using the formula of Williams[35] over the age periods indicated, and thus represent and average absorption rate of that period taking into account the change in Mn content and root dry matter over that period. Data represent a U.S. variety 'Herta' (open symbols)[31] and the average of two Australian barley genotypes: a released cultivar 'Weeah' and a breeders line WI2585 (closed symbols).[54] $pMn^{2+} = -\log(Mn$ activity). Standard errors of the mean are shown as vertical lines.

Mn contents over an extended period of time (usually 2 to 7 days) it measures net accumulation rather than short-term (unidirectional) flux. Thus, these data demonstrate that at low levels of Mn^{2+} activity Mn is actually lost from root tissues to the "free" ligand of the nutrient solution. Moreover, consistent with the low absorption rates found in young plants, this loss of Mn is also more marked in young plants. It is likely that the initial loss of Mn and the slow increase in Mn absorption could have serious detrimental, and possibly unrealized, effects on growth especially in plants grown on soils low in Mn or in plants derived from seeds low in Mn.

The effect of pH on Mn nutrition can be quite complicated, as high pH makes Mn less available in soils yet low pH inhibits Mn absorption from solution because of H^+ competition.[24] However, even in solution culture experiments it is difficult to separate the effects of pH on Mn absorption per se from its effects on Mn availability because of changes in chelation of Mn (and other nutrients) with changes in pH. In a recent experiment, Huang et al.[34] used a chelate buffering technique to maintain a constant activity of Mn (and other micronutrient cations) over a range of pH levels. They showed that increasing pH (without decreasing Mn activity) increased the absorption rate of Mn by barley plants, thus supporting the suggestions that, in solution, H^+ ions inhibit Mn absorption. Indeed the absorption rate was increased from a rate which could not supply adequate Mn at pH 6.00 to one which could at pH 7.7. This implies that the critical level of Mn supply cited above may depend on the pH of the solution. A lowering of the critical level of Mn supply with increasing pH would be beneficial to plants growing in calcareous soils, in which Mn availability is greatly depressed.

Zn Absorption

The arguments and inferences outlined above concerning previous studies on Mn absorption also apply to studies on Zn absorption. Thus, similar to Mn, absorption of Zn from solution culture has been studied, but at concentrations generally regarded as adequate to luxurious (see Kochian[2]). In recent studies where Zn has been supplied at levels which range from deficient to adequate, the absorption rate of Zn showed a curved response to Zn^{2+} activity (Figure 3). At low concentrations of Zn^{2+} it appears as though Zn is also lost from plants to the solution.[36] The most recent estimate[37] of the critical value for Zn^{2+} activity in solution for a plant to meet its needs is about $10^{-10.5}$ M. In a similar experiment with the same barley variety, this value was confirmed.[54] This is some 100 times lower than that required for Mn. Thus in the case of Zn, over the range of supply that is most likely to affect dry matter through Zn nutritional status, the rate of Zn absorption is not linearly related to Zn supply. This would indicate that there is some fundamental difference in the processes of Mn and Zn absorption by barley roots.

Implications from Genotypic Differences in Mn Nutrition

Barley genotypes exist that differ in their ability to grow on soils with low Mn availability. Graham[38] put forward several hypotheses which might explain

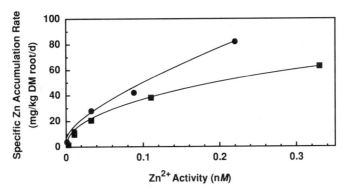

Figure 3. Specific accumulation rate of Zn by 'Herta' barley grown at various levels of Zn^{2+} in a chelate-buffered nutrient solution in two similar experiments. Accumulation rates were calculated using the formula of Williams.[35] The upper line is from the data of Welch and Norvell[36] (calculated from plants harvested at 10 and 17 days old) and the lower line from unpublished data[54] (calculated from plants harvested at 12 and 17 days old).

this differential Mn "efficiency". Among them, he suggested Mn absorption rates as a possibility. However, using a chelate-buffered nutrient solution at pH 6.00, similar to that described by Webb et al.,[31] we have not been able to demonstrate genotypic differences in absorption rates which could account for the differences in their efficiency in the field (Figure 4). This would suggest that the differences in efficiency might lie in some other mechanism. Graham[38] also suggested other hypotheses, such as differing abilities to acquire Mn rather than just the process of absorption, which could explain genotypic differences in tolerance to soils low in Mn. Such possibilities could include differences in root morphology or release of Mn mobilizing compounds. Whether differences exist in root morphology or in root exudates that could account for differences in efficiency, has not been established. However, it has been shown that when plants are grown in a mixed culture system in soil where Mn-efficient plants are grown in close proximity with inefficient plants, there is no advantage imparted to the inefficient variety (Figure 5). The results with mixed culture experiments would suggest that Mn acquisition by cereals is not enhanced by the release of some Mn mobilizing compound as has been shown for Fe acquisition in cereals,[39] even though Mn absorption can be enhanced by phytosiderophores produced in response to Fe deficiency[26] and would seem a likely mechanism.[24] Indeed, we have one experiment which has shown in solution culture that plants returned to Mn-adequate solutions after a period of Mn deprivation did not have an enhanced absorption rate of Mn but rather, a depressed rate (Figure 6). This implies that barley roots do not respond to Mn deficiency by increasing their ability to take up Mn. This is in contrast with results from studies of other nutrient deficiencies such as P and Fe, which show enhanced absorption rates of the deprived nutrient when plants are returned to an adequate supply of that nutrient.[40-42]

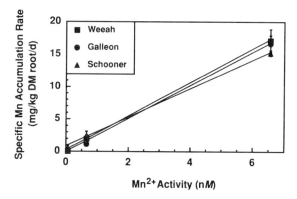

Figure 4.   Specific accumulation rate of Mn by three genotypes of barley known to differ in their ability to survive on Mn-impoverished soils in the field. Plants were grown at various levels of Mn^{2+} in a chelate-buffered nutrient solution. Accumulation rates calculated using the formula of Williams[35] for plants harvested at 19 and 23 days after imbibing seed.[56] Experimental conditions were almost identical to those described by Webb et al.[31] Standard errors of the mean are shown as vertical lines.

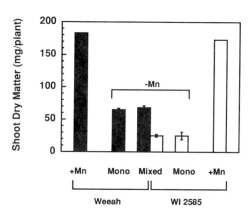

Figure 5.   Growth of two genotypes of barley known to differ in their ability to survive on Mn- impoverished soils in the field. Plants were grown in a controlled environment cabinet in a Mn-deficient soil in either mixed- or monoculture. Except for Mn, all other essential nutrients were supplied. In the field, Weeah survives better on these Mn-deficient soils and is thus regarded as "efficient". By contrast, WI2585 is regarded as inefficient. Potential growth in the presence of added Mn is shown by the "+Mn" treatments. Data from Ref. 57. Standard errors of the mean are shown as vertical lines.

TRANSLOCATION OF MN AND ZN

As with absorption of Mn and Zn, the biochemical aspects of Mn and Zn translocation have been reviewed quite recently.[1,2] Loneragan[43] has also reviewed the redistribution of Mn within plants. Thus, this section will concentrate on the consequences of the limited nature of Mn and Zn translocation, especially to growing roots.

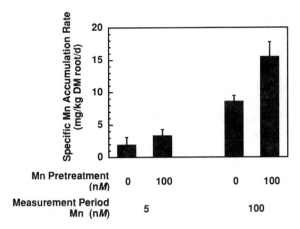

Figure 6. Effect of Mn pretreatment on subsequent specific accumulation rate of Mn by barley grown in nutrient solution. Accumulation rates were calculated using the formula of Williams[35] over the 7-day "measurement period" from day 28 to day 35.[58] Standard errors of the mean are shown as vertical lines.

A practical consequence of adding fertilizer only to the surface zone (top 10 cm) of cultivated land is that roots growing in the deeper zones may encounter areas of low nutrient availability. If a nutrient which has low availability in the deeper rooting zones has restricted mobility to the root system from other parts of the plant, then the roots in that zone may suffer deficiency while the rest of the plant appears healthy. This is particularly important if the nutrient also has limited mobility in the soil. Although there have been a substantial number of experiments studying translocation of micronutrient metals to roots, many of these have involved foliar applications, including the much-cited work of Bukovac and Wittwer.[44] As Loneragan[43] has pointed out for Mn, there is no guarantee that a foliar-applied nutrient will behave in the same way as an endogenous nutrient. Thus, in our work we have used "split-root systems" that allow supply of micronutrient metals to only a portion of the root system. This approach has allowed us to study the movement of micronutrient metals from an adequately supplied root to an inadequately supplied root in an otherwise healthy plant.

Mn Translocation

If Mn is supplied to only part of a root system, sufficient Mn may not be translocated to the part of the root system not supplied with Mn to support the growth of that part of the root system. In an early experiment with barley, Henkens and Jongman[45] demonstrated that only a small amount of radioactive Mn was translocated from one root system to the other, even though a substantial amount was translocated to the shoot. However, it is difficult to estimate how much this small amount of radioactive Mn might represent in total Mn. In a more recent experiment with subterranean clover, Nable and Loneragan[46] showed that the root not supplied with Mn produced only 50% of the dry mass (DM) of the root supplied with Mn (Table 1) and that the Mn concentration in that root fell to 12 mg/kg DM compared to 109 mg/kg DM in the root supplied with Mn. In a similar but more extensive experiment with wheat,[55] the concentration of Mn in the root not supplied with Mn fell rapidly to a level which would be considered deficient in shoots (Table 1). Although the depression in dry matter of the root not supplied with Mn may not have been significant, it is supportive of the results of Nable and Loneragan cited above.

Clearly, unless Mn is supplied to the entire root system, regions of the root will become deficient and may suffer restricted growth. This will obviously have serious implications on absorption of water and other nutrients from this zone. Under conditions not as favorable as those of the experimental conditions, restricted absorption of nutrients and water might affect ultimate yield even though shoots could appear healthy.

Zn Translocation

Translocation of Zn from a root zone of adequate supply to one of inadequate supply is not sufficient to maintain all root functions and root growth. Webb and Loneragan[47] have shown that P absorption was depressed in the root system not supplied with Zn, even though enough Zn was absorbed elsewhere to produce healthy shoots. Loneragan et al.[48] have similarly shown that the root not supplied with Zn could not acidify its bathing solution to the same extent as the one supplied with Zn (Figure 7). In both experiments, the root supplied with Zn behaved in a fashion similar to the control in which both root systems were supplied with Zn. Furthermore, the root not supplied with Zn behaved in a fashion similar to the control in which neither root system was supplied with Zn. Although the relevance of these experiments to field conditions may be argued, it is clear that when Zn is omitted from the immediate environment of a root, that root does not function in the same way as a root supplied with Zn.

At the time of harvest, root weights from the side not supplied with Zn were the same as those supplied with Zn, although Zn concentrations had fallen to levels considered deficient. Calculations of Zn movement indicate that the rate of translocation would not be adequate to maintain growth for an extended period. That this would eventually have an effect on root growth is confirmed by the experiments of Robson and Snowball,[49] in which the dry weight of the root not supplied with Zn was significantly less than the root supplied with Zn. These results indicate that if a particular root zone is not adequately supplied with Zn, that root will not function to its full potential.

The practical consequence of this has been demonstrated in a pot experiment in which the soil in the surface zone was supplied with Zn, but the deeper zone was either supplied or not supplied with Zn.[50] When a Zn-inefficient variety of wheat was grown in the treatment without Zn in the deeper zone, its water use was depressed by 12%, its maturity delayed by 10 days, and grain

Table 1.   Effect of Omitting Mn from One Root System on Dry Matter and Mn Concentration in Roots and Shoots

	Mn in Solution[a]			
Solution A	+		+	
Solution B	+		0	
Clover				
Dry matter (mg·plant⁻¹)				
Shoot	343	(31)	361	(27)
Root A	45	(3)	49	(1)
Root B	51	(2)	25	(3)
Mn concentration (mg·kg DM⁻¹)				
Shoot	35	(3)	22	(2)
Root A	101	(9)	109	(12)
Root B	117	(1)	12	(1)
Wheat				
Dry matter (mg·plant⁻¹)				
Shoot	3749	(156)	3785	(102)
Root A	501	(50)	518	(23)
Root B	507	(30)	471	(21)
Mn concentration (mg·kg DM⁻¹)				
Youngest emerged leaf blade	41	(3)	23	(2)
Root A	42	(5)	41	(0)
Root B	42	(1)	7	(0)

Note: Plants were grown in nutrient solutions with their root systems divided between two pots. Results represent the means of four (clover) or three (wheat) replicates. Standard errors of the means are shown in parentheses. Data for clover from Ref. 46 and for wheat from Ref. 55.

[a] Manganese supply was 0.5 μM for clover and 1.0 μM for wheat.

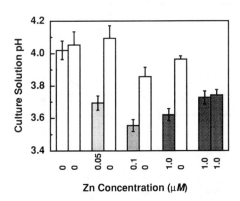

Figure 7. Effect of localized Zn treatment on the ability of that part of the root to acidify its rhizosphere when grown in nutrient solution in the presence of an ammonium based nitrogen source. The data from the 0.0 and 1.0,1.0 treatments serve as a control. Standard errors of the mean are shown as vertical lines.[48]

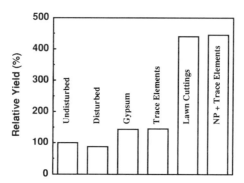

Figure 8. Effect of incorporating mineral and organic fertilizers in the subsoil on the grain yield of wheat. Prior to sowing, surface soil was removed and subsoil modified as indicated then replaced before the surface soil was returned. The area was then sown as a part of the farmer's normal paddock operations, including application of the farmer's standard fertilizer. The yield by normal cultural practices is indicated by the "undisturbed" plot. The control plot for soil amendments is represented by the "disturbed" plot, as it was excavated and mixed in the same way as other treatments but received no extra fertilization.[51]

yield reduced by 25%. The surprising feature of these results is that only 20% of the total root mass was in the treatment zone. That is, 80% of the root was adequately supplied with Zn, yet the immediate nutrition of only 20% of the root system had these substantial effects on plant development and physiology. The above differences were not found in the Zn-efficient variety. However, the concentration of Zn in the grain was halved (as it was in the inefficient variety) when only 20% of the root mass was in the nil Zn treatment zone. The results of these experiments show that even a small amount of root, when not adequately supplied with Zn, can have substantial and important effects on aspects of plant growth and physiology.

General Conclusions Regarding Translocation of Mn and Zn

As these effects of both Mn and Zn are likely to occur in plants that otherwise appear healthy, there may be real problems in agricultural environments that only receive fertilizer to the surface soil layers. The potential for enhancing production by modifying the deeper soil layers has been demonstrated by Graham[51] in farming environments (Figure 8). Whether similar responses to individual nutrients can also be attained in soils responsive to a particular nutrient is yet to be tested. However, it is likely that there is enormous potential for improvement in soils with low availability of Zn and Mn at depth because of the restricted mobility of Mn and Zn to roots.

Localized deficiency in micronutrients, especially Mn and Zn, may also make the roots more susceptible to disease.[8,15,52]

CONCLUSIONS

With the availability of new techniques such as chelate buffering, it is now possible to create experimental conditions that allow researchers to maintain constant levels of micronutrient metals in small-volume nutrient solutions. To date, these solutions have been used to grow plants at low levels of supply without having to be concerned with nutrient depletion. So far, absorption rates of nutrients have only been calculated using the approximation formula of Williams.[35] This formula calculates the average absorption rate over (usually) a reasonably long period of time, such as several days. While this has advantages in that it averages out diurnal fluctuations, it does not have the same sensitivity as short-term experiments with isotopes. Thus, it should now be possible to conduct short-term absorption experiments using radioactive tracers in such a way that kinetic parameters can be determined with concentrations of free ions more closely representing the ranges of deficiency and adequacy.

REFERENCES

1. Clarkson, D.T., The uptake and translocation of manganese by plant roots, in *Manganese in Soils and Plants,* Graham, R.D., Hannam, R.J., and Uren, N.C., Eds., Kluwer Academic, Dordrecht, 1988, 101.
2. Kochian, L.V., Mechanisms of micronutrient uptake and translocation in plants, in *Micronutrients in Agriculture,* 2nd ed., Mortvedt, J.J., Cox, F.R., Shuman, L.M., and Welch, R.M., Eds., Soil Science Society of America, Madison, WI, 1991, 229.
3. Marschner, H., *Mineral Nutrition of Higher Plants,* Academic Press, London, 1986.
4. Campbell, L.C. and Nable, R.O., Physiological functions of manganese in plants, in *Manganese in Soils and Plants,* Graham, R.D., Hannam, R.J., and Uren, N.C., Eds., Kluwer Academic, Dordrecht, 1988, 139.
5. Römheld, V. and Marschner, H., Function of micronutrients in plants, in *Micronutrients in Agriculture,* 2nd ed., Mortvedt, J.J., Cox, F.R., Shuman, L.M., and Welch, R.M., Eds., Soil Science Society of America, Madison, WI, 1991, 297.
6. Sevilla, F., Lopez-Gorge, J., Gomez, M., and Del Rio, L.A., Manganese superoxide dismutase from higher plants. Purification of a new Mn-containing enzyme, *Planta,* 150, 153, 1980.
7. Burnell, J.N., The biochemistry of manganese in plants, in *Manganese in Soils and Plants,* Graham, R.D., Hannam, R.J., and Uren, N.C., Eds., Kluwer Academic, Dordrecht, 1988, 125.
8. Graham, R.D. and Webb, M.J., Micronutrients and disease resistance and tolerance in plants, in *Micronutrients in Agriculture,* 2nd ed., Mortvedt, J.J., Cox, F.R., Shuman, L.M., and Welch, R.M., Eds., Soil Science Society of America, Madison, WI, 1991, 329.
9. Jyung, W.H., Ehmann, A., Schlender, K.K., and Scala, J., Zinc nutrition and starch metabolism in *Phaseolus vugaris* L., *Plant Physiol,* 55, 414, 1975.

10. Skoog, F., Relationships between zinc and auxin in the growth of higher plants, *Am. J. Bot.,* 27, 939, 1940.

11. Tsui, C., The role of zinc in auxin synthesis in the tomato plant, *Am. J. Bot.,* 35, 172, 1948.

12. Cakmak, I., Marschner, H., and Bangerth, F., Effect of Zn nutritional status on growth, protein metabolism and levels of indole-3-acetic acid and other phyto-hormones in bean (*Phaseolus vugaris* L.), *J. Exp. Bot.,* 40, 405, 1989.

13. Welch, R.M., Webb, M.J., and Loneragan, J.F., Zinc in membrane function and its role in phosphorus toxicity, in *Plant Nutrition 1982,* Proc. Ninth Int. Plant Nutr. Colloq., Scaife, A., Ed., Commonwealth Agricultural Bureaux, Slough, U.K., 1982, 710.

14. Cakmak, I. and Marschner, H., Increase in membrane permeability and exuda-tion in roots of zinc deficient plants, *J. Plant Physiol.,* 132, 356, 1988.

15. Sparrow, D.H. and Graham, R.D., Susceptibility of zinc-deficient wheat plants to colonization by *Fusarium graminearum* Schw. Group 1, *Plant Soil,* 112, 261, 1988.

16. Thongbai, P., Hannam, R.J., Graham, R.D., and Webb, M.J., Interaction between zinc nutritional status of cereals and *Rhizoctonia* root rot severity. I. Field observations, *Plant Soil,* 153, 207, 1993.

17. Thongbai, P., Graham, R.D., Neate, S.M., and Webb, M.J., Interaction between zinc nutritional status of cereals and *Rhizoctonia* root rot severity. II. Effect of Zn on disease severity of wheat under controlled conditions, *Plant Soil,* 153, 215, 1993.

18. Gilkes, R.J. and McKenzie, R.M., Geochemistry of manganese in soil, in *Man-ganese in Soils and Plants,* Graham, R.D., Hannam, R.J., and Uren, N.C., Eds., Kluwer Academic, Dordrecht, 1988, 23.

19. Norvell, W.A., Inorganic reactions of manganese in soils, in *Manganese in Soils and Plants,* Graham, R.D., Hannam, R.J., and Uren, N.C., Eds., Kluwer Aca-demic, Dordrecht, 1988, 37.

20. Bartlett, R.J., Manganese redox reactions and organic interactions in soils, in *Manganese in Soils and Plants,* Graham, R.D., Hannam, R.J., and Uren, N.C., Eds., Kluwer Academic, Dordrecht, 1988, 59.

21. Hughes, N.P. and Williams, R.J.P., An introduction to manganese biological chemistry, in *Manganese in Soils and Plants,* Graham, R.D., Hannam, R.J., and Uren, N.C., Eds., Kluwer Academic, Dordrecht, 1988, 7.

22. Shuman, L.M., Chemical forms of micronutrients in soils, in *Micronutrients in Agriculture,* 2nd ed., Mortvedt, J.J., Cox, F.R., Shuman, L.M., and Welch, R.M., Eds., Soil Science Society of America, Madison, WI, 1991, 113.

23. Ghiorse, W.C., The biology of manganese transforming microorganisms in soil, in *Manganese in Soils and Plants,* Graham, R.D., Hannam, R.J., and Uren, N.C., Eds., Kluwer Academic, Dordrecht, 1988, 75.

24. Marschner, H., Mechanisms of manganese acquisition by roots from soils, in *Manganese in Soils and Plants,* Graham, R.D., Hannam, R.J., and Uren, N.C., Eds., Kluwer Academic, Dordrecht, 1988, 191.

25. Lindsay, W.L., Inorganic equilibria affecting micronutrients in soils, in *Micronutrients in Agriculture,* 2nd ed., Mortvedt, J.J., Cox, F.R., Shuman, L.M., and Welch, R.M., Eds., Soil Science Society of America, Madison, WI, 1991, 89.

26. Marschner, H., Treeby, M., and Römheld, V., Role of root-induced changes in the rhizosphere for iron acquisition in higher plants, *Z. Pflanzenernäehr. Bodenkd.*, 152, 197, 1989.

27. Bowen, J.E., Absorption of copper, zinc and manganese by sugar cane tissue, *Plant Physiol.*, 44, 255, 1969.

28. Bowen, J.E., Kinetics of active uptake of boron, zinc, copper, and manganese in barley and sugarcane, *J. Plant Nutr.*, 3, 215, 1981.

29. Landi, S. and Fagioli, F., Efficiency of manganese and copper uptake by excised roots of maize genotypes, *J. Plant Nutr.*, 6, 957, 1983.

30. Maas, E.V., Moore, D.P., and Mason, B.J., Manganese absorption by excised barley roots, *Plant Physiol.*, 43, 527, 1968.

31. Webb, M.J., Norvell, W.A., Welch, R.M., and Graham, R.D., Using a chelate-buffered nutrient solution to establish the critical solution activity of Mn^{2+} required by barley (*Hordeum vulgare* L.), *Plant Soil*, 153, 195, 1993.

32. Parker, D.R., Chaney, R.L., and Norvell, W.A., Chemical equilibrium models: applications to plant nutrition research, in *Chemical Equilibrium and Reaction Models*, Loeppert, R.H., Ed., Soil Science Society of America, Madison, WI, in press.

33. Norvell, W.A., Reactions of metal chelates in soils and nutrient solutions, in *Micronutrients in Agriculture*, 2nd ed., Mortvedt, J.J., Cox, F.R., Shuman, L.M., and Welch, R.M., Eds., Soil Science Society of America, Madison, WI, 1991, 187.

34. Huang, C.Y., Webb, M.J., and Graham, R.D., Effect of pH on Mn absorption among barley genotypes in a chelate-buffered nutrient solution, in *Plant Nutrition — From Genetic Engineering to Field Practice, Proc. XIIth Int. Plant Nutr. Colloq.*, Barrow, N.J., Ed., Kluwer Academic, Dordrecht, 1993, 653.

35. Williams, R.F., The effects of phosphorus on the rates of intake of phosphorus and nitrogen and upon certain aspects of phosphorus metabolism in graminaceous plants, *Aust. J. Sci. Res. (B)*, 1, 333, 1948.

36. Welch, R.M. and Norvell, W.A., Growth and nutrient uptake by barley (*Hordeum vulgare* L. cv Herta): Studies using a HEDTA-buffered nutrient solution technique. II. Role of zinc in uptake and leakage of mineral nutrients by roots, *Plant Physiol.*, 101, 627, 1993.

37. Norvell, W.A. and Welch, R.M., Growth and nutrient uptake by barley (*Hordeum vulgare* L. cv Herta): Studies using a HEDTA-buffered nutrient solution technique. I. Zinc ion requirements, *Plant Physiol.*, 101, 619, 1993.

38. Graham, R.D., Genotypic differences in tolerance to manganese deficiency, in *Manganese in Soils and Plants*, Graham, R.D., Hannam, R.J., and Uren, N.C., Eds., Kluwer Academic, Dordrecht, 1988, 261.

39. Zhang, F., Römheld, V., and Marschner, H., Effect of zinc deficiency in wheat on the release of zinc and iron mobilizing root exudates, *Z. Pflanzenernäehr. Bodenkd.*, 152, 205, 1989.

40. Clarkson, D.T., Sanderson, J., and Scattergood, C. B., Influence of phosphate-stress on phosphate absorption and translocation by various parts of the root system of *Hordeum vulgare* L. (barley), *Planta*, 139, 47, 1978.

41. Clarkson, D.T. and Sanderson, J., Sites of absorption and translocation of iron in barley roots. Tracer and microautoradiographic studies, *Plant Physiol.*, 61, 731, 1978.

42. Römheld, V. and Marschner, H., Mobilization of iron in the rhizosphere of different plant species, *Adv. Plant Nutr.*, 2, 155, 1986.
43. Loneragan, J.F., Distribution and movement of manganese in plants, in *Manganese in Soils and Plants*, Graham, R.D., Hannam, R.J., and Uren, N.C., Eds., Kluwer Academic, Dordrecht, 1988, 113.
44. Bukovac, M.J. and Wittwer, S.H., Absorption and mobility of foliar applied nutrients, *Plant Physiol.*, 32, 428, 1957.
45. Henkens, Ch.H. and Jongman, E., The movement of manganese in the plant and the practical consequences, *Neth. J. Agric. Sci.*, 13, 392, 1965.
46. Nable, R.O. and Loneragan, J.F., Translocation of manganese in subterranean clover (*Trifolium subterraneum* L. cv. Seaton Park). II. Effects of leaf senescence and of restricting supply of manganese to part of a split root system, *Aust. J. Plant Physiol.*, 11, 113, 1984.
47. Webb, M.J. and Loneragan, J.F., Zinc translocation to wheat roots and its implications for a phosphorus/zinc interaction in wheat plants, *J. Plant Nutr.*, 13, 1499, 1990.
48. Loneragan, J.F., Kirk, G.J., and Webb, M.J., Translocation and function of zinc in roots, *J. Plant Nutr.*, 10, 1247, 1987.
49. Robson, A.D. and Snowball, K., The effect of chlorsulfuron on the uptake and utilization of copper and zinc in wheat, *Aust. J. Agric. Res.*, 41, 19, 1990.
50. Nable, R.O. and Webb, M.J., Further evidence that zinc is required throughout the root zone for optimal development, grain production and water-use in wheat, *Plant Soil*, 150, 297, 1993.
51. Graham, R.D., Breeding wheats for tolerance to micronutrient-deficient soils. Present status and priorities, in *Wheat for Nontraditional, Warm Areas*, Saunders, D.A., Ed., CIMMYT, Mexico, D.F., 1991, 315.
52. Wilhelm, N.S., Graham, R.D., and Rovira, A.D., Application of different sources of manganese sulfate decreases take-all (*Gaeumannomyces graminis* var *tritici*) of wheat grown in a manganese deficient soil, *Aust. J. Agric. Res.*, 39, 1, 1988.
53. Carroll, M.D., Asher, C.J., and Loneragan, J.F., unpublished results.
54. Webb, M.J., Norvell, W.A., Welch, R.M., and Graham, R.D., unpublished results.
55. Loneragan, J.F. and Webb, M.J., unpublished results.
56. Huang, C.Y., Webb, M.J., and Graham, R.D., Mn efficiency is expressed in barley in soil system but not in solution culture, *J. Plant Nutr.*, 17(1), in press, 1994.
57. Dinkelaker, B.E. and Graham, R.D., unpublished results.
58. Webb, M.J. and Graham, R.D., unpublished results.

CHAPTER **14**

Modeling of Iron Availability in the Plant Rhizosphere

David E. Crowley and Dirk Gries

INTRODUCTION

Nutritional competition for iron is an important phenomenon with wide-ranging consequences for microbial ecology and plant adaptation to iron-limiting soils. Given the complexity of the rhizosphere, studies on iron competition are usually approached only in simplified experimental systems that allow dissection of specific components of the below-ground system. In order to utilize data from such experiments, conceptual models are important for providing a framework to integrate experimental results,[1] and have proven useful for identifying biological, chemical, and kinetic factors that influence iron competition.[2-4] After several decades of research, there is now an extensive data base on parameters related to plant and microbial iron acquisition. Using this information, it has become possible to begin simulation modeling of processes that govern iron availability and test current hypotheses that describe plant-microbial interactions in the rhizosphere.

Since the advent of rhizosphere modeling in the 1970s,[5,6] there has been considerable progress in mathematical modeling of factors that influence plant-microbial interactions, mineral nutrient flux,[7] and microbial population dynamics in the rhizosphere.[8] Current models have focused primarily on carbon, but there are undoubtedly important links between carbon and iron that have yet to be explored.[9] Based on chemostat experiments that date back to early research by Winogradsky, it has been well established that microorganisms adjust their growth rates in relation to the carbon supply. However, apart from siderophore production, there is little information on how microbes adjust their growth in response to iron deprivation over a range of carbon supply levels. As we build on the framework of earlier models, simulation models on iron will need to consider the importance of carbon supply, its relevance to iron competition mediated by biosynthetic chelate production, and degradation in different root zones.[9,10]

In this chapter, we have compiled pertinent data related to iron nutritional competition in the rhizosphere and explore the use of simulation modeling to describe the relationships between plant demand for iron, plant growth rates, and depletion of iron in the rhizosphere. We also describe methodology developed to study iron competition, and the composition of plant and microbially produced chelates in an artificial system. While a mathematical simulation inherently oversimplifies complex processes, we have found our initial models to be useful for examining components of competition and for conducting sensitivity analyses of various parameters such as root exudation rates and release patterns, initial microbial population densities, and microbial growth rates. With further development, this approach should allow us to determine which factors are most important in governing competition for iron, and to better understand the complex interdependence of plants and microorganisms in iron-limiting soils.

PLANT UTILIZATION OF BIOSYNTHETIC CHELATES

The predominant effect of the plant in generating a rhizosphere is the deposition of organic material that provides substrate for microbial growth. Another effect that is less commonly considered is the depletion of nutrients that are taken up by the roots to support plant growth. Generally, soil microorganisms are considered to be more competitive than plants for macronutrients and to have first access to nutrients that cycle in the soil system. For example, it has been well established that microorganisms can decrease plant availability of nitrogen and phosphorus by immobilization of these nutrients in the microbial biomass under high carbon conditions. Similar effects can be hypothesized for nutrients such as iron that are required at low quantities, but which have poor availability at neutral to alkaline pH. However, there are unique differences between the macronutrients and iron in that both plants and microorganisms have highly efficient mechanisms for responding to iron limitation.

Among the chemical, physical, and biological factors potentially affecting nutritional competition, inducible mechanisms for responding to iron deficiencies have been the most extensively investigated. Research on plant iron nutrition has resulted in the division of plants into two mechanistic strategies. The first, employed by almost all vascular plants, involves an increase in proton ATPase activity and induction of an iron-chelate reductase, which together lower pH and increase the solubility of inorganic iron external to the plant root.[11-13] The second strategy, employed exclusively by *Poaceae* (graminaceous) species, involves the use of phytosiderophores for iron transport.[14] Representative plants within either strategy group also have been shown to use a variety of microbial siderophores for iron acquisition.[15] Recent studies show that there are high numbers of siderophore-producing bacteria in the rhizosphere of both iron-efficient and iron-inefficient grasses.[16] The ecological relevance of microbial siderophores in plant nutrition remains controversial, however, since the quantities and types of microbial siderophores in the rhizosphere, as well as the mechanisms by which plants acquire iron from these compounds, are not yet well understood.

A paradox in determining the relative importance of plant and microbial siderophores in plant nutrition is that either type of chelate can be shown to be effective or ineffective, depending on the experimental conditions. This has led to some interesting controversy among laboratories investigating these chelators. The most severe problem with phytosiderophores is that they are extremely labile with respect to degradation and are ineffective for mobilizing iron in hydroponic culture solutions inoculated with soil microorganisms. Similarly, *in situ*-produced microbial siderophores do not appear to be effective under these conditions either, possibly because they are not produced in sufficient quantities or else are also rapidly degraded in the hydroponic media. The primary experiments in support of microbial siderophores show that when representative compounds are supplied at the same concentrations used to maintain plants with synthetic chelates, they are generally quite effective for supplying iron. However, because phytosiderophores are taken up by grasses at much faster rates than microbial siderophores, and are transported by means of a specific transport system, it has been argued that microbial siderophores are unimportant for directly supplying iron to grasses. Other less restrictive hypotheses are that phytosiderophores are produced by grasses as necessary to augment iron provided by microbial siderophores,[4] or that they are released into the cortex tissue to mobilize iron that has been deposited into the apoplast by microbial siderophores.[17]

Very high concentrations of siderophores, ranging from 100 to 230 μM or higher, are produced by common root-colonizing microbes when cultured in low iron media.[10] The most common bacteria isolated from grass roots belong to the genus *Pseudomonas,* which comprise up to 50% of the population isolated on nonselective media.[10,18] Unfortunately, nonselective media detect only 1 to 10% of the total population determined by direct counting, so many potentially important strains remain unidentified. Siderophore production is generally correlated with the relative degree of iron stress such that as iron falls below 1 μM concentration, large quantities of siderophores are produced very rapidly. Concentrations of pyoverdin, a siderophore produced by *Pseudomonas putida,* have been measured at 300 μM in 48-h cultures.[19] Another example is the root-associated, nitrogen-fixing bacterium, *Azotobacter vinelandii,* that requires 25 μM iron for optimal growth.[10] Under iron-stress conditions imposed during steady state growth in a chemostat, this bacterium produces 2 μmol of siderophore per milligram dry weight of cells — an amount greater than the cell biomass. When produced at microsites in the rhizosphere, these quantities represent extremely high concentrations that could have either beneficial or detrimental effects on iron availability, depending on the ability of the plant to use the siderophore as an iron source.

Recently, it has been shown that microorganisms can have many anomalous effects on plant iron uptake that confound studies on plant iron uptake mechanisms.[20,21] Consequently, to study plant mechanisms, and not simply the net response of a randomly inoculated plant-microbial system, it is necessary to separate out microbial effects by use of axenic culture.[20] Using such a system, it has been shown that plants such as cucumber and oat can take up iron from the siderophore, ferrioxamine B (FOB), by what appears to be a constitutive transport mechanism. In cucumber, this seems to involve a facilitated diffusion mechanism for uptake of the intact chelate that discriminates against the deferrated complex.[22] Early studies with oat suggested the operation of an iron-stress inducible system, but upon further examination it was demonstrated that the inducible component may have been caused by microbial uptake of the siderophore. More recent studies conducted under axenic conditions show that iron-stressed oat and corn only slightly increase their abilities to utilize FOB.[20,23]

Differences in the abilities of iron-efficient Coker 227 and inefficient TAM 0-312 oat to use FOB under nonaxenic conditions also suggest a plant-mediated mechanism is responsible for uptake of iron from the siderophore. Since it is unlikely that root-colonizing microorganisms differentially affect these two cultivars, there must be differences either in the ability of iron-efficient Coker to obtain iron directly from FOB, or in the importance of phytosiderophore-mediated uptake of iron that has been deposited in the root apoplast. The latter is somewhat controversial, and is supported by the fact that Coker produces phytosiderophores in response to iron stress, whereas TAM does not.[24,25] Conversely, studies with wheat and corn suggest indirect acquisition is not likely, as the microbial siderophore FOB does not release significant quantities of iron into the apoplast, as has been found for synthetic chelators that have lower stability constants than most microbial siderophores.[17,26]

Another approach that has been used for preventing microbial effects on plant iron uptake has been to supply antibiotics in nutrient solutions during short-term uptake studies.[27] In the presence of microorganisms, an analog of FOB was found to be reductively deferrated at locations behind the root tip and at sites of lateral root emergence in both cotton and maize. Addition of antibiotics suppressed the deferration of the siderophore, which suggested that microorganisms on the root surface were causing iron reduction. However, whether microbial activity indirectly resulted in supplying iron to the plants was not determined. Under nonaxenic conditions, physiological concentrations of the siderophore were adequate to supply iron for normal growth over a 3-week period. Studies directly comparing axenic and inoculated plants are still necessary to determine to what extent use of ferrioxamine B is mediated by a plant mechanism or by various microbial processes that could include reductive release of iron or degradative destruction of the chelator.

Analyses of plant growth rates and root:shoot ratios in relation to iron demand suggest that, at least for some plants, iron uptake rates from ecologically relevant concentrations of FOB would be sufficient to support growth. For example, growth analysis data for oat show that depending on the growth stage, an iron uptake rate of 9 to 36 nM g^{-1} h^{-1} would provide optimal quantities of iron (100 μg g^{-1} dry wt).[28] Extrapolating from measurements of iron uptake rates from different siderophores, this requirement may be satisfied by 5 to 10 μM microbial siderophore or 1 μM phytosiderophore. Another example is provided by nonaxenic sorghum, which takes up iron from the *Pseudomonas putida* siderophore, pseudobactin, at a rate of 25 nmol g^{-1} h^{-1} when supplied at 10 μM concentration in short-term studies. Although this is some 100-fold less than the uptake rate for phytosiderophores, long-term experiments also confirmed that 10 μM pseudobactin provided sufficient iron for plant growth.[29] This is a concentration that could reasonably occur in the plant rhizosphere. Thus, this type of comparison may be more useful in evaluating the efficacy of phytosiderophores and microbial siderophores than comparisons based strictly on differences in uptake rates.

To characterize specific microbial effects and their net importance in plant iron nutrition, processes governing the efficacy of different chelators will need to be better examined in gnotobiotic experiments with defined cultures of microorganisms. In these studies, hydroponic culture systems have limited utility since the rhizosphere effect is eliminated by rapid diffusion of exudates and siderophores that become diluted in the culture medium. Moreover, experimental results from hydroponic culture are open to misinterpretation, since they eliminate any temporal and spatial effects of plant and microbial activity in different root zones. New experimental systems are needed to study the rhizosphere with minimal introduction of artifacts, yet which still allow the study of treatment effects at the microsite level. At this stage of research, insight into iron competition may also be obtained by integrating empirical data from previous research into static and dynamic models of the rhizosphere.

SIMULATION MODELING OF PLANT AND MICROBIAL DEMAND FOR IRON

To examine interactions between plants and microorganisms in plant iron nutrition, we have expanded upon a recent conceptual model describing iron competition (Figure 1).[4] This model, which was developed for graminaceous plants, particularly emphasizes the importance of plant growth for initiating nutritional competition. As iron is taken up by the plant the soluble iron pool becomes depleted, with concomitant production of phytosiderophores and siderophores when iron falls below critical levels required for

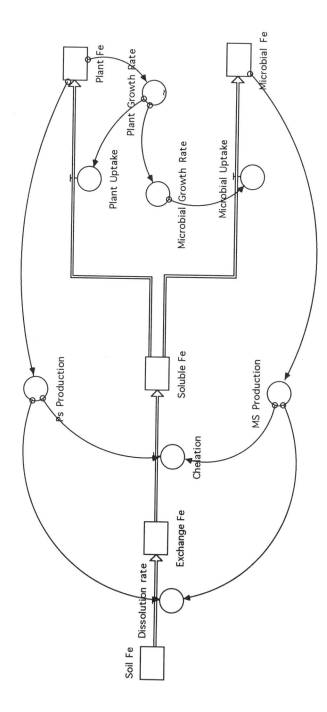

Figure 1.   Conceptual model of iron competition in the rhizosphere in which plant growth establishes the initial demand for iron that depletes the rhizosphere and thereafter results in the induction of phytosiderophore and microbial siderophore production.

growth. Competition between microorganisms occurs as microbial populations grow and produce siderophores. Depending on the relative demand for iron and the efficacy of phytosiderophores, the plant may also enter into competition with microorganisms as microbial siderophores accumulate and phytosidero-phores are degraded and utilized as substrate for microbial growth. The relative degree to which this occurs is dependent on root growth rates and root carbon exudation patterns that separate zones of phytosiderophore release, microbial growth, and siderophore production.

In determining whether plant uptake of iron can deplete the rhizosphere and establish conditions for nutritional competition, it is necessary first to calculate the relative demands for iron by the plant and microbial components of the system. This can be achieved either with a static model using a set of assumed values, or by computer simulation modeling which allows a better exploration of sensitivity to specific parameters such as plant root and microbial growth rates, total extractable iron, and the dimensions of the rhizosphere volume from which iron is extracted. Using simple calculations similar to those performed through simulation modeling, the first step in examining iron depletion is to determine the amount of iron available in the rhizosphere and the quantity necessary to support plant growth. Based on previous calculations,[30] a typical barley root is assumed to measure 0.5 mm in diameter and to have a root length of 8.4 cm mg^{-1} dry wt. The resulting rhizosphere cylinder, a 0.5-mm shell around the root, is calculated to be approximately 100 mm^3 in volume, and in a soil of bulk density of 1.6 to contain 0.16 g of soil. If we next assume a shoot to root ratio of 2:1 and an iron requirement of 100 μg g^{-1} tissue, such that the plant Fe demand will be 300 ng mg^{-1} root that must be supplied from the rhizocylinder. In a typical iron-limiting soil containing 2 μg g^{-1} soil DTPA extractable iron, the rhizocylinder volume calculated above would contain 320 ng of extractable iron, which provides just iron than that needed for plant growth. If the DTPA extractable iron is barely less than this quantity, or if the plant is inefficient in acquiring this quantity of iron, the plant undergoes increasing iron stress and will boost phytosiderophore production to increase its rate of iron assimilation.

Using similar calculations, the microbial iron demand can be calculated from measurements of the soil microbial biomass. Actual measurements of the biomass vary, depending on the experimental systems in which they have been measured. In soil, the microbial biomass has been estimated to range from 0.3 to 6 mg g^{-1} soil,[31] which at an average microbial iron content of 50 μg g^{-1} translates to 15 to 300 ng of Fe demand per gram of soil. In the rhizosphere volume of 100 mm^3 calculated above, the soil bacterial biomass would require 2.4 to 48 ng of iron, which is considerably less than the iron theoretically required by the plant. A similar value can be obtained from empirical measure-ments in sand culture, which show bacteria directly supported by root exudates have a biomass of 36% of the root dry wt.[32] Using this value, the microbial biomass associated with 1 mg of root tissue requires about 18 ng of iron. The correct value probably lies somewhere between these two estimates, since the actively growing microbial biomass involved in iron acquisition is derived both from utilization of root exudates, and by growth of microorganisms on native organic matter after they have been primed by the growth on root exudates.[33]

Given the approximate tenfold greater iron requirement of the plant, the above calculations strongly suggest that the plant represents the major sink for iron which is responsible for depleting the rhizosphere. Even in very low iron soil, where there is the greatest potential for competition, a reduction in plant growth as a result of iron deficiency would result in less rhizodeposition and a concomitant reduction in microbial growth. On the other hand, it must also be considered that microorganisms are capable of producing extremely high quantities of siderophore in relation to their biomass. In effect, this establishes an artificial demand for iron that could be competitive with the plant. For example, maximum phytosiderophore production rates by plants are measured at 50 μmol g^{-1} dry root weight during a 2-h peak production period, or approximately 1 mg of phytosiderophore. In comparison, *Pseudomonas fluorescens* produces 1 g g^{-1} dry weight or more of pyoverdin in overnight cultures,[34] or 1000 times more siderophore per unit biomass than the daily amount produced by a grass such as barley. Another example is the *Azotobacter vinelandii* strain cited earlier, which produces 2 μmol mg^{-1} dry weight of siderophore. Using a static model, production of this quantity of siderophore translates to a solution concentration of 21 mM at 25% soil moisture. This quantity of chelate could sequester 39 μg of iron, or 180 times the DTPA extractable level of 216 ng calculated to be available in the rhizosphere of 1 mg of root tissue. This compares with a 7.4 mM concentration calculated for phytosiderophore production by barley root tips under axenic conditions. Based on these types of calculations, the rhizosphere pool of microbial siderophores represents a significant sink for iron that could significantly affect iron availability in the rhizosphere.

KINETIC FACTORS RELATED TO IRON MOBILIZATION

In addition to iron demand, there are several kinetic factors that must be considered in iron depletion and establishment of competition for iron in the rhizosphere.[35] Calculations used by the simulation model in Figure 1 are based on soluble iron determined by DTPA extraction.[36] This value is an empirical measurement of the soluble iron concentration after shaking soil for 3 h in a 50-mM solution of DTPA. Compared to *in situ* chelation by biosynthetic chelates, the DTPA extraction is a harsh treatment that mobilizes almost all available iron. Shaking during the extraction procedure eliminates diffusion processes and exposes mineral surfaces that may not normally be accessible to iron chelators. Moreover, the use of a 50-mmol DTPA concentration for extraction probably overestimates iron that can be mobilized by phytosiderophores and microbial siderophores which are produced and utilized at much lower concentrations. Even if all of the DTPA extractable iron could be extracted from the rhizocylinder by biosynthetic chelates, mobilization and uptake of iron by the plant roots will be rate-limited by diffusion, as well as by slow chelation reaction kinetics. For synthetic chelators, slow reaction kinetics have been found to decrease soluble iron concentrations by two orders of magnitude below levels predicted by simple equilibrium models. Although the reaction rate kinetics for siderophores can probably be predicted by stability constants and kinetic modeling data,[35] the simple models become much more complex with chelate mixtures such as those which might occur in the rhizosphere. For this reason, simulation models of siderophore and phytosiderophore competition will need to be validated with empirical measurements of ferrated siderophores and phytosiderophores in rhizosphere soil water extracts.

Relative rates of iron mobilization have been compared for synthetic chelates, phytosiderophores from barley, and the microbial siderophore, ferrioxamine B.[37-39] These experiments are generally quite favorable for supporting an important role of phytosiderophores in iron mobilization, but also indicate limitations. In one study,[37] supply of 10 μM phytosiderophore increased the mobilization of iron from a radiolabeled soil by a factor of 10- to 20-fold, but was inhibited by high NaCl concentrations (20 to 80 mM) and was decreased when amorphous iron mineral surfaces were occluded by reaction with phosphate. Phytosiderophores also mobilize large quantities of zinc, manganese, and copper which can comprise 80 to 90% of the total phytosiderophore-chelated metal pool in soils, and lower their efficiency in chelating iron.[31,40] In comparison to phytosiderophores, the model siderophore, FOB, at the same concentration is somewhat more effective and specific for mobilization of iron. Microbial siderophores also have higher affinity for iron, which may allow them to strip iron from phytosiderophores and to competitively inhibit dissolution of mineral iron in a mixture of chelates.[4,41] For these reasons, as well as the fact that phytosiderophores are readily degraded, it can be argued that phytosiderophores would be poorly effective in direct competition with most of the microbial siderophores that have been characterized to date.

SPATIAL AND TEMPORAL FACTORS RELATED TO COMPETITION

One important process that is hypothesized to ameliorate direct plant-microbial competition is spatial separation of phytosiderophore and microbial siderophore production in different root zones.[30] From observations of root growth and microbial colonization, it has been found that as a root grows through soil, the tip and the zone of elongation are relatively sterile for a period of time prior to colonization by soil microorganisms.[42,43] After inoculation of the apex by bacteria carried on the root cap, or by contact with microcolonies and various inocula on soil particle surfaces, rapid colonization of the root surface takes place over a period of 3 to 4 days. This initial exponential growth phase is followed by a leveling off of microbial growth and a subsequent slow growth phase of expanding microcolonies.[44] This separation of phytosiderophore production from high microbial populations is critical, not only for allowing the plant time to extract available iron before it can be taken up by the microbial populations, but also for allowing phytosiderophores to function prior to bio-degradation. The separation of zones of microbial and phytosiderophore accumulation suggests that current ideas about the role of microbially deposited iron precipitates for supplying iron to phytosiderophores may not be highly relevant to plant nutrition in a natural system.

Another factor that may contribute to the efficacy of phytosiderophores is the diurnal production cycle that results in spiking of high concentrations of phytosiderophores as the root advances through the soil. One advantage to this cycle is that iron dissolution reactions from mineral surfaces are concentration-dependent such that the chelates are more effective in mobilizing iron when present at high concentrations.[45] A second advantage is that secretion of high pulse concentrations may allow temporary accumulation of phytosiderophores in the presence of phytosiderophore-degrader organisms. Lastly, production of excess phytosiderophore is beneficial for preventing the extracellular loss of iron which dissociates from phytosiderophores and nonspecifically adsorbs to negatively charged phospholipids on the root plasmalemma.[46]

Empirical data for phytosiderophore release show that pulsing of phytosiderophores results in extremely high concentrations of chelator in proportion to the available iron present in the rhizosphere. Typical quantities of phytosiderophore produced by *Poaceae* under iron-stress conditions range from 1 to 20 μmol g^{-1} root per 2-h daily cycle. If this quantity is released by the root tips, which are assumed to comprise 5% of the root system,[30] the phytosiderophore release is 20 to 400 nM mg^{-1} root which is secreted into 160 mg of rhizocylinder soil containing 5.7 nM iron. Even after considering zinc, copper, and manganese complexation, the 4- to 80-fold ratio of phytosiderophore to iron should be effective for sequestering available iron prior to chelate degradation. To test the effect of diurnal pulsing on phytosiderophore accumulation, we have constructed a computer simulation model (Figure 2) that examines microbial growth behind the root tip when a given quantity of phytosiderophore is secreted diurnally or continuously at a low concentration. The model assumes that, in either instance, the same total of amount of phytosiderophore is released into the rhizosphere. Using estimates provided by Darrah,[47] some of the carbon that is consumed by the microbial population is returned to the soluble carbon pool by decomposition of the necromass generated by microbial death. The carbon conversion efficiency is assumed to be 60% with a maintenance carbon respiration of 7.6 μg carbon per milligram of biomass per hour.

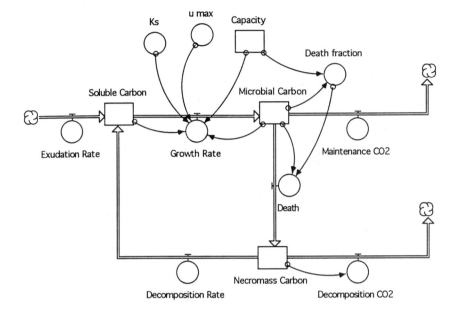

Figure 2. Computer simulation model (STELLA Software Program) used to predict phytosiderophore accumulation patterns and microbial growth dynamics with either pulsed release or constant release of phytosiderophore from plant roots. Microbial biomass C parameters: μ max, maximum microbial growth rate; Ks, affinity constant for phytosiderophore utilization; capacity, maximum attainable biomass carbon; death fraction, proportion of population converted to necromass per hour.
Assumed values: μmax = 1–16; Ks = 0.45; initial microbial biomass C = 0.05 mg; capacity = 4 mg g^{-1} soil; initial soluble carbon = 0; initial necromass carbon = 0; necromass decomposition rate = 0.05; exudation rate = PULSE (1 mg C, initial pulse @ 12 h, pulse interval 24 h) or = CONSTANT 0.042 h^{-1}; maintenance CO_2 = 0.0076 mg C per milligram of biomass.
Equations: Microbial Growth Rate = Microbial Carbon * (μmax * Soluble Carbon/Soluble Carbon + Ks) * (1-(Microbial Carbon/Capacity))
Capacity (t) = Capacity (t – dt)
Soluble Carbon (t) = Soluble Carbon (t – dt) + (Exudation Rate + Decomposition Rate – Growth Rate) * dt
Necromass Carbon (t) = Necromass Carbon (t – dt) + (Death – Decomposition Rate – Decomposition CO_2) * dt
Death = Microbial Carbon * Death fraction
Death fraction = (Microbial Carbon/Capacity) * 0.05
Decomposition CO_2 = 0.4 * Necromass Carbon

Results of the computer simulation clearly show the advantage of pulsed release of phytosiderophore with respect to degradation by microorganisms (Figure 3A). Using model parameters that assume an initial bacterial population biomass of 50 µg carbon per gram of soil,[47] and a bacterial generation time of 4 h, microbial growth proceeds rapidly with log phase growth over 2 days; a value in agreement with direct observations by Bowen.[48] With a pulsed release, the phytosiderophore level on day 1 peaks very quickly and is then

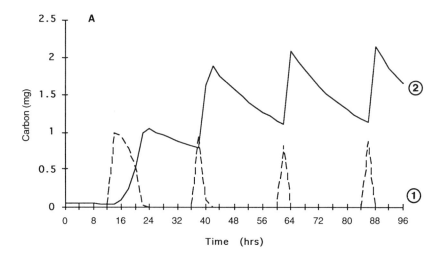

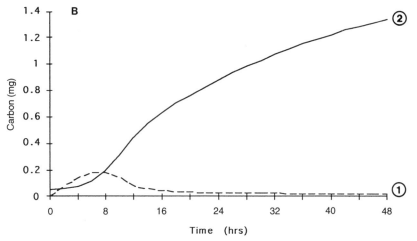

Figure 3.   Predicted soluble carbon accumulation (1,---) and microbial biomass generation (2,—)
over a 7-day (168 h) interval with either pulsed (A) or constant release (B) of
phytosiderophore from plant roots. Model parameters and equations used to generate these
simulations are as described in Figure 2.

consumed over a period of 16 h by the rapidly growing bacterial biomass. At
the second pulse interval, 24 h later, the peak reaches a similar high concentration, but the effective period in solution is shorter as indicated by the peak
width over time. On each of the succeeding days, the pulsed release pattern
allows a period of about 8 h during which the phytosiderophore can mobilize
iron. A simulation run with phytosiderophore released at a constant rate reveals
an entirely different pattern (Figure 3B). In contrast to the pulsed release that

is effective at each pulse interval, a constant low release of phytosiderophore results in a lower initial concentration of phytosiderophores, followed by rapid degradation, and no net accumulation at any time point thereafter. In this case, phytosiderophores would only be effective behind actively growing root tips that can grow fast enough to temporarily escape colonization.

A sensitivity analysis examining the accumulation of phytosiderophore in the presence of a degrader population shows that maximum microbial growth rate (μ max) is the most important factor controlling the longevity of each pulse. At a microbial generation time of 2 h, or twice as fast in Figure 3, the duration of the first pulse is only 4 h, as opposed to 12 h, and 2 h for each subsequent pulse event, as opposed to 8 h. With a constant release, the phytosiderophore reaches a peak level at 8 h and obtains a concentration that is only half that obtained with a slower generation time. The phytosiderophore level then diminishes to essentially zero by 12 h and thereafter is consumed as rapidly as it is produced. On the other extreme, at very slow growth rates with a generation time of 16 h (μ max = 0.0625), the microbial population can not keep up with either pulsed or constant release, which leads to continual accumulation of the phytosiderophore. Although some rhizosphere microorganisms grow at these slow rates,[9] the latter is not likely to occur since all available evidence shows phytosiderophores do not accumulate in the presence of established phytosiderophore-degrader populations.

Our simulation model results for individual roots with either pulsed or continuous release are in good agreement with hydroponic data that show phytosiderophores released by model plants, such as sorghum or barley, are subject to extremely rapid degradation after only a few days of production in nonsterile media.[49] Hydroponic systems contrast with soil in that after selective enrichment for phytosiderophore degraders, there is no spatial separation of the degrader populations from the sites of phytosiderophore release. As documented for corn grown under iron-stress conditions in nonsterile hydroponics, even extracellular apoplastic iron appears to be poorly mobilized by phytosiderophores in hydroponic culture.[26] Phytosiderophore cross-feeding experiments that demonstrate there is relatively poor cross-feeding between iron-efficient and inefficient cultivars in hydroponic media could also be adversely affected by this problem.[24] Given the extreme effect of microbial degradation, phytosiderophore uptake experiments with hydroponically grown plants are best conducted with axenic plants, or at a minimum, after a chemical treatment to surface sterilize the roots.

Calculations of phytosiderophore release under iron-stress conditions show that phytosiderophores are a major source of carbon for microbial growth, particularly at the root tip. Each molecule of mugineic acid contains 12 carbon atoms, which at a release rate of 10 μM g^{-1} root per day represents an input of 2.4 μM (28 μg) carbon per milligram of root per day. If this quantity is concentrated at the root tips that are assumed to comprise 5% of the root system, the carbon release of 560 µg mg^{-1} root during even a single pulse far exceeds that required to generate the microbial biomass normally measured in the rhizosphere. Using empirical approaches such as the multicompartment root box[50] and direct observation of microbial populations, it should be possible to determine whether phytosiderophore release results in a higher microbial population density behind the root tip than occurs under iron-sufficient growth. Also, since the microbial growth curve is strongly influenced by the carbon release pattern, it is important to know whether phytosiderophores comprise the bulk of the carbon released in the zone behind the root tip or only a portion of the total root exudate. In order to produce a microbial biomass of 300 µg mg^{-1} root at the root apices, approximately 300 µg of carbon are required at a conversion efficiency of 60%. In studies of barley under iron-sufficient conditions, total exudate release ranges from 20 to 130 µg mg^{-1} of root tissue,[51] which is much less than the amount calculated above for phytosiderophore release. Unfortunately, the localization of nonphytosiderophore carbon release may not be easy to document with any certainty since it appears to be variable depending on individual root growth rates and time of day. As a generalization, slow-growing roots have been found to release more carbon per unit biomass than fast growing roots, and exudation rates for all roots are approximately twofold higher during the day than at night.[51]

PLANT GROWTH AND IRON DEMAND

The importance of plant demand for iron in establishing nutritional competition has not yet been examined quantitatively, but is likely to be quite variable for plant species that have different growth rates, iron demand, and mechanisms for responding to iron limitation. Several studies have now shown that the relative iron uptake efficiency in *Poaceae* grasses is proportional to the release rate of phytosiderophores.[41,43] However, iron stress also must be defined in relation to plant growth rates, iron demand, and the root to shoot ratio for different plant species. Shoot iron content is quite variable for different plants, ranging from 1 to 7 µM g^{-1} tissue (50 to 350 ppm). To determine the quantity that is actually required for growth, it is necessary to measure the relative growth rates of individual plants grown over a range of iron concentrations under otherwise nutrient sufficient conditions. Another approach may involve the use of chelator buffering of hydroponic nutrient solutions to determine the concentration of free soluble Fe and corresponding internal concentrations of iron that are required for normal growth. If the plant has a very low growth rate, the amount of iron taken up per day per unit root weight is much less than that which would be required for growth of a rapidly growing plant species. This leads to a lower flux density of iron per unit surface area for slow-growing plants, and conversely, a high flux density of nutrient transfer from bulk soil for fast-growing plants. For this reason, it is not surprising that corn and sorghum, which have extremely high rates of biomass production, are among the most iron-inefficient plants even though they produce phytosiderophores. As other plant species and cultivars are subjected to evaluation for iron-uptake efficiency, realistic comparisons between species and ecotypes should be based on phytosiderophore production in relation to the relative growth rate, and not simply on phytosiderophore production quantities per gram of root.

Root growth patterns may have great importance in determining the relative iron uptake efficiency of grasses as well as for establishing iron nutritional competition in the rhizosphere. Since release of phytosiderophores by grasses occurs behind the root tips, the number of growing tips present on the root system may be proportional to the iron-uptake efficiency of the plant and the amount of iron that is depleted from the rhizosphere. Other plant responses might include a shift in the root:shoot ratio such that more of the plant biomass is allocated for nutrient acquisition. In recent studies to examine this phenomenon, selected calcicole and calcifuge grasses native to central Europe were examined to determine the importance of these responses for iron efficiency.[52] Calcicole plants were found to rely on phytosiderophore production and did not significantly change the root surface area per unit biomass, whereas calcifuge species increased their root surface area, but were not as efficient in producing phytosiderophores. The greater tolerance of calcicole species to iron-limiting conditions suggests that increased production of phytosiderophores may be a more effective response than allocating increased carbon to the root system. At the same time, by not increasing the root surface area, iron depletion may be more severe in the rhizosphere of calcicole plants, leading to greater microbial competition and production of microbial siderophores. Whether microbial siderophores contribute to the iron nutrition of highly efficient grasses native to calcareous soils remains to be determined.

We have frequently observed that under conditions of iron stress in hydroponic culture, grasses will change their branching pattern such that there are greater numbers of root tips and lateral branching on iron-stressed plants. As predicted from our model, differences in the relative growth rates of first, second, and third order laterals may be important for governing the efficacy of phytosiderophores. Roots with slow-growing determinant growth patterns may not have a sufficient growth rate to escape microbial colonization at the apex. Under these circumstances, phytosiderophores might be rapidly consumed and have only a low, intermittent accumulation in the rhizosphere where they would be in competition with microbial siderophores. On the other hand, if microbial siderophores are important in contributing to plant iron acquisition, it is likely that slow-growing laterals with high exudate production would result in maximal production of microbial siderophores that provide the predominant form of available iron in these root zones. Another area of high microbial activity where siderophores are likely to be produced at high concentrations will include breaks in the root endodermis, particularly at sites of lateral root emergence.[27] Since these sites provide a direct entry into the root apoplast that is continuous with the xylem, apoplasmic transport of microbial siderophores to the shoot may represent yet another important process for plant iron nutrition that has not yet been well investigated.

RELATIONSHIPS BETWEEN PLANT GROWTH AND MICROBIAL COMPETITION

The same relationship between growth rate and iron demand also holds true for microorganisms, and may help to explain the development of different microbial communities on plants. Since most, if not all of the carbon in the rhizosphere is derived from the plant, competition between microorganisms will depend not only on the relative growth rates of two organisms that are competing, but also on the plant growth rate that establishes the rate of carbon release. With rapidly growing plants, there will be depletion of iron and concurrent high deposition of carbon that establishes conditions for particularly intense competition for iron. Conversely, microorganisms that colonize slower growing plant species, or that are secondary colonizers of older root tissue, would be expected to have lower growth rates and less immediate demand for iron needed to utilize low amounts of carbon released by the roots. Under low carbon conditions, iron that is taken up may be replenished by diffusion rather than by induced production of siderophores. These rules of competition have consequences not only for basic knowledge on microbial interactions, but also for applications in which microorganisms are introduced into soils for various purposes such as biocontrol of root disease, or to promote plant growth by production of growth regulators.

During the primary colonization of new root tissue, microbial competitiveness is a function of the initial population and differences in generation time for various microbial species, as well as possible amensal effects such as release of antibiotics. Subsequently, as the rhizosphere community matures, secondary colonization on older roots is characterized by slow growth in a crowded, increasingly oligotrophic environment. Based on the distinct transition that occurs along the root, strategies for iron competition may change in specific root zones depending on the carbon supply, long-term alterations in the soil chemical environment, and the differential production of siderophores by specific populations. One of the current challenges in rhizosphere biology is to better understand the factors that are important in determining the species composition of microorganisms on plant roots, and how the microbial community can be manipulated by agricultural practices. Increasing evidence suggests that siderophore production plays a central role in mediating intermicrobial competition, which can result in both amensal or commensal relationships between microorganisms, depending on their ability to utilize different siderophore types.[53]

One of the best examples of the interaction between carbon and iron in microbial competition is that documented to occur between plant beneficial strains of *Pseudomonas fluorescens* and the root pathogen, *Fusarium* sp.[29,54] Under high carbon conditions, *P. fluorescens* must produce its siderophore to inhibit growth of the pathogen, whereas under low carbon conditions, less siderophore is produced and competition becomes increasingly based on differences in relative growth rates with respect to carbon. In this particular research, siderophore production was found to be reduced at lower growth rates, and was not directly proportional to the generation time, i.e., siderophore production decreased faster than the rate of cell division. Further studies using chemostats are necessary to better examine this phenomenon over the range of carbon and iron levels that occur in the rhizosphere. If only negligible levels of siderophores are produced along older parts of the roots that release low amounts of carbon, the importance of microbial siderophores in plant nutrition or disease suppression may be diminished or restricted to certain root zones.

Apart from disease suppression, other general effects of pseudomonads on plant growth are highly variable and poorly understood. Surveys of dozens of strains of *Pseudomonas* sp. on winter wheat have shown that many strains are inhibitory to root growth,[55] whereas certain strains of *P. cepacia*, *P. fluorescens*, and *P. putida* are beneficial for increasing plant height, biomass, and numbers of tillers.[56] Yet another aspect of *Pseudomonas* siderophores in iron nutritional competition is the effect of low iron on the survival and activity of symbiotic nitrogen-fixing bacteria. Under conditions of low iron availability, siderophores produced by fluorescent pseudomonads have been shown to inhibit root nodulation by rhizobium,[57] and to cause changes in nodule occupancy by competing strains.[58] Thus, in order to optimize the microbial community for plant growth and survival of highly effective nitrogen-fixing bacteria, it will be necessary to better understand factors affecting the balance between beneficial and detrimental populations of *Pseudomonas* sp. and how this is affected by iron.

SIDEROPHORE COMPOSITION OF THE RHIZOSPHERE

Presently, bioassay techniques have been developed to detect a variety of siderophores in soils,[2,59] but quantitative data for siderophore production rates and net accumulation in microsites are severely limited. In previous attempts to quantify siderophore concentrations, sampling methods have involved collection of soil along the entire root length. Undoubtedly, however, the concentrations of different siderophores and net composition of the mixture will be highly variable for different root zones. With regard to patterns of siderophore utilization, it is particularly interesting that most microorganisms have an ability to utilize a number of siderophores as iron sources, but that some microbes also specialize in the production of highly unique siderophores that are strain specific. In the case of pseudomonads, which are among the most aggressive root colonizers, the dual strategies of employing unique siderophores for nutritional competition, while also utilizing more generic siderophores, seems to be particularly well refined. Possibly, during the evolution of strains that are adapted for intense competition, pseudomonads have undergone selection for an ability to produce unique compounds that are capable of sequestering iron to the detriment of competitors. Further behind the root tip, slower growing organisms such as *Arthrobacter* and *Streptomyces* might have little need for unique siderophores and instead rely on widely produced and utilized siderophores, such as the ferrioxamines and ferrichromes. One of the immediate needs for better understanding nutritional competition are methods for detecting and quantifying specific types of siderophores in different root zones.

In the past few years, considerable progress has been achieved in the development of high performance liquid chromatography (HPLC) techniques for detection of phytosiderophores and microbial siderophores, although there are not yet any published methods for simultaneous detection of both types of compounds.[60,61] For rhizosphere competition studies it is also important to be able to detect whether the biosynthetic chelates are chelated with iron or other trace metals. This problem is especially critical using published methods for detection of phytosiderophores, which cannot be used to differentiate between the ferrated and deferrated forms, nor used preparatively since the phytosiderophore is derivatized with OPA for detection by UV fluorescence. In an attempt to solve this problem, we have been investigating the use of a polymeric HPLC column (Dionex OmniPac PCX 500) that combines both cation exchange and reverse phase retention properties in conjunction with UV and flow cell radioisotope detection. Chromatogram peak identification is accomplished by comparison of elution times with purified standards and is further facilitated by analysis of peak purity using a diode array detector and post-run computer analysis. Using purified siderophore standards obtained from microbial cultures, our methods routinely allow rapid separation and quantification of mixtures of a variety of different siderophores and certain phytosiderophores (Figure 4). Additionally, with an autosampler, our preliminary data indicate these procedures can be used for kinetic studies in which radiolabeled solid phase iron solubilization rates and metal exchange rates are monitored over time for defined siderophore mixtures. The minimum detection limits (less than 1 ng of siderophore) with the UV and isotope detectors are well below physiological concentrations that would be expected in rhizosphere water extracts.

To study the siderophore composition of the rhizosphere using HPLC techniques we have devised an artificial system that allows us to inoculate plant

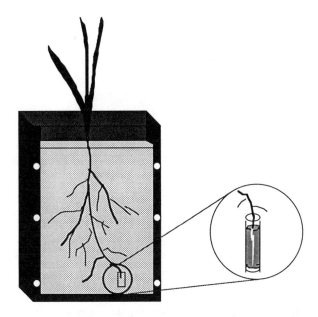

Figure 4.   Diagram of root box and rhizotube assembly used for sampling of rhizosphere exudates and siderophores after inoculation with selected microorganisms. Rhizotube assembly may consist either of a glass tube, or a microfuge spin filter which allows displacement extraction of rhizosphere solution by centrifugation.

roots with specific microorganisms under defined conditions that can be replicated. This system employs a root box with a removable side that allows access to the roots (Figure 5). To work with individual root tips, microfuge spin filter tubes are prepared with agarose media containing radioactive iron and a suspension of bacteria that can be added at different inoculum density levels. The tubes are inserted into the container media underneath an actively growing root tip that is directed toward the top of the tube. Because of the small contained volume, it is possible to use high levels of radioisotopes without contaminating large quantities of soil, and several tubes can be inserted into any one root box, allowing good replication of experimental treatments. Subsequently, the root box is closed and the plant is placed back into a controlled environment growth chamber for 1 to 2 days. During this time the roots of plants such as corn or barley grow well over 1 cm/d into the microfuge tube media and become colonized by the bacteria. After removal of the tube, by excision of the root, the microfuge spin filter is placed in a receptacle tube and spun at 10,000 rpm for 10 min to extract the water solution containing plant and microbial chelates from the agarose media. This solution is then injected into the HPLC for analysis.

Using this system we have detected pyoverdin production from a *Pseudomonas putida* isolate originally obtained from peanut roots after inoculation onto corn roots. Interestingly, to our knowledge, this is the first time that a microbial

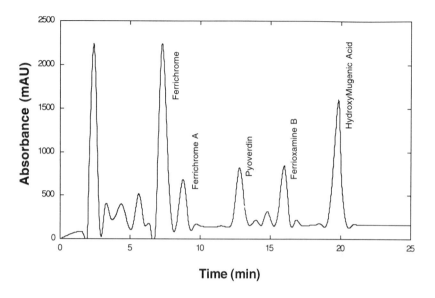

Figure 5. High performance liquid chromatography chromatogram of a mixture of ferrated microbial siderophores and the phytosiderophore, [^{55}Fe]-mugineic acid, after separation by reverse phase chromatography on a polymeric, cationic Dionex PCX 500 column. Eluant: 1% acetonitrile 99% H$_2$0 for 0–5 min, with linear gradient to 70% acetonitrile:0.01 M KCl over 5–20 min. Peak detection by UV absorption at 205 nm, with radioisotope peak confirmation by flow radiometry.

siderophore has been directly extracted from a microbial culture on plant roots. Using the radioisotope detection system, we can also detect the secretion of phytosiderophores into the agar media. This methodology shows great promise for examining specific microbial effects on phytosiderophore degradation and accumulation, as well as for detecting microbial siderophore concentrations that may be significant for plant iron nutrition. Use of the spin tubes is of course not limited to agarose media, but can also be used for sand or soil media containing different levels of iron or microbes, or amended to set different initial pH or organic matter levels. Microorganisms recovered from the soil can also be plated to determine population numbers of specific microorganisms on selective media, and the root section above the culture tube can be analyzed for radioactive iron content to determine whether a specific inoculation was beneficial or detrimental to plant iron uptake.

SUMMARY

Nutritional competition for iron has been established as an important factor governing microbial populations and plant-microbial interactions in both aquatic and terrestrial ecosystems.[53,62] However, apart from iron transport mechanisms that are utilized by individual organisms, very little is known yet about processes that govern iron availability and competition in natural systems, particularly in relation to carbon supply and microbial growth rates. Both beneficial and detrimental affects of microorganisms on plant iron uptake have been documented, including increased availability of iron to plants that utilize microbial siderophores, and in the case of grasses, the detrimental effects of microorganisms that degrade phytosiderophores for use as a substrate for growth. Better characterization of the role of microorganisms in plant growth and nutrition remains one of the most difficult and challenging problems in terrestrial ecology. Over the past decades, many clever experimental systems have been devised to study plant microbial interactions.[58-61] Nevertheless, in accordance with the principle of indeterminancy, every technique that has been devised to study the rhizosphere in some way disrupts the system.[63-66] Since any addition of microorganisms to soil or plant cultures alters the population balance and distribution of microorganisms in the microbial community, the best approach for minimally disruptive experiments would be with intact soil cores that are converted into microcosms. Recognizing the constraints on any one experimental approach, clever combinations of methodologies in conjunction with simulation modeling will be required to study the complex interdependence of plants and microorganisms that has evolved during adaptation to different soil chemical conditions.

Presently, very little is known about soil factors that affect plant species composition in grasslands or the structure of microbial communities that are associated with plant roots. In this regard, the evolution of calcicole and calcifuge plant species poses some particularly interesting questions, and suggest that it may be possible to link ecophysiological processes in the rhizosphere to plant adaptation to calcareous soils. At the other end of the spectrum, a better understanding of nutritional competition in the rhizosphere may allow us to manipulate the survival and activity of microorganisms that are introduced into soils for biocontrol of root disease, nitrogen fixation, and other purposes. Previously, most research on microbial dynamics in the rhizosphere has focused on the total microbial community rather than on individual populations. This will probably continue to be the case until better tools are available to track specific microbial populations in soils and changes that occur in the vicinity of plant roots with respect to iron and other nutrients. Carefully designed artificial systems provide many opportunities for studying plant microbial interactions under gnotobiotic conditions. The use of molecular biology techniques, such as the bioluminescence reporter gene,[67,68] which could be employed for tracking iron-regulated microbial activity, also offers a promising tool for studying the effects of nutrient limitation on individual microbial populations and how competition for nutrients may be affected by plant growth. With these new approaches, it will be interesting to see how our perspective of the rhizosphere changes in the coming decade.

REFERENCES

1. Rovira, A. D., Rhizosphere research — 85 years of progress and frustration, in *The Rhizosphere and Plant Growth,* Keister, D. L. and Cregan, P. B., Eds., Kluwer Academic, Boston, 1991, 3.

2. Bossier, P. and Verstraete, W., Ecology of Arthrobacter JG-9-detectable hydroxmate siderophores in soils, *Soil Biol. Biochem.,* 18, 487, 1986.

3. Buyer, J. S. and Sikora, L. J., Rhizosphere interactions and siderophores, in *The Rhizosphere and Plant Growth,* Keister, D. L. and Cregan, P. B., Eds., Kluwer Academic, Boston, 1991, 263.

4. Crowley, D. E., Wang, Y. C., Reid, C. P. P., and Szaniszlo, P. J., Mechanisms of iron acquisition from siderophores by microorganisms and plants, *Plant Soil,* 130, 179, 1991.

5. Newman, E. I. and Watson, A., Microbial abundance in the rhizosphere: a computer model, *Plant Soil,* 48, 17, 1977.

6. Bowen, G. D. and Rovira, A. D., Are modelling approaches useful in rhizosphere biology?, *Bull. Ecol. Res. Comm.,* 17, 443, 1973.

7. Cushman, J. H., Nutrient transport inside and outside the root rhizosphere: generalized model (Nonlinear root boundary conditions), *Soil Sci.,* 138, 164, 1984.

8. Lynch, J. M. and Whipps, J. M., Substrate flow in the rhizosphere, in *The Rhizosphere and Plant Growth,* Keister, D. L. and Cregan, P. B., Eds., Kluwer Academic, Boston, 1991, 15.

9. Christensen, H., Funck, J. D., and Kjoller, A., Growth rate of rhizosphere bacteria measured directly by the tritiated thymidine incorporation technique, *Soil Biol. Biochem.,* 21, 113, 1989.

10. Alexander, D. B. and Zuberer, D. A., Use of chrome azurol S reagents to evaluate siderophore production by rhizosphere bacteria, *Biol. Fertil. Soils,* 12, 39, 1991.

11. Bienfait, H. F., Lubberding, H. J., Heutink, P., Linder, L., Visser, J., Kaptein, R., and Dijkstra, K., Rhizosphere acidification by iron deficient bean plants: the role of trace amounts of divalent metal ions, *Plant Physiol.,* 90, 359, 1989.

12. Bienfait, H. F., Mechanism of Fe-efficiency reactions of higher plants, *J. Plant Nutr.,* 11, 605, 1988.

13. Zocchi, G. and Cocucci, S., Fe uptake mechanism in Fe-efficient cucumber roots, *Plant Physiol.,* 92, 908, 1990.

14. Nomoto, K., Sugiura, Y., and Takagi, S., Mugineic acids, studies on phytosiderophores, in *Iron Transport in Animals, Plants, and Microorganisms,* Winklemann, G., Van der Helm, D. and Neilands, J. B., Eds., VCH Publishers, Weinheim, Germany, 1987, 401.

15. Jenkinson, D. S. and Ladd, J. N., Microbial biomass in soil: measurement and turnover, in *Soil Biochemistry,* Paul, E. A. and Ladd, J. N., Eds., Marcel Dekker, New York, 1981, 415.

16. Alexander, D. B. and Zuberer, D. A., Siderophore-producing bacteria isolated from roots of iron-efficient and inefficient grasses, in *The Rhizosphere and Plant Growth,* Keister, D. L. and Cregan, P. B., Eds., Kluwer Academic, Beltsville, MD, 1991, 308.

17. Zhang, F., Römheld, V., and Marschner, H., Role of the root apoplasm for iron acquisition by wheat plants, *Plant Physiol.,* 97, 1302, 1991.

18. Kleeberger, A., Castorph, H., and Klingmuller, W., The rhizosphere microflora of wheat and barley with special reference to gram-negative bacteria (*Enterobacter agglomerans, Pseudomonas fluorescens*), *Arch. Microbiol.*, 136, 306, 1983.

19. Meyer, J. M., Halle, F., Hohnadel, D., Lemanceau, P., and Ratefiarivelo, H., Siderophores of *Pseudomonas* — biological properties, in *Iron Transport in Animals, Plants, and Microorganisms*, Winklemann, G., Van der Helm, D., and Neilands, J. B., Eds., VCH Publishers, Weinheim, Germany, 1987, 370.

20. Crowley, D. E., Römheld, V., Marschner, H., and Szaniszlo, P. J., Root-microbial effects on plant iron uptake from siderophores and phytosiderophores, *Plant Soil*, 142, 1, 1992.

21. Mozafar, A., Duss, F., and Oertli, J. J., Effect of *Pseudomonas fluorescens* on the root exudates of two tomato mutants differently sensitive to Fe chlorosis, *Plant Soil*, 144, 167, 1992.

22. Wang, Y. C., Brown, H. N., Crowley, D. E., and Szaniszlo, P. J., Evidence for direct utilization of a siderophore, ferrioxamine B, in axenically-grown cucumber, *Plant Cell Environ.*, 16, 579, 1993.

23. Wang, Y. C., Crowley, D. E., and Szaniszlo, P. J., unpublished data.

24. Brown, J. C., Jolley, V. D., and Mel Lytle, C., Comparative evaluation of iron solubilizing substances (phytosiderophores) released by oat and corn. Iron-efficient and iron inefficient plants, in *Iron Nutrition and Interactions in Plants*, Chen, Y. and Hadar, Y., Eds., Kluwer Academic, Boston, 1991, 189.

25. Jolley, V. D. and Brown, J. C., Iron efficient and inefficient oats. I. Differences in phytosiderophore release, *J. Plant Nutr.*, 12, 423, 1989.

26. Bienfait, H. F., Van den Briel, W., and Mesland-Mul, N. T., Free space iron pool in roots: generation and mobilization, *Plant Physiol.*, 78, 596, 1985.

27. Bar-Ness, E., Hadar, Y., Shanzer, A., and Libman, J., Iron uptake by plants from microbial siderophores. A study with 7-nitrobenz-2 oxa-1,3-diazole-desferrioxamine as fluorescent ferrioxamine B analog, *Plant Physiol.*, 99, 1329, 1992.

28. Crowley, D. E. and Szaniszlo, P. J., Iron uptake by oat from phytosiderophores and ferrioxamine B in long term axenic culture, 3rd Int. Conf. Iron Transport, Storage, and Metabolism, Austin, TX, July 1990.

29. Elad, Y. and Baker, R., The role of competition for iron and carbon in suppression of chlamydospore germination of *Fusarium* spp. by *Pseudomonas* spp, *Phytopathology*, 75, 1053, 1985.

30. Römheld, V., The role of phytosiderophores for acquisition of iron and other micronutrients in graminaceous species. An ecological approach, *Plant Soil*, 130, 127, 1991.

31. Treeby, M., Marschner, H., and Romheld, V., Mobilization of iron and other micronutrient cations from a calcareous soil by plant borne, microbial, and synthetic chelators, *Plant Soil*, 114, 217, 1989.

32. Whipps, J. M. and Lynch, J. M., Substrate flow and utilization in the rhizosphere of cereals (*Triticum aestivam*, wheat, *Hordeum vulgare*, barley), *New Phytol.*, 95, 605, 1983.

33. Crowley, D. E., Reid, C. P. P., and Szaniszlo, P. J., Microbial siderophores as iron sources for plants, in *Iron Transport in Animals, Plants, and Microorganisms*, Winkelmann, G., Van der Helm, D., and Neilands, J. B., Eds., VCH Publishers, Weinheim, Germany, 1987.

34. Meyer, J. M. and Abdallah, M. A., The fluorescent pigment of *Pseudomonas fluorescens:* biosynthesis, purification, and physicochemical properties, *J. Gen. Microbiol.,* 107, 319, 1978.
35. Erich, M. S., Duxbury, J. M., Bouldin, D. R., and Cary, E., The influence of organic complexing agents on iron mobility in a simulated rhizosphere, *Soil Sci. Soc. Am. J.,* 51, 1207, 1987.
36. Takagi, S., Kamei, S., and Yu, M. H., Efficiency of iron extraction from soil by plantborne, microbial, and synthetic chelators, *J. Plant Nutr.,* 11, 643, 1988.
37. Awad, F., Romheld, V., and Marschner, H., Mobilization of ferric iron from a calcareous soil by plant-borne chelators (phytosiderophores), *J. Plant Nutr.,* 11, 6, 1988.
38. Fekete, F. A., Spence, J. T., and Emery, T., Siderophores produced by nitrogen-fixing *Azotobacter vinelandii* OP in iron-limited continuous culture, *Appl. Environ. Microbiol.,* 46, 1297, 1983.
39. Lindsay, W. L. and Norvell, W. A., Development of a DTPA soil test for zinc, iron, manganese, and copper, *Soil Sci. Soc. Am. J.,* 42, 421, 1978.
40. Zhang, F. S., Treeby, M., Römheld, V., and Marschner, H., Mobilization of iron by phytosiderophores as affected by other micronutrients, in *Iron Nutrition and Interactions in Plants,* Chen, Y. and Hadar, Y., Eds., Kluwer Academic, Boston, 1992, 205.
41. Bar-Ness, E., Chen, Y., Hadar, Y., Marschner, H., and Romheld, V., Siderophores of *Pseudomonas putida* as an iron source for dicot and monocot plants, in *Iron Nutrition and Interactions in Plants,* Chen, Y. and Hadar, Y., Eds., Kluwer Academic, Boston, 1992, 271.
42. Parke, J. L., Root colonization by indigenous and introduced microorganisms, in *The Rhizosphere and Plant Growth,* Keister, D. L. and Cregan, P. B., Eds., Kluwer Academic, Boston, 1991, 33.
43. Rovira, A. D. and Campbell, R., Scanning electron microscopy of microorganisms on the roots of wheat, *Microbiol. Ecol.,* 1, 15, 1974.
44. Bowen, G. D., Microbial dynamics in the rhizosphere. Possible strategies in managing rhizosphere populations, in *The Rhizosphere and Plant Growth,* Keister, D. L. and Cregan, P. B., Eds., Kluwer Academic, Boston, 1991, 25.
45. Stumm, W. and Wieland, E., Dissolution of oxide and silicate minerals: rates depend on surface speciation, in *Aquatic Chemical Kinetics,* Stumm, W., Ed., Wiley-Interscience, New York, 1990, 367.
46. Mihashi, S., Mori, S., and Nishizawa, N., Enhancement of ferric-mugineic acid uptake by iron deficient barley roots in the presence of excess free mugineic acid in the medium, in *Iron Nutrition and Interactions in Plants,* Chen, Y. and Hadar, Y., Eds., Kluwer Academic, Dordrecht, 1991, 167.
47. Darrah, P. R., Models of the rhizosphere. I. Microbial population dynamics around a root releasing soluble and insoluble carbon, *Plant Soil,* 133, 187, 1991.
48. Bowen, G. D., Integrated and experimental approaches to the growth of organisms around roots, in *Soil Borne Plant Pathogens,* Schippers, B. and Gams, W., Eds., Academic Press, London, 1979, 209.
49. Römheld, V. and Marschner, H., Genotypical differences among graminaceous species in release of phytosiderophores, *Plant Soil,* 123, 147, 1990.
50. Mihashi, S. and Mori, S., Characterization of mugineic acid Fe transporter in Fe-deficient barley roots using the multi-compartment transport box method, *Biol. Metals,* 2, 146, 1989.

51. Liljeroth, E., Baath, E., Mathiasson, I., and Lundborg, T., Root exudation and rhizoplane bacterial abundance of barley (*Hordeum vulgare* L.) in relation to nitrogen fertilization and root growth, *Plant Soil*, 127, 81, 1990.

52. Gries, D. and Runge, M., The ecological significance of iron mobilization in wildgrasses, *J. Plant Nutr.*, 15, 1727, 1992.

53. Bossier, P., Hofte, M., and Verstraete, W., Ecological significance of siderophores in soil, in *Advances in Microbial Ecology*, Marshall, K. C., Ed., Plenum Press, New York, 1988, 385.

54. Park, C. S., Paulitz, T. C., and Baker, R., Biocontrol of *Fusarium* wilt of cucumber resulting from interactions between *Pseudomonas putida* and nonpathogenic isolates of *Fusarium oxysporum*, *Phytopathology*, 78, 190, 1988.

55. Fredrickson, J. K. and Elliott, L. F., Colonization of winter wheat roots by inhibitory rhizobacteria, *Soil Sci. Soc. Am. J.*, 49, 1172, 1985.

56. Freitas, J. and Germida, J. J., Plant growth promoting rhizobacteria for winter wheat, *Can. J. Microbiol.*, 36, 265, 1990.

57. Fuhrmann, J. and Wollum, A. I., *In vitro* growth responses of *Bradyrhizobium japonicum* to soybean rhizosphere bacteria, *Soil Biol. Biochem.*, 21, 131, 1989.

58. Fuhrmann, J. and Wollum, A. I., Nodulation competition among *Bradyrhizobium japonicum* strains as influenced by rhizosphere bacteria and iron availability, *Biol. Fertil. Soils*, 7, 108, 1989.

59. Nelson, M. N., Cooper, C. R., Crowley, D. E., Reid, C. P. P., and Szaniszlo, P. J., An *Escherichia coli* bioassay of individual siderophores in soil, *J. Plant Nutr.*, 11, 915, 1988.

60. Konetschny-Rapp, S., Huschka, H., Winkelmann, G., and Jung, G., High performance liquid chromatography of siderophores from fungi, *Biol. Metals*, 1, 9, 1988.

61. Kawai, S., Sato, Y., Takagi, S., and Nomoto, K., Separation and determination of mugineic acid and its analogs by high-performance liquid chromatography, *J. Chromatogr.*, 391, 325, 1987.

62. Murphy, T. P., Lean, D. R. S., and Nalewajako, C., Blue-green algae: the excretion of iron-selective chelators enables them to dominate other algae, *Science*, 192, 900, 1976.

63. Billes, G., A technique for the axenic culture of plants for the study of the rhizosphere effect, *Oecol. Plant.*, 13, 83, 1978.

64. Helal, H. M. and Sauerbeck, D. R., Method to study turnover processes in soil layers of different proximity to roots, *Soil Biol. Biochem.*, 15, 223, 1983.

65. Polonenko, D. R. and Mayfield, C. I., A direct observation technique for studies on rhizoplane and rhizosphere colonization, *Plant Soil*, 51, 405, 1979.

66. Martens, R., Apparatus to study the quantitative relationships between root exudates and microbial populations in the rhizosphere, *Soil Biol. Biochem.*, 14, 315, 1982.

67. Weger, L. A., Dunbar, P., Mahafee, W. F., Lugtenberg, B. J. J., and Sayler, G., Use of bioluminescence markers to detect *Pseudomonas* spp. in the rhizosphere, *Appl. Environ. Microbiol.*, 57, 3641, 1991.

68. Shaw, J. J., Dane, F., Geiger, D., and Kloepper, J. W., Use of bioluminescence for detection of genetically engineered microorganisms released into the environment, *Appl. Environ. Microbiol.*, 58, 267, 1992.

CHAPTER 15

Mechanisms of Iron Acquisition by Graminaceous (Strategy II) Plants

Satoshi Mori

INTRODUCTION

Discovery of Fe-Solubilizing Substances From Root Washings of Fe-Deficient Rice

Until recently, the susceptibility of rice cultivars to Fe deficiency has been poorly understood. In early work, Takagi[1] found that Fe deficiency in rice resulting from increased nutrient solution pH was more pronounced in plants grown in submerged culture than in a synthetic soil (Figure 1). The results from this research suggested that:

1. Under aerobic conditions, plants absorb Fe by use of root exudates that solubilize Fe at the root surface.
2. The root systems of rice produce less Fe-solubilizing substances than oat, which is less susceptible to Fe deficiency than rice.
3. The increased soil moisture that occurs in submerged culture enhances the leaching of the root exudates from the rhizosphere and consequently enhances Fe deficiency in rice.
4. For oat, the high rates of release of the Fe-solubilizing root exudates overcome Fe solubility problems associated with high pH.

However, in the presence of excess Cu^{2+}, where the Fe solubilization by the root exudates is competitively inhibited, Fe deficiency is induced at high pH.

Subsequent research by Takagi[2] proved the production of Fe-solubilizing root exudates. In these experiments, the roots of rice or oat were dipped into culture solutions with suspended $^{59}Fe(OH)_3$ gels. The gels were immediately adsorbed as a red/brown layer on the root surfaces. This Fe layer was tightly

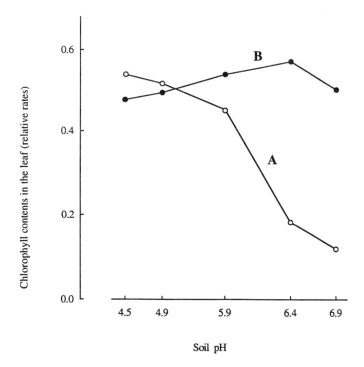

Figure 1. Effects of pH and water content in synthetic soils (made of sandy soils without organic matter) on the formation of Fe deficiency-induced chlorosis in rice seedlings; A: submerged culture; B: culture under 70% field capacity.

bound to the root surfaces and no leaching occurred when the roots were returned to Fe-free culture solutions. Time courses were subsequently taken to measure the levels of ^{59}Fe transported to the tops of the rice and oat plants. As shown in Figure 2A, the transport rates at pH 5 and 7 were very low, indicating that in water culture Fe absorption is negligible. A second set of plants was similarly exposed to the ^{59}Fe gels. However, instead of immediately placing the roots into the nutrient solutions, the plants were first kept moist by spraying with distilled water for 5 h, after which the plants were placed in nutrient solutions at pH 7.0. Results of this treatment (Figure 2A) showed a rapid increase in the ^{59}Fe content in the tops of the plants. Similar experiments with oats (Figure 2B) showed that at pH 5 and 7, and with the water-deficit treatment, the ^{59}Fe levels in the shoot tissue of the oat plants increased more rapidly than in rice.

These results indicated that the roots of both oat and the rice plants secreted Fe-solubilizing substances at the root's surface. These substances were concentrated at the root surface of the water-deficit plants. In contrast, the Fe-solubilizing compounds were diluted away from the plants whose roots were immediately placed in the nutrient solutions. As a consequence, little ^{59}Fe uptake occurred for these latter plants. For oats, the higher production of

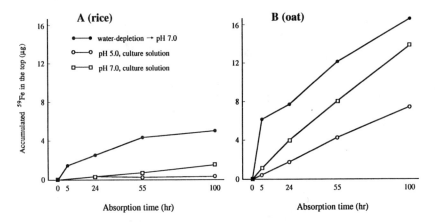

Figure 2.   Effect of pH of the culture solution and water-depletion treatment (mist culture) on the transport of ^{59}Fe from $^{59}Fe(OH)_3$ gels adsorbed on the root surface.

Fe-solubilizing substances was subsequently proven in experiments that compared the Fe-solubilizing activities of the root washings of oat and rice plants.[1,2]

Determination of the Chemical Structure of Mugineic Acid (MA)-Related Phytosiderophores

After the discovery of the Fe-solubilizing compounds, similar compounds were investigated by Tagaki from other graminaceous plants including: barley (*Hordeum vulgare* cv. Minorimugi[3] and Tochigi-golden melon),[4] oat *(Avena sativa)*,[5] rye *(Secale cereare)*,[6] and wheat *(Triticum aestivum)*.[4] Nomoto et al.[7] investigated the corresponding compounds from the beer barley (*Hordeum vulgare* var. distichom.). Chemical structures of mugineic acid (MA),[3] avenic acid (AVA),[5] 3-hydroxymugineic acid (HMA),[6] 3-epihydroxymugineic acid (epiHMA),[4] 2′-deoxymugineic acid (DMA),[4] and distichonic acid were subsequently determined.[7]

Strategy I and Strategy II

Characterization of phytosiderophore production and the corresponding structural details of these compounds significantly contributed to the understanding of the different Fe uptake mechanisms that occur in dicots and monocots. Proposals made by Takagi,[8] and subsequently Römheld and Marschner,[9] outlined two operative mechanisms: Strategy I for a number of Fe deficiency-tolerant dicots, and Strategy II for certain graminaceous monocots.

The remaining portion of this chapter will summarize the properties of the family of mugineic acid-related compounds produced by the graminaceous monocots. The emphasis of this chapter will center on the pertinent portions of Figure 3, mainly the biosynthesis, secretion, chelation, degradation, and absorption of phytosiderophore compounds.

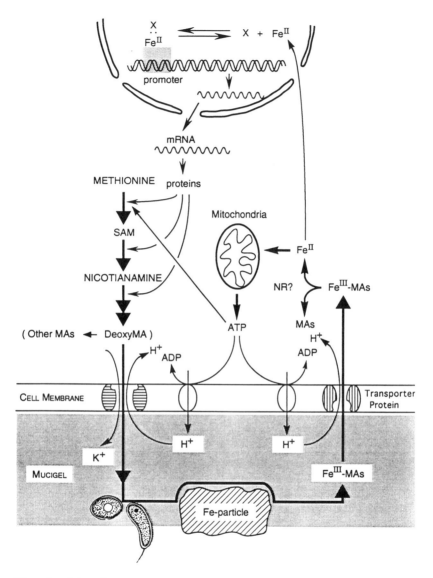

Figure 3.   Schematic representation of the dynamics of phytosiderophores; biosynthesis, secretion, chelation, degradation, and absorption.

BIOSYNTHESIS

Methionine — The Main Precursor of Mugineic Acid (MA) and Related Phytosiderophores

Initial investigations into the biosynthesis of MAs required developing improved techniques for the quantification and chromatographic separation of

MAs. Originally, the levels of MAs were determined as the amounts of total Fe-solubilizing substances. Early methods of MA qualitative analysis involved separation by either paper electrophoresis or paper chromatography. The separated MAs were detected by sequential exposure of chromatograms to $FeCl_3$ and NH_3. Paper chromatograms were then washed with water, and treated with o-phenanthroline and sunlight. The MAs were detected as white spots on a pink background of Fe^{2+}(o-phenanthroline)$_3$.

These early techniques were not useful, however, for rapid MA analysis. In 1985, Kawai and co-workers,[11] and subsequently Mori et al.,[10,12] developed separations of MAs by HPLC, with the latter group implementing the use of a radioanalyzer[13] for the detection of ^{14}C-labeled MAs and MA precursors. Use of this system ultimately led to the identification of methionine as the sole amino acid precursor of MA.

Analysis of the MAs in Fe-deficient barley showed MAs in both the roots and shoots. The levels were highest in the roots. MAs were also detected in the xylem sap and, for rice, also in the phloem.[14] To locate the site of biosynthesis of MAs in corn, Fe-deficient roots were analyzed in sequential 1-cm portions from the root meristems. The highest levels of MAs were obtained within 1 cm of the root tip. In subsequent experiments with barley, (cv. Ehimehadaka No. 1) this portion of the roots, under semiaseptic conditions, was treated with antibiotics and then exposed to ^{14}C-amino acids and ^{14}C-organic acids.

Results of these treatments showed that, in barley only, methionine was highly incorporated within 5 min into DMA, and subsequently MA and epiHMA. In rice, only DMA labeling occurred (Figure 4). In contrast, experiments with oat showed wide variations in the types of phytosiderophores produced, depending on the plant cultivar. Only AVA was produced in certain cultivars, whereas MA, DMA, and AVA occurred in others. Similar to rice, only DMA labeling occurred in wheat (cv. Haruhikari) and corn (cv. Kou No.50).[10] In barley, MA production seemed to occur only in the root tissue. Treatment of leaf discs with ^{14}C-methionine resulted in no incorporation of radioactivity into MAs.[15]

Based on the chemical structures of MAs, three methionine molecules were thought to be incorporated into each molecule of MA. To examine this hypothesis, ^{13}C-methionine was supplied to intact roots of Fe-deficient barley (cv. Minorimugi). Subsequent isolation and ^{13}C-NMR analysis of the MA and DMA showed that all three carboxylic groups of both compounds were labeled with ^{13}C; which confirmed this hypothesis.[16] Because most graminaceous plants contain DMA, this compound was thought to be the first product among MAs. However, for oat it is still unknown whether DMA is a precursor of AVA, or if there are other pathways from methionine to AVA (Figure 5).

General Existence of Nicotianamine in Plants

The next step in elucidating the related biosynthetic pathways was to identify the intermediate between methionine and DMA or AVA. Inspection of

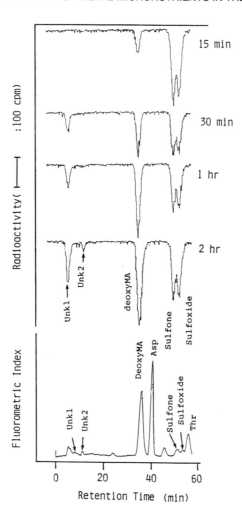

Figure 4.   Time course of metabolic incorporation on 1-^{14}C-methionine into DMA in Fe-deficient rice root cells. (From Mori, S. and Nishizawa, N., *Plant Cell Physiol.*, 28, 1081, 1987.)

natural products in the plant kingdom showed that nicotianamine (NA) had a closely analogous structure to DMA. NA was found in *Nicotiana,* and its chemical structure was determined by Noma et al.[17] This compound was shown to occur in 11 species of Solanaceae, *Fagus sylvatica,* and *Zea mays.*[18] Budesinsky et al.[19] found NA in *Beta vulgaris, Medicago sativa,* and *Lycopersicon esculentum.* Fushiya et al.[20] identified small amounts of NA in the seedlings of oat and rice. These findings suggest that NA generally exists, not only in dicots, but also in monocots. In addition, NA may have a strong relation with MAs in its metabolism in monocots. It was also thought that DMA could possibly form from the NA molecule through deamination and

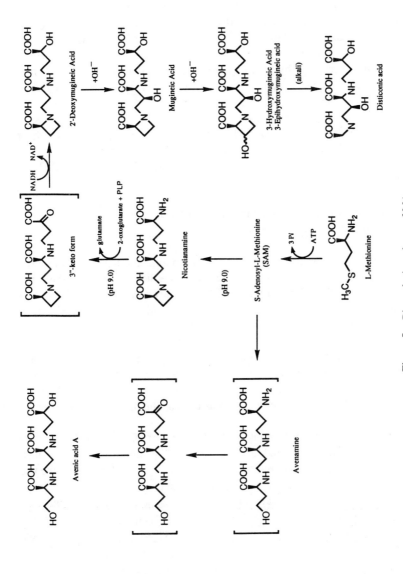

Figure 5. Biosynthetic pathway of MAs.

hydroxylation. This proposal suggested that NA might be a precursor of MAs. If so, then NA would be labeled with ^{14}C when Fe-deficient barley roots were fed ^{14}C-methionine. However, we could detect no labeled NA from barley root tissue extracts through analysis using the HPLC-radioanalyzer. Moreover, the amount of NA in barley appears to be very low in the roots, regardless of the Fe status of the plants. Therefore, further research is needed to determine if NA is a precursor of MA, and to establish whether NA turnover in the roots is too rapid to allow for NA detection.

Subsequent research by Kawai et al.[21] suggested that NA was an intermediate metabolite between methionine and MAs. When the roots of Fe-deficient barley were treated with ^{14}C-methionine and 1 mM aminooxyacetate (as an inhibitor of the transamination reactions), ^{14}C-NA accumulated in the tissue extract with no formation of labeled MAs. Yet, in spite of a great deal of effort to make ^{14}C-labeled NA by the in vitro system developed by Shojima et al.,[22] the incorporation of labeled NA into MAs by exogenous supply of labeled NA to the Fe-deficient barley root tips has not yet been confirmed.

Establishment of the Cell-Free System For the Biosynthesis of Nicotianamine

Since 1988, Shojima et al.[22] have been researching a cell-free system for the biosynthesis of MAs. The initial system used the homogenate of whole Fe-deficient roots supplemented with ATP, NADH, and NADPH. However, in this system no radioactivity was incorporated into MAs from ^{14}C-methionine. This failure to incorporate ^{14}C suggested that inhibitory compounds from basal portions of the roots prevented labeling, and that the enzymes responsible for these reactions in root tips were diluted by including basal portions of the roots. To prevent these interferences, the system was modified to include homogenates prepared only from 1 cm root-tip segments. This resulted in occasional detection of ^{14}C-NA when ^{14}C-methionine was used as the precursor. However, no ^{14}C-MAs were obtained.

To improve the reproducibility in the formation of ^{14}C-labeled NA, the cell-free system was further altered to use only roots of severely Fe-deficient plants. Changes were also made to remove possible polyphenol inhibitors by the use of the polyphenol adsorbant, PVP360, and elution through a Sephadex column. Other changes included the addition of the protease inhibitor, leupeptine (1.4 µM). Finally, the pH of the cell-free system was optimized.[22,23] With the addition of several changes implemented by Higuchi et al., the current cell-free system is listed below.

Extraction Buffer For Sample Preparation

Taps buffer pH 8.5, 200 mM; MgCl$_2$, 5 mM; KCl, 5 mM; EDTA, 1 mM; DTT, 30 mM; polychlar-AT, 5%(w/v); glycerol, 5%(v/v); leupeptine, 1.4 uM.

Gel Filtration Buffer

Taps buffer pH 8.5, 50 mM; MgCl$_2$, 5 mM; KCl, 5 mM; EDTA, 1 mM; DTT, 10 mM.

Sample Preparation

At 4°C, 1-cm root tips (3 g FW) are collected from Fe-deficient barley, homogenized with a mortar and pestle, and filtered through 3 layers of 70-μm nylon filter. The filtrate is centrifuged at 160 × G for 5 min. The supernatant is then eluted with the gel filtration buffer through an extensively washed Sephadex-G25 (medium) column (2.7 × 50 cm). The first peak, monitored at 280 nm, is collected (60 ml). Leupeptine (1.4 μM) is added, and the sample is concentrated with Centriprep (2 kDa cutoff, ADVANTEC) to a final volume of 10 ml. This sample is termed the crude enzyme.

Reaction Conditions

Radiolabeled methionine supplied as 0.3 MBq of l-[1-^{14}C]methionine (2.0 GBq/nmol) is added to 1 ml of crude enzyme. Final concentrations of ATP and leupeptine are added as 10 mM and 1.4 μM, respectively. Reactions are run for 1 to 2 h at 25°C in a constant temperature shaker waterbath (90 rpm). The reactions are started with the addition of ^{14}C-methionine. The reactions are stopped with the addition of 1/40 (v/v) 5 N HCl. The reaction solution is centrifuged at 1500 × G for 20 min, and the supernatant is filtered through a 0.45-μm membrane filter. The levels of radiolabeled MAs are either analyzed by the HPLC-radioanalyzer, or the peaks are collected off the HPLC and analyzed by a liquid scintillation counter.

By the use of this system, it was determined that the biosynthesis of NA from methionine was strongly induced during Fe deficiency. It was also determined that an alkaline pH (pH 9.0) was optimum for this biosynthesis. This high pH optimum suggested that the site of NA biosynthesis should be sequestered in the root cells. Nishizawa and Mori[24] have found vesicles which originate from the rough endoplasmic reticulum in the Fe-deficient barley root cells before sunrise. These vesicles were specifically labeled with ^{14}C when ^{14}C-methionine was supplied to the roots after sunset.[25] We speculate that NA may be produced in those vesicles.

S-Adenosyl Methionine (SAM) as a Precursor of Nicotianamine

In the cell-free system used to study NA biosynthesis, ATP, Mg^{2+}, and K$^+$ are included. These chemical species are essential in the production of S-adenosylmethionine (SAM) from methionine.[26] The author speculated that SAM might be a more immediate precursor than methionine in the biosynthesis of NA. To examine this hypothesis, experiments were conducted in which 1-^{14}C-SAM (50 µCi) was supplied *in vivo* to Fe-deficient barley root tips. This treatment however, failed to produce detectable levels of labeled NA. Subsequently, 1-^{14}C-SAM and 1-^{14}C-methionine and ATP were separately supplied to the cell-free system. The results of these latter experiments showed that ATP was required for the reaction and that SAM was incorporated into NA faster than methionine. This proved that SAM is a direct precursor of NA. Furthermore, the enzyme activity (NA synthetase) was strongly induced in Fe-deficient barley, and the optimal pH for the activity was 9.0.[23] No cofactors other than K^+, Mg^{2+}, and ATP were needed for the NA biosynthesis.

Finally, although the biosynthesis of NA involves the combination of three molecules of SAM, the dimeric intermediate of two SAM molecules has not yet been detected.

Biosynthetic Pathway Between NA and DMA

From an organic synthetic approach, the conversion of NA to DMA would first require a deamination of NA, followed by hydroxylation. Based on a possible biosynthetic pathway, this conversion could involve a 3″-keto species generated after deamination (Figure 5). A H^+ could subsequently be supplied by a proton donor. To prove the occurrence of this proposed biosynthetic pathway for DMA, Shojima et al.[23] synthesized a large quantity of ^{14}C-NA, which was then supplied to the cell-free systems containing 2-oxoglutarate (or pyruvate or oxalacetate) and NADH (or NADPH). DMA was successfully synthesized in these reactions. The existence of the hypothetical 3″-keto acid (the compound in brackets in Figure 5) was indirectly proved by Ohata et al.

Nicotianamine was supplied in the cell-free system[22] which contains 2-oxoglutarate and PLP, as an amino group acceptor and cofactor, respectively. At the end of the enzyme reaction 10 mM NaBH$_4$ was added. This treatment resulted in the chemical reduction of an unidentified 3″-keto form to DMA and the amount of DMA could be measured by HPLC. By this method the activity of nicotianamine aminotransferase, which converts nicotianamine to the 3″-keto form, was measured. The nicotianamine aminotransferase was induced in the root tips of barley by Fe deficiency. Fe-sufficient barley roots had almost negligible amounts of this enzyme activity.

Chromosomal Location of the Genes Which Code For the Enzymes Responsible For the Hydroxylation of DMA to MA or MA to HMA

Subsequent research on the biosynthesis of MA was directed towards detecting the occurrence of the enzyme(s) responsible for the hydroxylation reactions involved with AVA, MA, HMA, and epiHMA biosynthesis. To directly establish the biosynthetic pathway from DMA to MA, or from MA to epiHMA, ^{14}C-DMA and ^{14}C-MA were supplied exogenously to the tips of Fe-deficient barley roots.[10] The roots and the solution cultures were subsequently analyzed for MA synthesis from DMA, or epiHMA synthesis from MA. The analysis, however, failed to establish the above pathways. This is possibly due to the failure of the absorbed MAs to be translocated to the site of the biosynthesis of MAs in the root tissue.[24] Alternatively, the supplied ^{14}C-MAs could have been enzymatically degraded via other unknown pathways.

These findings indicated that alternate approaches in studying the hydroxylation enzymes were needed. One such alternate approach was to determine the chromosomal location of the genes encoding for the enzyme(s) responsible for the hydroxylation reactions. To accomplish this, two chromosomal addition lines were used. These included: wheat (cv. Chinese Spring)-barley (cv. Betzes) addition lines,[27] and the wheat (cv. Chinese Spring)-rye (cv. Imperial) addition lines.[28] These lines are illustrated in Tables 1 and 2.[27,28] For example, Add-1 has chromosome 1 of Betzes barley added to the chromosomes of Chinese Spring wheat. After Fe deficiency occurred in all of the plant strains, root washings were collected and analyzed for MAs. Table 1 shows that Chinese Spring wheat secreted only DMA, while Betzes barley produced DMA, MA, and HMA. Among the addition lines, only Add-4 secreted MA. This suggested that the gene for the hydroxylase responsible for the conversion of DMA to MA is located on chromosome 4 of Betzes barley. Table 2 shows that Chinese Spring wheat secreted DMA, while Imperial rye secreted DMA, MA, and HMA. Among the addition lines, only R-5 secreted DMA, MA, and HMA. This suggested that both genes for the hydroxylases responsible for the conversion of DMA to MA, and MA to HMA, are located on chromosome 5 of Imperial rye. Most surveys for genes coding for morphological traits and enzyme production in cereals have located these genes on homologous chromosomes in the tribe Triticeae. However, in the case of the genes responsible for MA synthesis from DMA, one is located on chromosome 4 of Betzes barley, and another is on chromosome 5R of Imperial rye. There are other examples of genes located on nonhomologous chromosomes of barley and rye, i.e., leaf esterase on chromosomes 3 and 7 in barley, and 3R, 4R, 5R, and 6R in rye. Other genes, of perhaps more fundamental importance, such as the ones coding for the rRNA, are located with tandem duplication on the nucleolus organizing region of chromosomes 6 and 7 in barley, and 1R in rye.[28]

Figure 5 summarizes the current findings on the MA biosynthetic pathways from methionine to MA. The biosynthetic pathway from methionine to AVA, as shown in Figure 5, has yet to be established.

SECRETION OF MAS

Circadian Rhythm

Figure 6A shows the pattern of MA secretion for Fe-deficient barley cultured under 14 h of light followed by 10 h of darkness. The secretion peak was

Table 1. **Secreted Amount of MAs by Betzes Barley Chromosome Addition Lines, Chinese Spring Wheat and Betzes Barley**

Lines	Secreted MAs (μg/plant/day)		
	epiHMA	MA	DMA
Chinese Spring			7900
Betzes	1990	430	1227
Add-1			923
Add-2			250
Add-3			197
Add-4		1257	5707
Add-6			6500
Add-7			2357

Note: Add-5 is not obtained because of male sterility.

Table 2. **Secreted Amount of MAs By Imperial Rye Chromosome Addition Lines, Chinese Spring Wheat and Imperial Rye**

Lines	Secreted MAs (μg/plant/day)		
	HMA	MA	DMA
Chinese Spring			2800
Imperial	170	46	120
R-1			1133
R-2			1500
R-3			1400
R-4			4200
R-5	33	263	4366
R-6			1300
R-7			433

sharpest just after the initial illumination. In contrast to this, under continuous darkness the amount of MAs produced decreased and the peak of production was less well defined. Finally, MA production ceased (Figure 6B).[30] These results indicate that the production has a circadian cycle, and also that the transport of the photosynthetic substrates from the upper portions of the plants to the roots is necessary. Moreover, the production of well-defined circadian peaks appears to be closely tied to the temperature cycle. Under continuous light, but different day and night temperatures, the well-defined peak of MA release occurs (Figure 7A), whereas under constant temperature poorly defined peaks of MA release occur (Figure 7B). The cycles in Figure 7C show that with the standard light cycles, but with reverse temperature conditions, the peak of MA release shifts from the morning hours to the evening.[30]

These observations suggest that the initiation and regulation of MA release is not due to the light signal, but rather to the synchronous temperature increase with the illumination. The dominant factor that acts as a trigger for the initia-

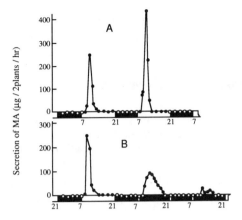

Figure 6. Secretion pattern of mugineic acid from Fe-deficient barley roots. Culture conditions
— A: 14 h light/10 h dark, B: after 14 h light/10 h dark, dark was continued for 2 days.
(From Takagi, S., in *Physiology of Metal Relating Compounds,* 1991, 5 [in Japanese].
With permission from Hakuyusha Press.)

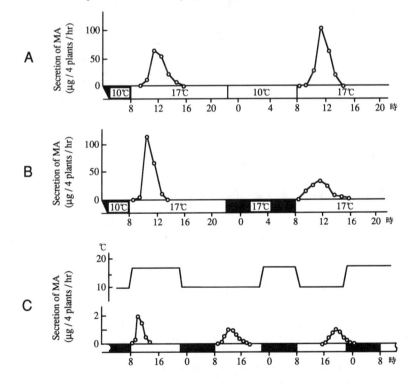

Figure 7. Secretion pattern of mugineic acid from Fe-deficient barley roots under three different
culture conditions; A: continuous light and usual temperature change, B: 14 h light
under constant temperature, C: 14 h light/10 h dark under reversed temperature
conditions. (From Takagi, S., in *Physiology of Metal Relating Compounds,* 1991, 5 [in
Japanese]. With permission from Hakuyusha Press.)

tion of MA release is the time of the temperature increase. In general, the greater the difference in temperature, the sharper the peak of release of the MAs.[30]

Energy Dependence of MA Release

Immediately after light exposure for Fe-deficient barley, the ATPase inhibitor, DCCD (dicyclohexylcarbodiimide, at 0, 10, 50, and 100 μM) was exogenously supplied for 6 h in deionized water in which barley roots were dipped for collecting root washings. The amounts of MAs in both roots and root washings were analyzed by HPLC. As shown in Figure 8, DCCD inhibited the release of MA by the roots. This suggests that the MA secretion is ATP dependent.[31]

Synchronous Release of MAs With K+

Takagi constructed the continuous collecting systems for root washings from single barley (cv. Minorimugi) plants.[30] The secreted root washings were immediately treated with thymol to avoid degradation of these root washings by root surface bacteria. The molar correlations between MA and the other ions in the root washings were determined. Among the released ions, only K^+ had a high equimolar correlation with MA (r = 0.971, Figure 9). This suggested that MA was released as K^+MA^- or that MA was cotransported with K^+ through the action of a H^+,K^+-ATPase.

CHELATION AND SOLUBILIZATION OF FE

Chemical Structure of Fe³⁺-Mugineic Acid

Because of the difficulty of crystalizing Fe^{3+}-MA, Co^{3+}-MA was used for X-ray analysis by Mino et al.[32] The molecular structure and properties of Fe^{3+}-MA are described in detail by Nomoto et al.[8] and Sugiura et al.[33]

Chelating Constant of Fe³⁺-Mugineic Acid

The value of the log stability constant (18.1) of the Fe^{3+} chelate of MA was initially proposed by Nomoto et al.[8] This value however, has been considered too low relative to the values reported for other microbial Fe^{3+} siderophore chelates. With a log stability constant of 18.1, it was likely that MAs would be unable to compete with the much more stable microbial siderophores in the rhizosphere. A low stability constant could result in the induction of Fe deficiency in the *Gramineae* under high pH conditions in the rhizosphere. Consequently, Murakami et al.[34] reexamined the value of the stability constant of Fe^{3+}-MA. They estimated the stability constant value to be between $10^{32.5}$

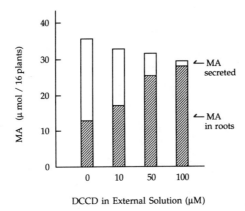

Figure 8.   Effect of DCCD on the secretion of MA from the roots of Fe-deficient barley.

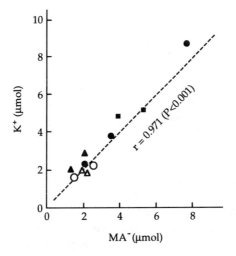

Figure 9.   Correlation between MA⁻ and K⁺ in the root washings of Fe-deficient barley (cv. Minorimugi). (From Takagi, S., in *Physiology of Metal Relating Compounds,* 1991, 5 [in Japanese]. With permission from Hakuyusha Press.)

and $10^{33.3}$. However, these values were still in question because ^{59}Fe was found to transfer from ^{59}FeMA to FOB (log K_s 30.6).[55] This indicates that the stability constant of FeMA is less than that of FeFOB. Jurkevitch et al.[35] compared the Fe^{3+} chelate of pseudobactin St3 and MA. After 10 h, the Fe from FeMA was nearly completely transferred to pseudobactin St3 (Figure 10).

From these two examples, it can be inferred that the microbial Fe^{3+}-siderophores have higher stabilities than Fe^{3+}MA. If this is the case, how then do plants overcome the competition by microbial siderophores? One possible means is that on a daily basis the Fe-deficient barley roots secrete MA in 1000-fold higher molar amounts than the amounts of Fe actually taken up by the

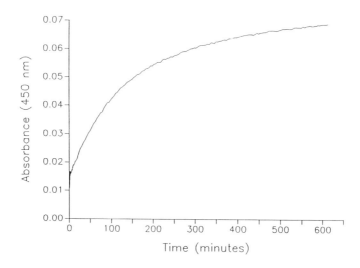

Figure 10. Ligand exchange between $^{55}Fe^{3+}MA$ and pseudobactin from strain St3 in BES [N,N-*bis*(2-hydroxyethyl)-2-aminoethanesulfonic acid] buffer (0.02 *M*, pH 7) at 30°C. (From Jurkevitch, E. et al., *BioMetals*, 6, 119, 1993.)

plants.[2] The concentrated levels of MA secreted at once with the circadian rhythm (Figure 6A) seems to be important in further avoiding the higher chelating ability of competing microbial siderophores.

Solubilization of Fe From Calcareous Soils

The Fe-solubilizing capacity of MA (0.1 to 5 μmol/g soil) was measured by Takagi for two calcareous soils (UT-2 from Utah and NM-2 from New Mexico).[30] There was less solubilization from NM-2 soil than from UT-2 soil (Figure 11). Of the three compounds, DTPA (diethylenetriaminepentaacetic acid), DFOB (deferrioxamine), and MA, MA showed the highest capacity to solubilize Fe from both soils. Hiratate et al.[36] synthesized four different iron oxide compounds and analyzed the levels of solubilized FeMA and the levels of MA adsorbed to the iron oxides between pH 3 and 11 at 4°C in the absence of light. For all of the materials tested, MA adsorption to the iron oxides was highest at low pH. With increasing pH the adsorption decreased, and at pH 10 adsorption levels were negligible. Between pH 3 and 7, MA adsorption to the iron oxide compounds increased in the order: ferrihydrite ≫ goethite > lepidocrite > hematite. MA solubilized less Fe from the crystalline iron oxides than from the amorphous forms. The amounts of solubilized Fe depended more on the amount of amorphous Fe than on the concentration of the MA. For ferrihydrite, the levels of solubilized Fe decreased at pH below 6 because of the increased adsorption of MA to ferrihydrite. On the other hand, at pH greater than 8 both the amounts of solubilized Fe and the levels of adsorbed MA were low.

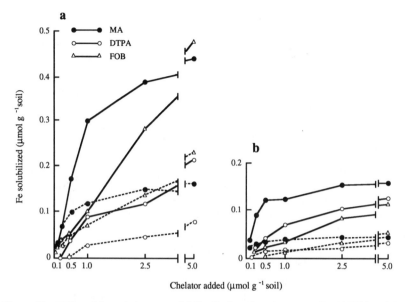

Figure 11. Ability of three chelators to solubilize Fe from the calcareous soils, (a) UT-2 and (b) NM-2; (– – –) 3 h incubation; (—)20 h incubation. (From Takagi, S., in *Physiology of Metal Relating Compounds,* 1991, 5 [in Japanese]. With permission from Hakuyusha Press.)

Consequently, in the case of ferrihydrite the level of solubilized Fe was highest between pH 6 and 8. From this, it is proposed that in Takagi's experiment, UT-2 soil had more amorphous Fe than NM-2.

Immobilization of Fe³⁺ By Phospholipids of Cell Membranes

During transport experiments using ^{55}FeMA and isolated membrane vesicles, Mihashi et al.[37] found ^{55}Fe bound to phospholipids of membrane vesicles. This observation raised questions as to how plants prevent this nonspecific binding at root cell membranes. One finding shows that when the ratio of MA to Fe was increased to 1000 in the multicompartmemt transport box experiment, the transport of ^{55}Fe was 32 times greater than at equimolar Fe to MA ratios. Large molar ratios of MA to Fe are achieved during the circadian cycle, and therefore, the cycle of MA release not only overcomes the competing Fe chelation by microbial siderophores and the degradation by microbes (see below), but also inhibits the Fe³⁺MA binding to the phospholipids of the plant root cell plasma membranes.[38]

DEGRADATION BY SOIL MICROBES

Contributions of microbial siderophores to Fe acquisition and translocation by graminaceous plants depend on culture conditions.[39] However the main compounds which solubilize Fe in the rhizosphere of grasses are the MAs. In general, microbial siderophores have higher stabilities with Fe^{3+} than the phytosiderophores. It is possible, therefore, that an increase in the number of microbes in the rhizosphere will result in the production of higher amounts of microbial siderophores which remove Fe from Fe^{3+}MA, thus causing depletion of Fe in the plant. In addition to this, MAs are subject to microbial degradation in the rhizosphere.[40]

In nutrient solutions, under axenic conditions, Fe-deficient maize released higher amounts of phytosiderophores (estimated as Fe-mobilizing compounds) relative to inoculated plants (Figure 12). The chlorophyll content of the youngest leaves was higher in the axenic plants. These results suggest that phytosiderophores released into nutrient solutions of inoculated plants are decomposed by bacteria; resulting in less Fe absorption and, hence, in the Fe deficiency of these plants.[41]

Ando soil was incubated with MA supplied at 24-h intervals with or without thymol to test the stability of MA in the soil.[42] As shown in Figure 13, thymol suppressed the bacterial degradation of MA.

Watanabe and Wada[40] separated several strains of bacteria from root surfaces of Fe-deficient water-cultured barley. The dominant bacteria (MA1) had high MA decomposition activity. This bacteria could grow in a medium containing MAs as the sole carbon source, and within a week decomposed the MAs in the following order: DMA > MA > epiHMA. This order of MAs was the same in the calcareous soil suspension cultures.[43] However, after several times of acclimating the cultures in soil, differences in the degradation rates between MAs disappeared. The bacteria MA1 could decompose Fe^{3+}MAs as well as MA.[43] This bacteria was classified as *Pseudomonas*,[40] but the exact taxonomical classification is still uncertain. Pure culture strains of *P. putida*, *P. saccharophilia*, and *Escherichia coli* showed no degradation activity. *P. fluorescens*, *P. aeruginosa*, *P. aureofaciens*, *P. stutzeri*, *P. fragi*, and *P. acidovorans* had only limited ability to degrade MA within a week's time.[56]

Consequently, the amounts of MAs in root washings fluctuate depending on the culture conditions, nutrient components,[44] variety of host plants,[45] and microbial composition of the rhizosphere. How does the plant avoid the attack of the MAs-decomposing bacteria? One solution is that the portion of the roots involved with MAs release and Fe^{3+}MAs absorption is near the root tip, where the bacterial population on the rhizoplane is less than on the basal portions of the root.[41] As a result, the released MAs effectively solubilize immobilized Fe in the rhizosphere at the root tips without competing with bacterial decomposition. After absorption of Fe^{3+}MAs during the day, residual (Fe^{3+})MAs in the rhizosphere are very likely decomposed during the night by the bacteria which have high population densities in the new basal portion of the roots.[46]

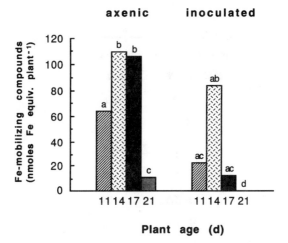

Figure 12. Fe-mobilizing compounds (phytosiderophores) in the solution of the limestone substrate at different plant ages. Statistical significance by Scheffe-test at $p = 0.01$ is shown by different letters. The concentration of phytosiderophores in the limestone substrate (calculated on plant basis) was slightly higher in axenic plants than in inoculated plants after 11 and 14 days (onset of Fe deficiency), and several times higher after 17 and 21 days. On day 21, there was a distinct decrease in phytosiderophores in the axenic treatments. (From von Wiren, N. et al., *Soil Biol. Biochem.*, 25, 371, 1993.)

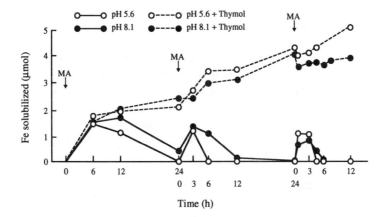

Figure 13. Time course of soil Fe solubilization when mugineic acid (MA) was added at 24 h-intervals. (From Takagi, S., in *Physiology of Metal Relating Compounds*, 1991, 5 [in Japanese]. With permission from Hakuyusha Press.)

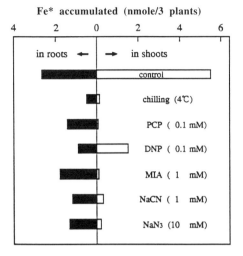

Figure 14. Effect of temperature and metabolic inhibitors on the absorption and transport of ^{59}Fe from [59]FeMA in the culture solution at pH 6.5.

ABSORPTION OF FE³⁺MAS BY PLANT ROOTS

Energy Dependence

Results of preliminary experiments by Takagi (Figure 14), showed that the absorption and translocation of Fe^{3+}MAs by rice roots is clearly energy dependent.[29,47]

Inhibition by BPDS

As previously discussed, Romheld and Marschner[9] tested bathophenanthrolinedisulfonic acid (BPDS), a strong Fe^{2+} chelator, as an inhibitor of Fe^{3+}MA absorption by roots of Fe-deficient graminaceous plants. They found, however, no inhibition by BPDS. This suggested that Fe^{3+}MA was directly absorbed by the plant roots by a "translocator" located in the root cell membranes. Their results were confirmed by the experiments done with the "multicompartment root box method" by our group.

Multicompartment Root Box Method

The characteristics of the transporter (translocator) were investigated by Mihashi and Mori[38] and Mihashi et al.[37] using the "multicompartment transport box method" which was developed by Kawasaki et al.[48] (Figure 15). In our case, [55]Fe was supplied in Compartment 2, and after 20 h of incubation at 25°C the root exudate from the cut end of the basal portion of the roots (Compartment 4) was

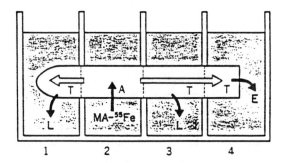

Figure 15. Multicompartment transport box method for the analysis of [55]FeMA transport. [55]FeMA supplied in compartment 2 is absorbed and accumulated (A), transported in the roots (T), and pumped out (E) into the solution of compartment 4. During transport some radioactivity leaked (L) from the root. There was high correlation ($r = 0.938$) between (T + L + E) and E. Therefore the E value was used as the representative value of "transport".

measured for radioactivity. [55]FeCl$_3$ was not transported in the Fe-sufficient roots, while it was rapidly transported in the Fe-deficient roots. [55]Fe from [55]Fe^{3+}MA was transported 32.2 times more from [55]FeCl$_3$ in the Fe-deficient roots. [55]Fe from [55]Fe^{3+}MA was transported 18.4 times more in the Fe-deficient roots than in the Fe-sufficient roots. BPDS did not inhibit the absorption of [55]Fe^{3+}MA in Fe-deficient barley roots. From the above results it is concluded that [55]Fe was absorbed as [55]Fe^{3+}MA, as earlier proposed by Römheld and Marschner,[9] and that during Fe deficiency, the transporter activity of Fe^{3+}MA is induced.

The transporter activity at the root tip (within 0.5 cm) was 10% that of the more basal portions of the roots. The effects of other inhibitors on the transport of [55]Fe^{3+}MA were the following: DCCD (100 μM) inhibited 84%; CCCP (5 μM) inhibited 92.9%; nigericin (5 μM) inhibited 50.4%; and NEM (100 μM) inhibited 55.2%. K$^+$ ion at 10 mM enhanced the transport of Fe^{3+}MA 2.5 times. This enhanced activity was completely canceled by 5 μM valinomycin. These results strongly suggest that the Fe^{3+}MA transport depends on H$^+$ or K$^+$ gradients produced by ATPase activity, and that the active site of the transporter molecule may have free sulfhydryl groups.[38] Whether K$^+$ cotransported with MA$^-$ into the rhizosphere from Fe-deficient roots has any close relation to Fe^{3+}MA transport into the roots is not known.

Plasma membranes purified from both Fe-deficient and Fe-sufficient barley roots were analyzed for differences in the protein composition. The compositional differences were analyzed by electrophoresis (SDS-PAGE) and silver staining. Three peptides (40, 28, and 14 kDa) in the plasma membranes of Fe-deficient plants occurred in higher amounts than in Fe-sufficient plants.[37] However, the differences detected for the barley roots were not as dramatic as those observed for similar experiments done with *E. coli*,[49] but rather were similar to those of fungal systems.[50]

FUTURE WORK

Much has been learned about phytosiderophore production during Fe deficiency by the graminaceous monocots. An important focus of future research is to clarify the mechanisms of phytosiderophore regulation by Fe. To study this, the genes coding for the enzymes of MAs biosynthesis and for the MAs transporter protein(s) need to be cloned.

With regard to the study of the genes that code for MA biosynthesis, the enzymes, nicotianamine synthetase and nicotianamine aminotransferase (Figure 5), are being purified by our group. These enzymes are already known to be induced in Fe-deficient barley roots. Nicotianamine aminotransferase and DMA-synthetase (3″-keto acid reductase) are the most important enzymes because the genes that code for these enzymes are considered to have been acquired during the evolution of graminaceous plants.[51] In contrast, the gene coding for the nicotianamine synthetase is known to exist both in dicots[52] as well as in monocots.[53]

The purification of the Fe^{3+}MA transporter protein(s) will require far more sophisticated means of detection and identification. A likely procedure for this is the development of a photoaffinity label for the transporter proteins with some radiolabeled inhibitor. For the development of an assay using plasma membrane vesicles for the transporter,[37] ^{3}H-labeled MAs of higher specific activity than ^{55}FeMAs or ^{59}FeMAs or ^{14}C-MAs[12] are essential, and will need to be produced through organic synthesis.

To study the functional genes involved with MA biosynthesis, the following sequence is generally followed: protein purification, amino acid sequence determination, DNA probe synthesis, cDNA cloning, genomic DNA cloning, and promotor regions (i.e., Fe-responsive element) characterization. For enzymes that have not been purified to homogeneity, differential hybridization is an alternate method to characterize the functional genes. We are currently cloning cDNAs in barley roots which are specifically expressed during Fe deficiency. By doing this we anticipate finding the genes for the biosynthesis of MAs and/or the proteins involved in the transport of the Fe^{3+}MAs.[54]

REFERENCES

1. Takagi, S., The absorption mechanism of heavy metals by the plants, especially on the secretion of iron-solubilizing substances and iron absorption from the plant roots, in *Studies on the Soil Science and Plant Nutrition in Modern Agriculture,* Yokendo Press, Tokyo, 1972, 66.
2. Takagi, S., Naturally occurring iron-chelating compounds in oat and rice root washings. I. Activity measurement and preliminary characterization, *Soil Sci. Plant Nutr.,* 22, 423, 1976.
3. Takemoto, T., Nomoto, K., Fushiya, S., Ouchi, R., Kusano, G., Hikino, H., Takagi, S., Matsuura, Y., and Kakudo, M., Structure of mugineic acid, a new amino acid possessing an iron-chelating activity from root washing of water-cultured *Hordeum vulgare* L., *Proc. Jpn. Acad.,* 54, B469, 1978.

4. Nomoto, K., Yoshioka, H., Arima, M., Takemoto, T., Fushiya, S., and Takagi, S., Structure of 2'-deoxymugineic acid, a novel amino acid possessing an iron chelating activity, *Chimia,* 35, 249, 1981.

5. Fushiya, S., Sato, Y., Nozoe, S., Nomoto, K., Takemoto, T., and Takagi, S., Avenic acid A, a new amino acid possessing an iron-chelating activity, *Tetrahedron Lett.,* 21, 3071, 1980.

6. Nomoto, K., Yoshioka, H., Takemoto, T., Fushiya, S., Nozoe, S., and Takagi, S., A new amino acid possessing Fe-chelating activity secreted from grasses, 22nd Symp. Chemistry of Natural Products, Fukuoka, Japan, 1979, 619.

7. Nomoto, K., Sugiura, Y., and Takagi, S., Mugineic acid, studies on phytosiderophores, in *Iron Transport in Microbes, Plants and Animals,* Winkelmann, G., van der Helm, D., and Neilands, J.B., Eds., VCH Publishers, Weinheim, Germany, 1987, chap. 22.

8. Takagi, S., Nomoto, K., and Takemoto, T., Physiological aspect of mugineic acid, a possible phytosiderophore of graminaceous plants, *J. Plant Nutr.,* 7, 469, 1984.

9. Römheld, V. and Marschner, H., Evidence for a specific uptake system for iron phytosiderophores in roots of grasses, *Plant Physiol.,* 80, 175, 1986.

10. Mori, S. and Nishizawa, N., Methionine as a dominant precursor of phytosiderophores in Graminaceous plants, *Plant Cell Physiol.,* 28, 1081, 1987.

11. Kawai, S., Sato, Y., Takagi, S., and Nomoto, K., Separation and determination of mugineic acid and its analogues by high performance liquid chromatography, *J. Chromatogr.,* 391, 325, 1987.

12. Mori, S., Nishizawa, N., Kawai, S., Sato, Y., and Takagi, S., Dynamic state of mugineic acid and analogous phytosiderophores in Fe-deficient barley, *J. Plant Nutr.,* 10, 1003, 1987.

13. Mori, S., A new continuous flow monitoring system for radioactive amino acids, *Agric. Biol. Chem.,* 45, 1881, 1981.

14. Mori, S., Nishizawa, N., Hayashi, H., Chino, M., Yoshimura, E., and Ishihara, J., Why are young rice plants highly susceptible to iron deficiency?, *Plant Soil,* 130, 143, 1991.

15. Mori, S., Ishihara, J., Hayashi, H., Chino, M., and Nishizawa, N.K., Dynamic state of phytosiderophores in Fe-deficient rice plants, Botanical Congress, Berlin, 1987, I-25b-2, 38.

16. Kawai, S., Ito, K., Takagi, S., Iwashita, T., and Nomoto, K., Studies on phytosiderophores: biosynthesis of mugineic acid and 2'-mugineic acid in *Hordeum vulgare* L var. Minorimugi, *Tetrahedron Lett.,* 29, 1053, 1988.

17. Noma, M., Noguchi, M., and Tamaki, E., A new amino acid, nicotianamine, from tobacco leaves, *Tetrahedron Lett.,* 22, 2017, 1971.

18. Noma, M. and Noguchi, M., Occurrence of nicotianamine in higher plants, *Phytochemistry,* 15, 1701, 1976.

19. Budesinsky, M., Budzikiewicz, H., Prochazka, Z., Ripperger, H., Romer, A., Scholz, G., and Schreiber, K., Nicotinamine, a possible phytosiderophore of general occurrence, *Phytochemistry,* 19, 2295, 1981.

20. Fushiya, S., Takahashi, K., Nakatsuyama, S., Sato, S., Nozoe, S., and Takagi, S., Occurrence of nicotianamine and avenic acid in *Avena sativa* and *Oryzae sativa, Phytochemistry,* 21, 1907, 1982.

21. Kawai, S., Ito, K., and Takagi, S., Biosynthetic pathway of mugineic acid in the roots of Fe-deficient barley, *Soil Sci. Plant Nutr.*, 34, 73, 1988 (in Japanese).

22. Shojima, S., Nishizawa, N.K., and Mori, S., Establishment of a cell-free system for the biosynthesis of nicotianamine, *Plant Cell Physiol.*, 30, 673, 1989.

23. Shojima, S., Nishizawa, N.K., Fushiya, S., Nozoe, S., Irifune, Y., and Mori, S., Biosynthesis of phytosiderophores — *in vitro* biosynthesis of 2'-deoxymugineic acid from L-methionine and nicotianamine, *Plant Physiol.*, 93, 1497, 1990.

24. Nishizawa, N. and Mori, S., The particular vesicle appearing in barley root cells and its relation to mugineic acid secretion, *J. Plant Nutr.*, 10, 1013, 1987.

25. Nishizawa, N.K., Shojima, S., and Mori, S., The ultrastructural studies on mugineic acid secretion by barley roots under the Fe-deficiency stress, in *Transactions of 14th ICSS*, Vol. IV-476-477, Kyoto, 1990.

26. Tabor, W. and Tabor, H., Methionine adenosyltransferase (*S*-adenosylmethionine synthetase) and *S*-adenosylmethionine decarboxylase, *Adv. Enzymol.*, 56, 251, 1984.

27. Mori, S. and Nishizawa, N., Identification of barley chromosome no. 4, possible encoder of genes of mugineic acid synthesis from 2'-deoxymugineic acid using wheat-barley addition lines, *Plant Cell Physiol.*, 30, 1057, 1989.

28. Mori, S., Nishizawa, N.K., and Fujigaki, J., Identification of rye chromosome 5R as a carrier of the genes for mugineic acid synthetase and 3-hydroxymugineic acid synthetase using wheat-rye addition lines, *Jpn. J. Genet.*, 65, 343, 1990.

29. Takagi, S., Difference of iron absorption mechanisms among the plants, in *Studies on Soil Science and Plant Nutrition*, Tohoku, Japan, 1984, 190 (in Japanese).

30. Takagi, S., Mugineic acid, in *Nutrition and Physiology of Metal Relating Compounds*, Hakuyusha Press, Tokyo, 1991, 5 (in Japanese).

31. Takagi, S., Secretion of mugineic acid and Fe-solubilization in the rhizosphere of barley roots, in *Studies on the Regulatory Factors in the Rhizosphere*, Report of the Monbusho Fund, 1988, 6 (in Japanese).

32. Mino, Y., Ishida, T., Ota, N., Inoue, M., Nomoto, K., Takemoto, T., Tanaka, H., and Sugiura, Y., Mugineic acid Fe(III) complex and its structurally analogous cobalt(III) complex. Characterization and implications for absorption and transport of iron in graminaceous plants, *J. Am. Chem. Soc.*, 105, 4671, 1983.

33. Sugiura, Y., Tanaka, H., Mino, Y., Ishida, T., Ota, N., Nomoto, K., Yoshioka, H., and Takemoto, T., Structure, properties and transport mechanism of iron(III) complex of mugineic acid, a possible phytosiderophore, *J. Am. Chem. Soc.*, 103, 6979, 1981.

34. Murakami, T., Ise, K., Hayakawa, M., Kamie, S., and Takagi, S., Stabilities of metal complexes of mugineic acids and their specific affinities for iron(III), *Chem. Lett.*, 2137, 1989.

35. Jurkevitch, E., Hadar, Y., Chen, Y., Chino, M., and Mori, S., Utilization of the phytosiderophore of mugineic acid as an iron source by rhizosphere fluorescent *Pseudomonas*, *BioMetals*, 6, 119, 1993.

36. Hiratate, S., Takagi, S., and Inoue, K., Conjugation of MA with synthetic Fe-compounds, *Soil Sci. Plant Nutr.*, 37, 25, 1991 (in Japanese).

37. Mihashi, S., Mori, S., and Nishizawa, N.K., Enhancement of mugineic acid-Fe^{3+} uptake by Fe-deficient barley roots with coexistence of excess free mugineic acid in the medium, *Plant Soil*, 130, 135, 1991.

38. Mihashi, S. and Mori, S., Characterization of mugineic acid-Fe transporter in Fe-deficient barley roots using the multicompartment transport box method, *Biol. Metals*, 2, 146, 1989.

39. Crowley, D.E., Römheld, V., Marschner, H., and Szaniszlo, P.J., Root-microbial effects on plant iron uptake from siderophores and phytosiderophores, *Plant Soil,* 142, 1, 1992.

40. Watanabe, S. and Wada, H., Mugineic acid decomposing bacteria isolated from rhizoplane of iron-deficient barley, Japan *Soil Sci. Plant Nutr.,* 60, 413, 1989.

41. von Wiren, N., Morel, J.L., Guckert, A., Römheld, V., and Marschner, H., Influence of microorganisms on iron acquisition in maize, *Soil Biol. Biochem.,* 25, 371, 1993.

42. Takagi, S., Mugineic acids as examples of root exudates which play an important role in nutrient uptake by plant roots, in *Phosphorus Nutrient of Grain Legumes in the Semi-Arid Tropics,* Johansen, C., Lee, K.K., and Sahrawat, K.L., Eds., Patancheru, India:ICRISAT, 1991, 77.

43. Watanabe, S., Matsumoto, S., and Wada, H., Microbial decomposition of mugineic acids. A new aspect of relation between plants and microorganisms, *Transactions of 14th ICSS,* Vol. III, Kyoto, Japan, 1990, 272.

44. Mori, S., Hachisuka, M., Kawai, S., Takagi, S., and Nishizawa, N.K., Peptides related to phytosiderophore secretion by Fe-deficient barley roots, *J. Plant Nutr.,* 111, 653, 1988.

45. Kawai, S., Takagi, S., and Sato, Y., Mugineic acid-family phytosiderophores in root-secretions of barley, corn, and sorghum varieties, *J. Plant Nutr.,* 11, 633, 1988.

46. Marschner, H., Römheld, V., and Kissel, M., Localization of phytosiderophore release and iron uptake along intact barley roots, *Physiol. Plant.,* 71, 157, 1987.

47. Takagi, S., Studies on the meaning of submerged culture of rice plant, Annu. Rep. Agric. Inst. Tohoku University, 18, 1, 1966 (in Japanese).

48. Kawasaki, T., Shimizu, G., and Moritsugu, M., Effects of high concentrations of sodium chloride and polyethylene glycol on the growth and ion absorption in plants. II. Multicompartment transport box experiment with excised roots of barley, *Plant Soil,* 75, 87, 1983.

49. Braun, V., Hantke, K., Katrin, E.-H., Koster, W., Pressler, U., Sauer, M., Schaffer, S., Schoffler, H., Staudenmaier, H., and Zimmermann, L., Iron transport systems in *Escherichia coli,* in *Iron Transport in Microbes, Plants, and Animals,* Winkelmann, G., van der Helm, D., and Neilands, J.B., Eds., VCH Publishers, Weinheim, Germany, 1987, chap. 2.

50. Huschka, H.G. and Winkelmann, G., Iron limitation and its effect on membrane proteins and siderophore transport in *Neurospora crassa, Biol. Metals,* 2, 108, 1989.

51. Römheld, V. and Marschner, H., Genotypical differences among graminaceous species in release of phytosiderophores and uptake of iron phytosiderophores, *Plant Soil,* 123, 147, 1990.

52. Shojima, S., Nishizawa, N.K., Fushiya, S., Nozoe, S., Kumashiro, T., Nagata, T., Ohota, T., and Mori, S., Biosynthesis of nicotianamine in the suspension-cultured cells of tobacco *(Nicotiana megalosiphon), Biol. Metals,* 2 142, 1989.

53. Nishizawa, N.K., Mori, S., Takahashi, S., and Ueda, T., Mugineic acid secretion by cultured barley cells derived from anther, *Protoplasma,* 148, 264, 1989.

54. Okumura, N., Nishizawa, N.K., Umehara, Y., and Mori, S., An iron deficiency specific cDNA from barley roots having two homologous cysteine-rich domains, *Plant Mol. Biol.,* 17, 531, 1991.

55. Takagi, S., personal communication.

56. Watanabe, S., personal communication.

CHAPTER 16

Genetically Controlled Uptake and Use of Iron by Plants

Von D. Jolley and John C. Brown

INTRODUCTION

Availability of nutrients in soils is a function of complex chemical, physical, and biological interactions. Under the extreme competition for nutrients found in soils, both higher plants and microorganisms have developed adaptive mechanisms that enhance their abilities to obtain essential nutrients, such as iron. Cells of higher organisms are fundamentally similar and the need for iron is universal. The major metabolic functions of iron are in reversible oxidation-reduction reaction systems and in formation of coordination complexes (chelates). Ferric forms predominate in both cells and soils, and their availability becomes essential to cell function. Iron-deficiency stress in a plant may induce biochemical reactions that increase availability of iron from the soil until sufficient iron has been absorbed by the plant, at which point the biochemical reactions are shut off. This cycle of reactions may be repeated under subsequent iron-deficiency stresses. Thus, the available supply of iron and the biochemical reactions induced are controlled by the plant as it interacts with the soil environment. These mechanisms are genetically controlled, but may be absent or uncontrolled in special cases. Without such adaptations, plants would likely not survive the harsh (iron) environment of calcareous soils.[1] The objective of this paper is to review current understanding of the adaptive mechanisms that control the availability and use of rhizosphere iron.

MECHANISMS OF SOLUBILIZING IRON FOR PLANT USE

Plant adaptation to iron-deficiency stress in dicotyledenous plants has been studied for over 30 years. Brown[2,3] and Brown and Jolley[4] have established that

0-87371-942-5/94/$0.00+$.50
© 1994 by CRC Press, Inc.

iron deficiency elicits biochemical activities that facilitate iron solubilization and absorption. The induced events in dicots and nongraminaceous monocots include: a release of hydrogen ions and "reductants" from roots, an enhanced reduction of ferric to ferrous iron at the plasmalemma of roots, and an accumulation of citrate in root tissues to accelerate transport of the solubilized iron to plant shoots and leaves. Grasses respond to iron-deficiency stress by producing the chelate, phytosiderophore (PS). In some cases grass roots have specialized uptake and transport mechanisms as well.[4-6] These mechanisms are genetically controlled and are primarily centered within and around the root. These two divergent adaptations of plants to obtain iron during iron-deficiency stress were categorized by Marschner et al.[6] as Strategy I (dicots and nongraminaceous monocots) and Strategy II (grasses).

Strategy I

Weiss[8] showed that PI-549-19-5-1 (now T203) soybean was susceptible to iron-deficiency stress, whereas Hawkeye (HA) soybean was not, and he coined the terms "iron inefficient" for T203 and "iron efficient" for Hawkeye. Using reciprocal grafts, Brown et al.[9] placed iron-inefficient T203 soybean tops on iron-efficient HA rootstocks and HA tops on T203 rootstocks. The iron-inefficient shoots grafted to the iron-efficient roots became iron efficient and remained green. In contrast, leaves of iron-efficient shoots grafted onto iron-inefficient roots turned chlorotic and eventually died. These findings were subsequently confirmed with iron-efficient and iron-inefficient tomato genotypes.[10] Thus, it was established that the controlling factors in absorption and transport of iron were associated with the root. As early as 1961, Brown et al.[11] proposed that the iron-efficient plants initiated some metabolic process that affected the solubility or availability of iron at the roots.

Differences in iron reductive capacity of T203 and HA soybean roots were established using ferricyanide-ferrichloride solutions as a source of ferric iron.[11] Significant reductive activity was present on the lateral roots of HA, but not of T203 soybean. This relationship was confirmed with studies on T3238FER (iron efficient) and T3238fer (iron inefficient) tomatoes. The former released hydrogen ions and reductants from their roots with a concomitant reduction of ferric to ferrous iron.[10] Marschner et al.[12] made similar observations using sunflower.

Some dicotyledonous plants initially respond to iron-deficiency stress by forming transfer cells in the epidermis of roots.[13] These cells release hydrogen ions and reductants, enhance iron absorption/translocation in roots, and degenerate when iron-deficiency stress is alleviated.[14] The iron-efficient T3238FER tomato develops transfer cells, whereas the iron-inefficient T3238fer does not.[15] The T3238fer tomato does not release hydrogen ions nor reductants, whereas the T3238FER does.[16] However, some iron-efficient plants proliferate root hairs rather than transfer cell development.[14,17]

Reductants released into the rhizosphere can reduce ferric to ferrous iron and thus solubilize it. The reductant released from tomato roots was identified as caffeic acid.[18] Barrett-Lennard et al.[19] and Sijmons and Bienfait[20] suggested that the quantity of reductants released from roots is insufficient to supply adequate ferrous iron to plants and thus inferred that reductants are not essential in making iron available to plants. However, something must deliver soluble iron to root surfaces, and the combination of hydrogen ion release and reductants would be very effective in solubilizing iron. Muskmelon does not release reductants during iron-deficiency stress response, but prolific hydrogen ion release appears to compensate for no reductant release.[21]

Reduction of ferric to ferrous iron is essential for iron absorption in Strategy I plants. Young lateral roots are the principal sites of reduction in both soybean and tomato.[16,17] The ferric iron ligand complex is split during the reduction step, with the resulting ferrous iron absorbed and the ligand remaining in the soil solution. Reduction is critical, and if BPDS, a strong ferrous chelator, is allowed to compete with the plant roots for the reduced iron, plants remain severely chlorotic because iron is not absorbed and translocated.

The iron-stress response mechanism in dicots seems to be a combination of biochemical activities.[22] There is a concomitant release of hydrogen ions and reductants and enhanced reduction of ferric to ferrous iron by roots in iron-efficient genotypes (Figure 1). Iron thus made available is transported within the xylem as ferric citrate. An accumulation of citrate is a part of this concert of events, apparently in preparation for transporting iron within the plant. In soybean and tomato, reduced iron appears to move through the protoxylem of the young lateral root to the metaxylem of the primary root, where it is oxidized to ferric iron, chelated by citrate, and transported to plant tops.[3]

Strategy II

The Strategy II response to iron-deficiency stress in grasses is the secretion of Fe(III) ligands similar in function to the microbial siderophores. Neilands[23] characterized siderophores as low molecular weight (500 to 1000 Da) ligands that are ferric-specific, biosynthetically produced, and iron-regulated. They function to supply iron to the cell via a specific uptake/transport system. In grasses, the ferric-specific ligands are exceptionally small molecules (300 to 400 Da). They are nonproteinogenic amino acids analogous to mugineic acid (MA) and have been collectively referred to as "mugineic acid-family phytosiderophores" (MAs).[24] These have been shown to be released in a diurnal pattern, with excretion initiation stimulated by light and with a 3- to 5-h release time. Subsequent reports suggest that iron deficiency stimulates accelerated synthesis and accumulation of MAs with release via an energy-dependent transport system attended by symport of equimolar potassium. Some have suggested that the periodicity of release of MAs is a strategy in grasses to avoid or minimize microbial degradation of MAs.[24,26]

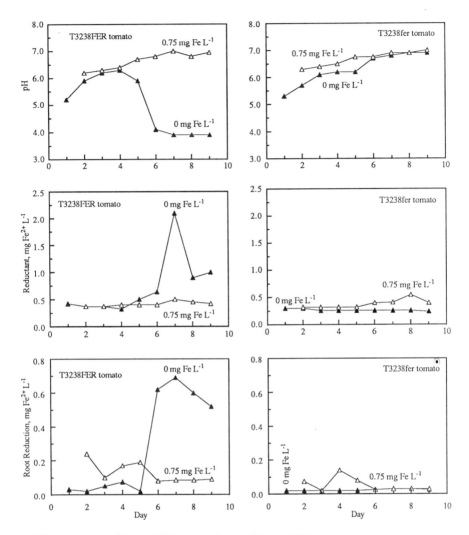

Figure 1. Iron-efficient (T3238FER) and iron-inefficient (T3238fer) tomato grown at two levels of iron in a modified Hoagland nutrient solution showing the response of T3238FER and the lack of response of T3238fer tomato to iron-deficiency stress. (From Camp, S. D., Jolley, V. D., and Brown, J. C., *J. Plant Nutr.*, 10, 423, 1987, with permission.)

Brown and co-workers[11,12] originally explained differences between iron-efficient and iron-inefficient cultivars of oat and corn based on the known Strategy I mechanisms. These differences included small differential abilities to exude hydrogen ions and reductants, to reduce iron at the roots, or to absorb other nutrients. Following the discovery by Takagi[5] in 1976 of an "iron solubilizing substance" released from roots of iron-stressed barley, understanding of the phytosiderophore (PS)-mediated uptake of iron by grasses has expanded

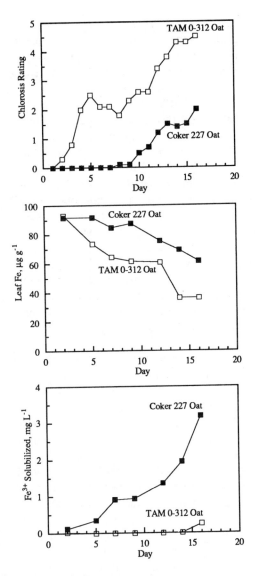

Figure 2. Chlorosis scores (0 = none and 5 = severe with necrosis), leaf iron content, and phytosiderophore release (ferric iron solubilized) over time for iron-efficient (Coker 227) and iron-inefficient (TAM 0-312) oat grown at 0.6 mg l^{-1} (no synthetic chelate) in a modified Hoagland nutrient solution. (Adapted from Jolley, V. D. and Brown, J. C., *J. Plant Nutr.*, 12, 423, 1989, with permission.)

dramatically. Release of PS provides a clearer explanation for these differences in some grasses (e.g., oats), but not in others (e.g., corn).

Jolley and Brown[29,30] found that iron-efficient Coker 227 and iron-inefficient TAM 0-312 oats differed in the quantity of PS released during iron-deficiency stress (Figure 2). However, this was not the explanation with WF9 and ys_1 corn, where little or no release of PS was observed despite wide divergence in iron-deficiency chlorosis development between the two cultivars.[31,32] Small differences in iron-reducing capacities were observed between WF9 and ys_1, but this small divergence is not sufficient to explain the diversity in ability to assimilate iron.[32] Römheld[26] reported that increased iron solubility in the rhizosphere of maize grown on calcareous soils is most likely caused by enhanced release of siderophores by microorganisms.

Quantity of PS has been proposed as an explanation for the observed iron efficiencies of various plant species. Römheld and Marschner[33] and Kawai et al.[34] ranked species according to rate of release of PS as: barley > wheat = rye > oat > corn > sorghum > rice. However, quantity of release of PS varies among genotypes within a genus in oat, barley, corn, and sorghum.[31,33,34]

The chemical structure of MAs was reported by Nomoto et al.[35] Since the initial elucidation of its structure, Mori and Nishizawa[36] and Kawai et al.[37] have proposed the following sequence of MA biosynthesis: L-methionine → S-adenosylmethionine → (polymerization of 2-aminobutyrate moieties) → nicotianamine → 2'-deoxymugineic acid (DMA) → mugineic acid (MA) → epihydroxymugineic acid (epiHMA).[37] Greater understanding of the biosynthetic pathway of MAs is likely to be achieved in the next few years.

Species and genotypes within species differ in the form of the PS (avenic acid, DMA, MA, or epiHMA) released.[33,34,36] The various forms probably differ in effectiveness at solubilizing iron, but only limited information is available on the solubility constants of these forms.[38,39] The stability constant of 18.1 proposed for mugineic acid is considerably below that of most microbial siderophores and commercial chelates. Consequently, some have questioned its effectiveness in solubilizing iron in soils.[40] Takagi[24] and Takagi and co-workers[39] reported that "despite their relatively small molecular size, MAs have a strong affinity for Fe(III) and provide octahedral ferric complexes (Fe(III)-MAs) of high stability, the putative formation constant of which is in the order of 10^{32}". A more recent report by Murakami et al.[39] suggested a formation constant for ferric-MA, -DMA and -epiHMA of around 33, which would be comparable to microbial siderophores.

Recent studies[24,26,41-43] have confirmed that MAs are effective in solubilizing iron from soils. Takagi reported[24] that MA (mugineic acid) and DMA (2'-deoxymugineic acid) showed highest solubilization of iron from soils of any of the synthetic chelates (EDDHA, DTPA, or EDTA) or siderophore (FOB) which were compared to it. He also found that solubilization efficiency of MA and DMA was less affected by pH of soil than other tested chelates. Interference by calcium and aluminum was also minimal with MA and DMA, which both had low affinity for calcium and aluminum compared to the synthetic chelates EDTA, or FOB. This reduced affinity may explain the high iron-solubilizing efficiency of MAs in high pH soils.

Another explanation for the effectiveness of the PS in making iron available to plants is an enhanced uptake system proposed by Marschner and Römheld.[44] Double-labeled iron-phytosiderophore complex supplied to iron-deficiency-stressed barley plants showed absorption by plants of both PS and iron. In contrast, iron chelated with EDDHA did not provide iron efficiently to barley. Mori and co-workers[45] suggested that rice plants absorbed iron-MAs, even though rice produced MA in low amounts, and that iron was also transported within the plant complexed with MA.

Clarkson and Sanderson[46] showed by tracer and microautoradiographic studies that rates of absorption and translocation of ^{59}Fe in intact seminal axes of barley roots were highest 1 to 4 cm from the root tip. High rates of uptake and translocation were restricted to a narrow band of maturing or recently matured cells in the root tip. These rates of uptake and translocation were seven- to tenfold higher in plants grown without iron than those grown with iron. This work was completed before the widespread knowledge of the existence of PS.

More recently, Marschner et al.[6] found that not only was there a 10- to 50-fold increase in the release of PS from iron-deficient compared to iron-sufficient barley plants, but that the release was limited to apical zones of both seminal and lateral roots. This iron-deficiency stimulation of release was coupled with an enhanced uptake of iron in these same apical root zones during iron-deficiency stress. Uptake of iron, provided as ferric EDDHA, was small and similar along the entire root surface. Uptake of iron, provided as iron-PS, was absorbed and translocated between 100 (iron-sufficient) and 1000 (iron-deficient plants) times higher than from EDDHA. Thus, not only is there an enhanced release of PS in grasses, but there is also a specific uptake and translocation system that is activated during iron-deficiency stress. Mori et al.[45] proposed a transporter protein involvement in the enhanced uptake of DMA and found the complexed ferric-DMA was transported in the xylem of iron-stressed rice plants.

Hopkins et al.[47-49] found that iron-deficiency-stressed WF9 maize, TAM 0-312 oat, 'Redland' B-line sorghum, fefe muskmelon, and T3238fer tomato were each able to absorb iron made available from PS released by iron-deficiency-stressed Coker 227 oat. However, ys_1 (ys_1/ys_1) maize was unable to procure iron in similar circumstances where grown in the presence of oat PS. In contrast, five soybean cultivars varying in iron efficiency did not get more iron when grown in combination with oat. Instead oat PS prevented iron uptake in relation to the iron-inefficiency of each soybean cultivar. Thus, soybean obtained iron from the iron-PS complex in relation to its ability to reduce iron at its roots. This supports the mechanism proposed for dicots, in which the ferric-chelate complex is split during iron absorption.[44] However, no explanation for uptake of iron from iron-PS in iron-inefficient dicots is yet identified. Variation in root reductive capacity and/or presence or absence of specific uptake sites may be responsible. Takagi et al.[25] had earlier found that although rice produces low amounts of PS, it was capable of rapid absorption and translocation of iron-PS supplied to it during iron-deficiency stress. This would suggest that rice has the enhanced uptake system without the ability to release much phytosiderophore. Likewise, TAM 0-312 lacks the ability to release large quantities of PS, but was able to obtain iron from Coker 227 oat PS.

GENETIC CONTROL

Iron must be made available to the cell in biological systems. As discussed above for plants classified as Strategy I and Strategy II, iron-deficiency stress may induce biochemical reactions in plants that facilitate iron uptake and that are subsequently shut off after sufficient iron has been made available.[3,4] These reactions may be induced repeatedly if iron-deficiency stress again occurs in the plant. The mechanisms are genetically managed[8,50-53] and may be absent[3,4] or uncontrollable in special cases.[54,55] Kneen and co-workers[54,55] recovered the mutant genotype E107 from seed of commercial pea (*Pisum sativum* L. cv "Sparkle") that had been treated for 1 h with 1% ethylmethane sulfonic acid (0.08 M). The iron concentration in the E107 mutant was over 50-fold greater than that in "Sparkle". Welch and LaRue[56] showed that the controlling mechanism for iron uptake in the mutant was located in the roots, and Grusak et al.[57] suggested that the mutant genotype functions as if it was constantly experiencing iron-deficiency stress, even when grown with a "normal" supply of iron. Iron uptake by the plant was out of control.

Uncontrolled uptake of iron by tumorous crown gall on sunflower stems was observed by Brown et al.[58] and Jolley et al.[59] The tumorous crown gall tissue of sunflower reduced more ferric to ferrous iron and contained higher iron concentrations than equivalent nontumorous stem tissue (Figure 3). Thus, an adequate supply of iron is needed for tumorous growth and the tumorous crown gall induces its own mechanisms for making iron available. One mechanism involved reduction of ferric to ferrous iron, but it seems uncontrolled by the plant.

Weinberg[60] suggested that iron might promote tumorous neoplastic growth because it is an essential nutrient for continuous cell proliferation. He also suggested that normal and neoplastic cells have a similar qualitative need for iron, but that continuous growth of the tumor cells would probably require an enhanced or diversified supply of iron.[61] Brown et al.[58] believe the tumorous crown gall on sunflower stems induces its own mechanism for making iron available to its cells that is not controlled by the plant. Tumorous neoplastic cells have been found that obtain iron from normal tissues despite vigorous attempts by the host to restrict access to iron.[61]

Genetic control of mechanisms affecting uptake and use of iron by normal and neoplastic cells denotes the important and essential role that iron plays in biochemistry.

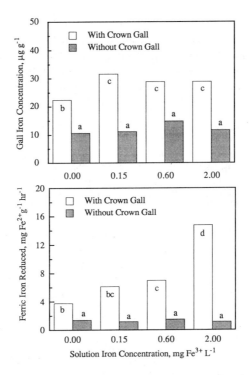

Figure 3. Ferric to ferrous iron reduction and iron concentration of sunflower stem tissues infected with or without crown gall grown at 0, 0.15, 0.6, and 2.0 mg l⁻¹. Columns with the same letter are not statistically significant at the 0.05 level; Duncan Waller K Ratio Test. (Adapted from Jolley, V. D., Terry, R. E., and Brown, J. C., *J. Plant Nutr.*, 14, 539, 1991, with permission.)

FACTORS INTERACTING WITH THE IRON-STRESS RESPONSE MECHANISMS

Other Nutrients

Significant interactions occur between micronutrients, and some of these are explained by their effects on the iron-stress response mechanism. Sanilac navy bean (a Strategy I plant) is more susceptible to zinc deficiency than Saginaw navy bean, and Sanilac accumulated iron under zinc-deficiency stress.[62] This enhanced uptake of iron in Sanilac was associated with an increased release of reductants and reduction of iron by roots of Sanilac under zinc-deficiency stress.[63] With adequate zinc, both genotypes reduced iron similarly. Thus, zinc-deficiency stress stimulated the initiation of the iron-stress response mechanism in Sanilac, but not Saginaw. Zinc deficiency symptoms in Sanilac were likely due to the zinc-iron imbalance caused by the increased uptake of iron.

Another interaction observed in the field has recently been explained by an effect of potassium on the iron-deficiency stress response mechanism in both Strategy I and II plants.[64-66] In the absence of potassium, both tomato and soybean were found to lack the ability to release hydrogen ions and reductants or to reduce iron at their roots.[64] Muskmelon did not reduce iron or release reductants from its roots, but did release hydrogen ions under a combination of iron and potassium stress.[65] Recently, PS release and iron uptake (Strategy II) were both observed to be inadequate in potassium-deficient oat compared to potassium-sufficient plants.[66] Thus, potassium is closely involved in maintaining an active iron-stress response mechanism in both Strategy I and II plants.

Soil copper, zinc, and manganese are known to be mobilized by PS released in response to iron deficiency,[41] yet their application to soils had only slight (copper) or no (zinc or manganese) inhibitory effects on iron mobilization by PS.[42] Zhang et al.[42] found that mobilization of iron from ferric hydroxide was not affected by additions of calcium chloride, magnesium sulfate, or manganese sulfate. Iron mobilization was slightly inhibited by zinc sulfate, and strongly inhibited by copper sulfate. Only added copper impaired iron mobilization in calcareous soils. However, copper is seldom high enough in soils to cause such impairment so PS should solubilize iron when released by plants in calcareous soils.

Zinc deficiency in several dicot and grass species increased root exudation of amino acids, sugars, and phenolics, but these exudates did not generally mobilize iron. Root exudates from zinc-deficiency-stressed wheat and barley increased mobilization of both zinc and iron from soils and synthetic resins. This mobilization was directly linked to release of phytosiderophore. The rate of iron uptake was not enhanced by zinc deficiency as it was under iron-deficiency stress. Römheld[26] suggested that the preferential transport system present in roots during iron-deficiency stress response is not induced during zinc or manganese stress. Ecologically, zinc and iron are both poorly available in well-aerated calcareous soils, and this may explain the release of PS during both iron and zinc deficiencies.[26]

Genotypes of several species differ in requirements for copper, boron, and molybdenum, or in tolerance to toxic levels of manganese or aluminum. These may eventually be explained by physiological mechanisms similar to iron or by nutrient interactions as with zinc/iron and potassium/iron. Relatively little is known about the potential mechanisms involved.

Ultraviolet Light

Light exposure and composition have been shown to have a notable effect on iron nutrition. In environmental growth chambers, cool white fluorescent (CWF) lamps in combination with incandescent lamps provide sufficient light (350 to 700 nm) to cultivate healthy plants. Low pressure sodium (LPS) lamps produce more light and radiation per unit of input energy than do CWF lamps, but their spectral output (560 to 610 nm) has no ultraviolet or blue light radiation.[69] Chlorophyll concentrations in lettuce and cotton were lower in leaves of plants grown under LPS than CWF light.[70] Leaf iron in cotton was generally lower in plants grown under LPS than CWF light.[71] Hydrogen ions were extruded by iron-deficiency-stressed roots grown under either LPS or CWF light, but reductants were extruded and ferric iron was reduced only by the roots of plants grown under CWF light.[72] Cotton plants grown in soil for 6 weeks under LPS light had disorganized grana similar to those found in iron-deficiency-stressed plants.[72] More iron was reduced in solutions exposed to CWF than to LPS light.[70] Perhaps UV light affects the reduction of ferric to ferrous iron in leaves and thus controls the availability of this iron to biological systems requiring iron in the plant.

Root Nodules and Iron-Deficiency Stress

Symbiotic dinitrogen fixation is associated with dicotyledonous legumes infected with *Bradyrhizobium* bacteria. The dinitrogen fixation process releases hydrogen ions into the rhizosphere and creates an environment for reduction of nitrogen within the root nodule. It also requires iron as part of the active enzymes within nodules. Both dinitrogen fixation and iron-deficiency stress response are chemically reducing processes, and the two processes could potentially enhance the activity of each other.[73] Iron-inefficient T203 soybean normally produces no iron-deficiency stress response. Iron-stressed, nodulated T203 plants developed an iron-stress response which resulted in less chlorosis than similarly stressed nonnodulated plants. Iron-stressed T203 soybeans were grown in a split-root system where one side of the roots was inoculated and the other was not. Iron deficiency response was limited to the inoculated side of the split-roots.[74] Similarly grown T203 soybean were inoculated with strains of *Bradyrhizobium* either effective or ineffective in fixing nitrogen, and the iron-stress response was limited to the roots inoculated with effective bacteria.[75] In every case where an iron-stress response was stimulated on nodulated soybean roots, enhanced iron uptake and leaf greening occurred. Thus, variable nodule activity in field soybean may be a major contributor to nonuniform field chlorosis observations.

SUMMARY

Numerous biochemical reactions in plants are dependent on an available supply of iron. Plants have adaptive mechanisms for obtaining sparingly available soil iron. Plant response to iron-deficiency stress is controlled genetically, resulting in variable uptake and use of iron. After sufficient iron is made available, biochemical responses to the stress halt, but will be repeated under subsequent iron-deficiency stress. These controlled reactions (in dicots and nongraminaceous monocots) include: release of hydrogen ions and reductants from roots, enhanced reduction of ferric to ferrous iron by the roots, accumulation of citrate in roots that chelate iron for transport, and (in grasses) release of a phytosiderophore that solubilizes iron in the growth medium to be available for enhanced uptake. These controlled reactions are not common to all plants and differ in the number and kind of activities dependent on the plant genotype.

Other factors interacting or affecting the genetically controlled mechanism of iron uptake are zinc- and potassium-deficiency stress, ultraviolet light, and symbiotic dinitrogen fixation associated with dicotyledonous legumes infected with *Bradyrhizobium* bacteria. Tumorous crown gall activates its own mechanism for making iron available to itself.

Micronutrients play a role in determining what biochemical reactions are functional in plants. Plant iron supply is dependent on iron made available to the plant through induced, genetically controlled biochemical reactions.

REFERENCES

1. Lindsay, W. L., Soil and plant relationships associated with iron deficiency with emphasis on nutrient interactions, *J. Plant Nutr.*, 7, 489, 1984.
2. Brown, J. C., Iron chlorosis in plants, *Adv. Agron.*, 13, 329, 1961.
3. Brown, J. C., Mechanism of iron uptake by plants, *Plant Cell Environ.*, 1, 249, 1978.
4. Brown, J. C. and Jolley, V. D., Plant metabolic responses to iron-deficiency stress, *BioScience,* 39, 546, 1989.
5. Takagi, S., Naturally occurring iron-chelating compounds in oat- and rice-root washings. I. Activity measurement and preliminary characterization, *Soil Sci. Plant Nutr.,* 22, 423, 1976.
6. Marschner, H., Römheld, V., and Kissel, M., Localization of phytosiderophore release and iron uptake along intact barley roots, *Physiol. Plant.,* 71, 157, 1987.
7. Marschner, H., Römheld, V., and Kissel, M., Different strategies in higher plants in mobilization and uptake of iron, *J. Plant Nutr.,* 9, 695, 1986.
8. Weiss, M. G., Inheritance and physiology of efficiency in iron utilization in soybeans, *Genetics,* 28, 253, 1943.
9. Brown, J. C., Holmes, R. S., and Tiffin, L. O., Iron chlorosis in soybeans as related to genotype of rootstock, *Soil Sci.,* 86, 75, 1958.
10. Brown, J. C., Chaney, R. L., and Ambler, J. E., A new tomato mutant inefficient in the transport of iron, *Physiol. Plant.,* 25, 48, 1971.
11. Brown, J. C., Holmes, R. S., and Tiffin, L. O., Iron chlorosis in soybeans as related to the genotype of rootstock. 3. Chlorosis susceptibility and reductive capacity at the root, *Soil Sci.,* 91, 127, 1961.
12. Marschner, H., Kalisch, A., and Römheld, V., Mechanism of iron uptake in different plant species, in *Plant Analysis and Fertilizer Problems,* Wehrmann, J., Ed., German Society of Plant Nutrition, Hanover, 1974, 273.
13. Kramer, D., Römheld, V., Lansberg, E., and Marschner, H., Induction of transfer-cell formation by iron deficiency in the root epidermis of *Helianthus annuus* L., *Planta,* 147, 335, 1980.

14. Landsberg, E. C., Regulation of iron-stress response by whole plant activity, *J. Plant Nutr.*, 7, 609, 1984.

15. Landsberg, E. C., Transfer cell formation in the root epidermis: a prerequisite for Fe-efficiency?, *J. Plant Nutr.*, 5, 415, 1982.

16. Brown, J. C. and Ambler, J. E., Iron stress response in tomato *(Lycopersicon esculentum)*. 1. Sites of Fe reduction, absorption and transport, *Physiol. Plant.*, 31, 221, 1974.

17. Ambler, J. E., Brown, J. C., and Gaugh, H. G., Sites of iron reduction in soybean plants, *Agron. J.*, 63, 95, 1971.

18. Olsen, R. A., Bennett, J. H., Blume, D., and Brown, J. C., Chemical aspects of the Fe stress response mechanism in tomatoes, *J. Plant Nutr.*, 3, 905, 1981.

19. Barrett-Lennard, E. G., Marschner, H., and Römheld, V., Mechanism of short term Fe (III) reduction by roots. Evidence against the role of secreted reductants, *Plant Physiol.*, 73, 893, 1983.

20. Sijmons, P. C. and Bienfait, H. F., Mechanism of iron reduction by roots of *Phaseolus vulgaris* L., *J. Plant Nutr.*, 7, 687, 1984.

21. Jolley, V. D., Brown, J. C., and Nugent, P. E., A genetically related response to iron deficiency stress in muskmelon, *Plant Soil*, 130, 87, 1991.

22. Camp, S. D., Jolley, V. D., and Brown, J. C., Comparative evaluation of factors involved in Fe stress response in tomato and soybean, *J. Plant Nutr.*, 10, 423, 1987.

23. Neilands, J. B., Iron absorption and transport in microorganisms, *Annu. Rev. Nutr.*, 1, 27, 1981.

24. Takagi, S., Mugineic acids as examples of root exudates which play an important role in nutrient uptake by plant roots, in *Phosphorous Nutrition of Grain Legumes in the Semi-Arid Tropics,* Johansen, C., Lee, K. K., and Sahrawat, K. L., Eds., ICRISAT, Patancheru, India, 1991, 77.

25. Takagi, S., Nomoto, K., and Takemoto, T., Physiological aspects of mugineic acid, a possible phytosiderophore of graminaceous plants, *J. Plant Nutr.*, 7, 469, 1984.

26. Römheld, V., The role of phytosiderophores in acquisition of iron and other micronutrients in graminaceous species. An ecological approach, *Plant Soil*, 130, 127, 1991.

27. Clark, R. B. and Brown, J. C., Internal root control of iron uptake and utilization in maize genotypes, *Plant Soil*, 40, 669, 1974.

28. Brown, J. C. and McDaniel, M. E., Factors associated with differential response of oat cultivars to iron stress, *Crop Sci.*, 18, 551, 1978.

29. Jolley, V. D. and Brown, J. C., Iron efficient and inefficient oats. I. Differences in phytosiderophore release, *J. Plant Nutr.*, 12, 423, 1989.

30. Jolley, V. D. and Brown, J. C., Iron inefficient and efficient oat cultivars. II. Characterization of phytosiderophore released in response to Fe deficiency stress, *J. Plant Nutr.*, 12, 923, 1989.

31. Brown, J. C., Jolley, V. D., and Lytle, C. M., Comparative evaluation of iron solubilizing substances (phytosiderophores) released by oats and corn. Iron-efficient and iron-inefficient plants, *Plant Soil*, 130, 157, 1991.

32. Lytle, C. M., Jolley, V. D., and Brown, J. C., Iron-efficient and iron-inefficient oats and corn respond differently to iron-deficiency stress, *Plant Soil*, 130, 165, 1991.

33. Römheld, V. and Marschner, H., Genotypical differences among graminaceous species in release of phytosiderophores and uptake of iron phytosiderophores, *Plant Soil*, 123, 147, 1990.

34. Kawai, S., Takagi, S., and Sato, Y., Mugineic acid-family phytosiderophores in root-secretions of barley, corn and sorghum varieties, *J. Plant Nutr.*, 11, 633, 1988.

35. Nomoto, K., Yashioka, H., Arima, M., Fushiya, S., Takagi, S., and Takemoto, T., Structure of 2'-deoxymugineic acid, a novel amino acid possessing an iron-chelating activity, *Chimia*, 7, 249, 1981.

36. Mori, S. and Nishizawa, N., Methionine as a dominant precursor of phytosiderophores in *Graminaceae* plants, *Plant Cell Physiol.*, 28, 1081, 1987.

37. Kawai, S., Itoh, K., Takagi, S., Iwashita, T., and Nomoto, K., Studies on phytosiderophores: biosynthesis of mugineic acid and 2'-deoxymugineic acid in *Hordeum vulgare* L. var. Minorimugi, *Tetrahedron Lett.*, 29, 1053, 1988.

38. Nomoto, K., Sugiura, Y., and Takagi, S., Mugineic acids, studies on phytosiderophores, in *Iron Transport in Microbes, Plants and Animals*, Winklemann, G., van der Helm, D., and Neilands, J. B., Eds., VCH Publishers, Weinheim, Germany, 1987, 401.

39. Murakami, T., Ise, K., Hayakawa, M., Kamei, S., and Takagi, S., Stabilities of metal complexes of mugineic acids and their specific affinities for iron (III), *Chem. Lett.*, 1989, 2137, 1989.

40. Crowley, D. E., Reid, C. P. P., and Szaniszlo, P. J., Microbial siderophores as iron sources for plants, in *Iron Transport in Microbes, Plants and Animals*, Winkelmann, G., van der Helm, D., and Neilands, J. B., Eds., VCH Publishers, Weinheim, Germany, 1987, 375.

41. Treeby, M., Marschner, H., and Römheld, V., Mobilization of iron and other micronutrient cations from a calcareous soil by plant-borne, microbial, and synthetic metal chelators, *Plant Soil*, 114, 217, 1989.

42. Zhang, F. S., Treeby, M., Römheld, V., and Marschner, H., Mobilization of iron by phytosiderophores as affected by other micronutrients, *Plant Soil*, 130, 173, 1991.

43. Awad, F., Römheld, V., and Marschner, H., Mobilization of ferric iron from a calcareous soil by plant-borne chelators (phytosiderophores), *J. Plant Nutr.*, 11, 701, 1988.

44. Römheld, V. and Marschner, H., Evidence for a specific uptake system for iron phytosiderophores in roots of grasses, *Plant Physiol.*, 80, 175, 1986.

45. Mori, S., Nishizawa, N., Hayashi, H., Chino, M., Yoshimura, E., and Ishihara, J., Why are young rice plants highly susceptible to iron deficiency?, *Plant Soil*, 130, 143, 1991.

46. Clarkson, D. T. and Sanderson, J., Sites of absorption and translocation of iron in barley roots. Tracer and microautoradiographic studies, *Plant Physiol.*, 61, 731, 1978.

47. Hopkins, B. G., Jolley, V. D., and Brown, J. C., Plant utilization of iron solubilized by oat phytosiderophore, *J. Plant Nutr.*, 15, 1599, 1992.

48. Hopkins, B. G., Jolley, V. D., and Brown, J. C., Variable inhibition of iron uptake by oat phytosiderophore in five soybean cultivars, *J. Plant Nutr.*, 15, 125, 1992.

49. Hopkins, B. G., Jolley, V. D., and Brown, J. C., Differential response of Fe-inefficient muskmelon, tomato, and soybean to phytosiderophore released by Coker 227 oat, *J. Plant Nutr.,* 15, 35, 1992.

50. McDaniel, M. E. and Brown, J. C., Differential iron chlorosis of oat cultivars — a review, *J. Plant Nutr.,* 5, 545, 1982.

51. Nugent, P. E. and Bhella, H. S., A new chlorotic mutant of muskmelon, *HortScience,* 23, 379, 1988.

52. Wann, E. V. and Hills, W. A., Inheritance of Fe transport in tomato stock T3238, *Tomato Genet. Coop. Rep.,* 22, 28, 1972.

53. Wann, E. V. and Hills, W. A., The genetics of boron and iron transport in tomato, *J. Hered.,* 64, 370, 1973.

54. Kneen, B. E. and LaRue, T. A., Induced symbiosis mutants of pea *(Pisum sativum)* and sweetclover *(Melilotus alba annua), Plant Sci.,* 58, 177, 1988.

55. Kneen, B. E., LaRue, T. A., Welch R. M., and Weeden, N. F., Pleiotropic effects of brz. A mutation in *Pisum sativum* (L.) cv. 'Sparkle' conditioning decreased nodulation and increased iron uptake and leaf necrosis, *Plant Physiol.,* 93, 717, 1990.

56. Welch, R. M. and LaRue, T. A., Physiological characteristics of Fe accumulation in the 'bronze' mutant *Pisum sativum* (L.) cv. 'Sparkle' E107 (brz brz), *Plant Physiol.,* 93, 723, 1990.

57. Grusak, M. A., Welch, R. M., and Kochian, L. V., Physiological characterization of a single-gene mutant of *Pisum sativum* exhibiting excess iron accumulation. I. Root iron reduction and iron uptake, *Plant Physiol.,* 93, 976, 1990.

58. Brown, J. C., Terry, R. E., Jolley, V. D., and Hopkins, B. G., Reduction of iron (Fe^{3+} to Fe^{2+}) by tumorous crown gall cells of sunflower, *J. Plant Nutr.,* 13, 1513, 1990.

59. Jolley, V. D., Terry, R. E., and Brown, J. C., Tumorous crown gall response to iron-deficiency stress in sunflower, *J. Plant Nutr.,* 14, 539, 1991.

60. Weinberg, E. D., Roles of iron in neoplasia promotion, prevention, and therapy, *Biol. Trace Element Res.,* 34, 123, 1992.

61. Weinberg, E. D., Cellular acquisition of iron and the iron-withholding defense against microbial and neoplastic invasion, in *Iron and Human Disease,* Lauffer, R. B., Ed., CRC Press, Boca Raton, FL, 1992.

62. Ellis, B. G., Responses and susceptibility in zinc deficiency — a symposium, *Crops Soils,* 18, 10, 1965.

63. Jolley, V. D. and Brown, J. C., Factors in iron-stress response mechanism enhanced by Zn-deficiency stress in Sanilac, but not Saginaw navy bean, *J. Plant Nutr.,* 14, 257, 1991.

64. Jolley, V. D., Brown, J. C., Blaylock, M. J., and Camp, S. D., A role for potassium in the use of iron by plants, *J. Plant Nutr.,* 11, 1159, 1988.

65. Hughes, D. F., Jolley, V. D., and Brown, J. C., Differential response of dicotyledonous plants to potassium-deficiency stress: Iron-stress response mechanism, *J. Plant Nutr.,* 13, 1405, 1990.

66. Hughes, D. F., Jolley, V. D., and Brown, J. C., Role for potassium in the iron-stress response mechanism of iron-efficient oat, *Soil Sci. Soc. Am. J.,* 56, 830, 1992.

67. Zhang, F. S., Römheld, V., and Marschner, H., Release of zinc mobilizing root exudates in different plant species as affected by zinc nutritional status, *J. Plant Nutr.,* 14, 675, 1991.

68. Zhang, F. S., Römheld, V., and Marschner, H., Effect of zinc deficiency in wheat on the release of zinc and iron mobilizing root exudates, *Z. Pflanzenernaehr. Bodenkd.*, 152, 205, 1989.

69. Cathey, H. M. and Campbell, L. E., Plant productivity: new approaches to efficient sources and environmental control, *Trans. ASAE*, 20, 360, 1977.

70. Brown, J. C., Cathey, H. M., Bennett, J. H., and Thimijan, R. W., Effect of light quality and temperature on Fe^{3+} reduction and chlorophyll concentration in plants, *Agron. J.*, 71, 1015, 1979.

71. Jolley, V. D., Brown, J. C., Pushnik, J. C., and Miller, G. W., Influence of ultra-violet (UV)-blue light radiation on the growth of cotton. I. Effect on iron nutrition and iron stress response, *J. Plant Nutr.*, 10, 333, 1987.

72. Pushnik, J. C., Miller, G. W., Jolley, V. D., Brown, J. C., Davis, T. D., and Barnes, A. M., Influences of ultra-violet (UV)-blue light radiation on growth of cotton. II. Photosynthesis, leaf anatomy, and iron reduction, *J. Plant Nutr.*, 10, 2283, 1987.

73. Soerensen, K. U., Terry, R. E., Jolley, V. D., Brown, J. C., and Vargas, M. E., The interaction of iron-stress response and root nodules in iron-efficient and -inefficient soybeans, *J. Plant Nutr.*, 11, 853, 1988.

74. Soerensen, K. U., Terry, R. E., Jolley, V. D., and Brown, J. C., Iron-stress response of inoculated and non-inoculated roots of an iron inefficient soybean cultivar in a split-root system, *J. Plant Nutr.*, 12, 437, 1989.

75. Terry, R. E. and Jolley, V. D., unpublished data, 1992.

CHAPTER 17

Mechanisms of Iron Uptake from Rhodotorulate-Iron by Tomato

G.W. Miller, S. Hasegawa, A. Shigematsu, and G.W. Welkie

INTRODUCTION

Microorganisms may develop an iron stress when grown under iron-deficiency conditions and secrete Fe^{3+}-specific chelators known as siderophores.[1,2] Some of these siderophores (such as rhodotorulic acid [RA]) have stability constants that can exceed 10^{30} strength.[3] This is comparable to the synthetic Fe chelate, EDDHA (ethylenediamine di-*o*-hydroxyphenylacetate). Tomato plants have been grown using rhodotorulate-^{59}Fe as the iron source. Iron-efficient plants were able to use the iron from this source for their growth, and it was concluded that plants in the soil may derive their iron, in part, from siderophores of microbial origin.[4]

The objective of this study was to determine if siderophore-Fe^{+3} from the yeast *Rhodotorula pilimanae* is first reduced at the root surface, releasing the ligand (RA), or if the whole complex (Fe-RA) is absorbed into the plant. Rhodotorulic acid forms a complex with Fe^{3+} involving 3 RA molecules and 2 atoms of Fe^{+3} with a molecular weight of 1410 Da.[5] Miller et al.[4] found that the ferrated rhodotorulic acid supplied Fe to Fe-efficient tomato. Miller et al.[6] used ^{59}Fe-rhodotorulate (^{59}Fe-RA) and ^{14}C-rhodotorulate-Fe (^{14}C-RA-Fe) in the growing medium of two iron-efficient tomato varieties. It was evident from autoradiograms and tissue sampling that ^{59}Fe and ^{14}C were abundant in roots, stems, and leaves. Unlike synthetic chelates, RA (or metabolized derivatives) was apparently absorbed by roots and translocated to the leaves.

The design of the experiment showing uptake of RA-Fe by tomato was not adequate to assign a definite mechanism for Fe uptake from rhodotorulate in tomato plants.[6] The ^{14}C from rhodotorulate was incorporated into the leaves at the 6-h treatment period. Whether this came from uptake of the metal complex,

0-87371-942-5/94/$0.00+$.50

uptake of the smaller RA molecule, or degradation products is uncertain. The results indicated that metabolism of RA took place since no increase in ^{14}C was noted in any tissue with longer treatment periods. Further experiments with RA labeled with ^{59}Fe and ^{14}C and the interaction of RA with Fe^{+3} at the root surface using microautoradiography are necessary to determine definite uptake mechanisms.

Shenker et al.[7] showed that a fungal siderophore (raphorin) could act as an Fe carrier at least as well as EDDHA can. Uptake of ^{55}Fe by tomato plants mediated from raphorin-Fe (0.5 μM) was almost as high as that from ^{55}Fe-EDTA that contained a tenfold higher concentration of the Fe-chelate. Bienfait[8] argued that due to their redox potential, microbial siderophores like FOB are not reduced at appreciable rates and thus do not have a significant role in the iron nutrition of plants. Later, Bienfait[9] proposed a strategy plants use to facilitate Fe uptake mediated by siderophores. Tiffin and Brown[10] measured the selective absorption of iron from iron chelates of Fe-EDDHA, EDTA, and DTPA. The relative uptake of these synthetic Fe chelates in the stem exudates from Fe-stressed and green soybeans established that ^{55}Fe was high and ^{14}C very low in Fe-stressed plants, compared to very low concentrations of both ^{14}C and ^{59}Fe in green plants. It was concluded that these chelates delivered Fe to the roots but were not involved with its translocation within the plant.

Iron-efficient plants may respond to Fe stress by increasing the reduction capacity of ferric compounds at the root surface.[11,12] Reduction of Fe^{+3}-chelates at the root surface would release Fe^{2+} for plant absorption and the ligand would remain at the root surface.

Leong and Neilands[13] suggested three possible mechanisms of transport of exogenous siderophores in microorganisms:

1. Donation of Fe to the cell membrane without penetration of the complex or ligand
2. Uptake of the complex followed by internal release of Fe
3. Dissociation of the complex with simultaneous uptake of both ligand and Fe

In *Rhodotorula pilimanae* the uptake of Fe from RA-Fe probably follows mechanisms[5] similar to the mechanism concluded to be operative in higher plants for synthetic chelating agents,[10] that is, release of Fe from the chelate and uptake of Fe into the cell.

The reduction of Fe from Fe-EDDHA ($Fe^{3+} \rightarrow Fe^{2+}$) in peanut showed that Fe was released at the root surface and then absorbed as the free Fe^{2+} form.[14] The uptake of Fe by a variety of different organisms from various siderophores, phytosiderophores, and synthetic chelating compounds follows different methods as shown in Table 1. In most of the organisms shown the Fe is taken up as the Fe^{3+}-siderophore complex, but it also is apparently reduced to Fe^{2+} by some organisms and then absorbed as the Fe^{2+} form.[15-17]

MATERIALS AND METHODS

Preparation of Rhodotorulic Acid

Rhodotorulic acid was obtained by growing *Rhodotorula pilimanae* under iron stress and then extracting the RA. Rhodotorulic acid ^{14}C was prepared by growing yeast (*R. pilimanae* strains, 68-274, 54-190) in an Fe-free culture containing media as described by Atkin et al.[18] After cultures were inoculated with *R. pilimanae*, 500 μCi of ^{14}C-sucrose (U-^{14}C, high specific activity, Amersham, 99.3% pure) was added to each 300-ml culture solution. After 2 h, an additional 10 g of sucrose (unlabeled) was added per liter. Cultures were monitored for RA concentration and harvested after about 14 days.[18] RA was isolated following the procedure of Atkins et al.,[18] then crystallized and recrystallized. After recrystallization, paper chromatograms showed only one radioactive peak characterized as ^{14}C-RA with a radiochemical purity of 97%. The melting point of this compound of 215 to 218°C was identical to that of commercial RA (Porphyrin Products Inc., Logan, UT). Unlabeled RA was prepared using the same procedures as listed above but with no added ^{14}C-sucrose.

Preparation of ^{14}C-RA, ^{14}C-RA-Fe, and RA-^{59}Fe

A solution containing 188 mg of ^{14}C-RA (0.547 mmol, specific activity 0.018 μ-Ci/mg, radioactive purity 97%) and 20 ml of 0.1 M $HClO_4$ was adjusted to pH 5.0 by adding 0.1 M NaOH. After adjusting the volume to 100 ml with distilled water, 15 ml of the solution was added to a 500-ml culture solution containing the growing plants (0.52 μCi ^{14}C per culture solution). A 10.7-ml solution containing 110 mg of ^{14}C-RA (0.320 mmol, specific activity 0.0265 μCi/mg, total radioactivity 2.92 μCi, radiochemical purity 97%), 20 mM $FeCl_3 \cdot 6H_2O$, and 0.1 N $ClHO_4$ (0.214 mmol $FeCl_3$; 12 mg Fe^{3+}) was adjusted to pH 5.0 using 0.1 M NaOH. After making up to a final volume of 100 ml with distilled water, 15 ml of the solution was added to a 500-ml culture solution containing the growing plants (0.44 μCi ^{14}C per culture solution and 3.6 ppm Fe^{3+}). A 5.35-ml solution containing 55.2 mg RA (0.160 mmol) in 0.1 N $ClHO_4$, 300 μl $^{59}FeCl_3$ dissolved in 0.1 N HCl (Amersham, 9.7 μg/ml Fe, 0.1 mCi/ml, total 30 μCi) and 20 mM $FeCl_3 \cdot 6H_2O$ (0.107 mmol, 6 mg Fe^{3+}) was adjusted to pH 5.0 using 0.1 M NaOH. After dilution with distilled water to a final volume of 100 ml, 15 ml of the solution was added to the 500-ml culture solutions containing the growing plants (4.5 μCi ^{59}Fe per culture solution, 1.8 ppm Fe^{3+}).

Tomato Culture

Fe-efficient (T3238FER) and -inefficient (T3238fer) tomato plants[4] were germinated in vermiculite and transferred to 500 ml nutrient solution (4 containers of

Table 1.   Form of Iron Uptake by Various Organisms

Species	Form of chelate (siderophore)	Iron uptake into organism	Ref.
Graminaceous, monocots	Mugineic acid-Fe ($MA-Fe^{3+}$)	$MA-Fe^{+3}$	15
Bacillus megaterium	Fe^{3+}-Schizokinen	Fe^{3+} Schizokinen	16
U. stilago sphaerogena	Ferrichrome	Ferrichrome	17
Aerobacter aerogenes	Fe^{3+} Aerobactin	Fe^{3+} Aerobactin	16
Dicots (peanut, tomato, etc.)	Fe^{3+}-EDTA Fe^{3+}-EDDHA	Fe^{2+}	14
Rhodotorula pilimanae	Fe^{3+}-Rhodotorulate (Fe^{3+}-RA)	Fe^{2+}	5

500 ml in a 4-1 container wrapped with aluminum foil). The plants were grown initially in 1/2 strength nutrient solution for 1 week and then transferred to full-strength nutrient solution containing no iron for treated plants[19] for about 15 days. Control plants of both varieties were grown with 3 ppm Fe chelated with HEDTA. The experimental design did not provide aseptic plant cultures.

The Fe-deficient plants were monitored daily for pH changes; when the pH decreased to below 5, Fe treatments were initiated as described in the previous section. Plants were harvested 3, 6, 24, and 48 h after addition of RA-^{14}C, ^{14}C-RA-Fe, or RA-[59]Fe. Tissue from the roots, stems, or leaves were extracted for radioactivity.

Extraction of Tissue

Immediately after treatment in the radioactive medium for 3, 6, 24, or 48 h, tomato plants were rinsed with water and separated into roots, stems, and leaves. Each part was weighed and freeze-dried under vacuum. After weighing and pulverizing, the dried plant material was added to the following solvents, successively: diethylether ($10 \ ml \times 3$), methanol ($10 \ ml \times 3$), and water ($10 \ ml \times 2$) and treated with an ultrasonic vibrator. In the extraction, insoluble residues were separated from the extract by centrifugation (3000 rpm, 10 min).

Measurement of Radioactivity

The tissue samples containing ^{14}C were dissolved using a Bray's scintillator and the radioactivity was measured with a liquid scintillation counter (Aloka LSC-1000). For the ^{59}Fe samples, a γ-counter (Aloka Auto Well Gamma System ARC-251) was used. Radioactivity of plant tissues was counted after decolorizing with alkaline solution and H_2O_2. A total dose of ^{59}Fe or ^{14}C was the amount of radioactivity added to the nutrient solution. The total dose added to the treatments containing ^{59}Fe was calculated using a γ-counter; those containing ^{14}C were calculated using the liquid scintillation counter.

Thin-Layer Chromatography (TLC)

TLC was carried out by spotting on a cellulose plate (Merck, Art. 5718) and *n*-butanol-acetic acid-water (3:1:1) was used as the solvent. Radioactivity on the plate was detected by a bio-imaging analyzer (Fuji Film Co. Ltd, FUJIX BAS 2000).

Autoradiography

Autoradiograms were prepared for each isotope experiment using the bio-imaging analyzer. Film plates were exposed for 48 h.

RESULTS

pH and Total Fe Content

The pH profiles of the plants growing under Fe-stress conditions are shown in Figure 1. The Fe-inefficient tomato T3238fer was incapable of modifying the pH of the nutrient solution and the plants remained chlorotic. Plants for the experiment were grown as outlined by Miller et al.[4] The T3238FER plants (Fe-efficient) depressed the nutrient solution pH to varying degrees in response to iron stress. The minus-Fe control plants reduced the pH of the nutrient solution to 3.4 on day 2, and the value remained relatively constant until the termination of the experiment. The nutrient solution of the Fe-HEDTA-treated plant showed a pH decrease to 4.6 on day 2, followed by a gradual increase, reaching approximately the original value of 6.7 on day 6.

The nutrient solution of the Fe-siderophore-exposed plants exhibited a rapid decline in pH to 3.6, followed by a rapid increase in pH of the solution. By day 4 the pH had returned to 6.8, and this increased only slightly to pH 6.9 by the termination of the experiment on day 6. The Fe-siderophore and Fe-HEDTA treatments resulted in the greening of the plants, while the minus-Fe control plants remained chlorotic. In Figure 2 the total iron contents of top leaves, bottom leaves, and stems of T3238FER and T3238fer plants for each treatment were analyzed separately at the termination of the experiment. It can be seen in Figure 2 that T3238fer showed slight response to either the Fe-HEDTA or Fe-siderophore treatment, whereas the T3238FER displayed a marked response. For example, the Fe concentration in the top leaves of T3238fer and T3238FER for the minus iron treatments were 40 and 47; Fe-HEDTA treatments, 65 and 140; and the Fe-RA treatments, 47 and 215 ppm Fe, respectively. Consequently only T3238FER (efficient) plants were used for further experiments.

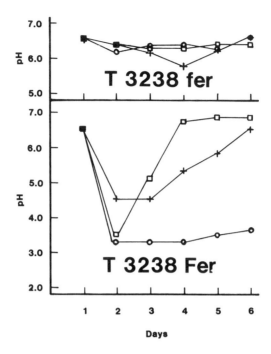

Figure 1.   Changes in the pH of the nutrient solutions of severely iron-stressed plants with iron resupplied; 0 = no Fe; □ = 0.5 ppm Fe as siderophore; + = 0.5 ppm Fe as HEDTA. (From Miller, G.W., Pushnik, J.C., Brown, J.C., Emery, T.E., Jolley, V.D., and Warnick, K.Y., *J. Plant Nutr.,* 8(3), 249, 1985, with permission.)

RA[59]-Fe Radioactivity, Chromatography, and Autoradiograms

The radioactivity in the tomato tissue after exposure to ^{59}Fe-RA for 6 to 48 h is shown in Table 2. Fe per gram of tissue in the roots was high after 6 h: 20.97×10^5 dpm/g (6.7% of total dose), and changed little after 24 h: 21.51×10^5 dpm/g (16.7% of total dose), and after 48 h: 17.05×10^5 dpm/g (12.6% of total dose). Although roots were rinsed prior to extraction and analyses, much of the Fe was probably adsorbed on the root surface. Stem tissue contained Fe at 3.11×10^5 dpm/g (2.8% total dose), 2.30×10^5 dpm/g (2.7% total dose), and 4.57×10^5 dpm/g (4.5% total dose) after 6, 24, and 48 h exposure, respectively. An increase in radioactivity from 6 to 48 h was evident, indicating an accumulation of Fe with time. Fe per gram of leaf tissue accumulated 4.84×10^5 dpm (3.9% total dose), 3.89×10^5 dpm (5.7% total dose) and 13.06×10^5 dpm (14.7% total dose) at 6, 24, and 48 h exposure, respectively. After extraction of tissue most of the radioactivity was found in the residue, with highest amounts in the roots of all treated tissues and in the leaves after 48 h. Some radioactivity was found in the water extract with traces in the ether extract, particularly in the roots.

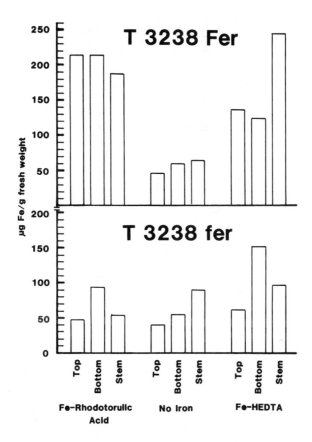

Figure 2. Iron content and distribution. Top leaves, bottom leaves, and stem tissue from severely iron-stressed T3238FER (efficient) and T3238fer (inefficient) tomato plants under various iron resupply treatments. (From Miller, G.W., Pushnik, J.C., Brown, J.C., Emery, T.E., Jolley, V.D., and Warnick, K.Y., *J. Plant Nutr.*, 8(3), 249, 1985, with permission.)

Autoradiograms of plants treated with RA-^{59}Fe are shown in Figure 3 after 3, 6, and 24 h. There was a marked change in the accumulation of Fe in the stems and leaves over this time treatment, with greatest accumulation in the stems and young leaves after the 6- and 24-h treatments. The roots showed very high radioactivity after 3 h due to the presence of RA-^{59}Fe in and on the root surface. This high radioactivity was also found after 6 and 24 h.

Thin-layer chromatograms of water and ether extracts from stem and leaves showed no indication of RA-^{59}Fe. The methanol extract from roots showed RA-^{59}Fe present at the 3-h treatment period, but this was probably adsorption from the solution.

Table 2. Radioactivity from ^{59}Fe-Rhodotorulate in Tomato Tissue after Different Time Exposures

Tissue	Treatment time (h)	Dry wt. (g)	Total tissue	% Total dose	dpm/g dry wt × 10⁻⁵			
					Ether extract	Methanol extract	Water extract	Residue
Root	6	0.318	20.97	(6.7)	0.54	0.04	3.51	11.52
	24	0.772	21.51	(16.7)	0.03	0.00	5.75	13.29
	48	0.736	17.05	(12.6)	0.02	0.00	4.34	11.66
Stem	6	0.883	3.11	(2.8)	0.02	0.00	0.09	5.52
	24	1.170	2.30	(2.7)	0.01	0.00	0.13	3.13
	48	0.736	4.57	(4.5)	0.01	0.00	0.20	3.83
Leaves	6	0.809	4.84	(3.9)	0.01	0.00	0.05	4.00
	24	1.448	3.89	(5.7)	0.00	0.00	0.22	4.26
	48	1.114	13.06	(14.7)	0.01	0.00	0.88	14.39

Note: Total radioactivity added to nutrient solution 9.923 × 10⁶ dpm as ^{59}Fe-rhodotorulate.

Figure 3.   Autoradiograms of tomato leaves and stems exposed to ^{59}Fe-RA after 3, 6, and 24 h and roots after 3 h.

^{14}C-RA-Fe Radioactivity, Chromatography, and Autoradiograms

Tomato plants were exposed to ^{14}C-RA-Fe in the growing medium. Radioactivity found in Fe-efficient tomato tissue from 6 to 48 h is shown in Table 3. Total radioactivity per gram of root was 21.43×10^4 dpm (11.4% total dose),

Table 3. Radioactivity from ^{14}C-Rhodotorulate-Fe in Tomato Tissue after Different Time Exposures

Tissue	Treatment time (h)	Dry wt. (g)	Total tissue	% Total dose	dpm/g dry wt $\times 10^{-4}$			
					Ether extract	Methanol extract	Water extract	Residue
Root	6	0.518	21.43	(11.4)	0.51	1.50	15.22	6.55
	24	0.360	46.77	(17.3)	0.62	1.58	12.58	30.40
	48	0.927	13.10	(12.5)	0.04	0.14	18.93	8.65
Stem	6	0.894	1.60	(1.5)	0.00	0.15	0.20	0.25
	24	0.457	1.74	(0.8)	0.11	0.53	0.55	1.10
	48	1.266	0.37	(0.5)	0.01	0.01	0.20	0.80
Leaves	6	1.056	0.93	(1.0)	0.00	0.00	0.05	0.23
	24	0.673	1.03	(0.7)	0.00	0.12	0.32	1.14
	48	1.460	0.23	(0.3)	0.00	0.00	0.31	0.68

Note: Total radioactivity added to nutrient solution (500 ml) 9.720 $\times 10^5$ dpm as ^{14}C-rhodotorulate-Fe.

46.77×10^4 dpm (17.3% total dose), and 13.10×10^4 dpm (12.5% total dose) at 6, 24, and 48 h, respectively. Adsorption at the root surface was likely, and no increased accumulation at 48 h was observed. Radioactivity per gram of stem was 1.60×10^4 dpm (1.5% total dose), 1.74×10^4 dpm (0.8% total dose), and 0.37×10^4 dpm (0.5% total dose) at 6, 24, and 48 h, respectively. Again, no noticeable increase with time took place. Radioactivity per gram of leaves was 0.93×10^4 dpm (1.0% of total dose), 1.03×10^4 dpm (0.7% total dose), and 0.23×10^4 dpm (0.3% total dose). Unlike the accumulation of ^{59}Fe found with RA-^{59}Fe, there was no accumulation of ^{14}C with ^{14}C-RA-Fe over time. The water extract and residue contained most of the radioactivity for the tomato tissue, but the ether extract for the roots showed some activity. Considerable activity in the methanol extract was found in extracts from roots and stems.

Autoradiograms of the tomato leaves and stems showed little difference in radioactivity after the 3- and 24-h treatment time (Figure 4), indicating no accumulation of ^{14}C over time from the ^{14}C-RA-Fe. Radioactivity of the roots was very high after 3 h, probably indicating high adsorption from the nutrient solution (Figure 4). Thin-layer chromatography from the water and ether extracts showed no indication of ^{14}C-RA-Fe or ^{14}C-RA. The methanol extract of the roots showed the presence of ^{14}C-RA and traces of ^{14}C-RA-Fe in the 3-, 6-, and 24-h treatment, but nothing was evident in the stems or leaves (Figure 5). This indicates that Fe may have been released on the root surface or in the root, leaving the free siderophore.

^{14}C-RA Radioactivity, Chromatography, and Autoradiography

Radioactivity in tomato (FER) treated with ^{14}C-RA and in tissue extracts is shown in Table 4. Total radioactivity in the roots was 23.00×10^4 dpm/g tissue (15.6% total dose), 37.07×10^4 dpm/g tissue (18.9% total dose), and 36.96×10^4 dpm/g tissue (19.7% total dose) at 6, 24, and 48 h respectively, after treatment initiation. During the same time frame, radioactivity in the stem was 1.32×10^4 dpm/g tissue, (0.3% total dose), 3.24×10^4 dpm/g tissue (1.8% total dose), and 1.21×10^4 dpm/g tissue (0.8% total dose); and radioactivity in the leaves was 0.30×10^4 dpm/g tissue (0.3% total dose), 1.09×10^4 dpm/g tissue (0.9% total dose), and 0.68×10^4 dpm/g tissue (0.5% total dose). No noticeable increase in radioactivity over time was found with any tissue. Radioactivity was highest in the residue followed by water extract, methanol extract, and ether extract.

Autoradiograms of tops from 3-, 6-, and 24-h exposure times showed little difference in intensity, similar to the ^{14}C-RA-Fe treatment (Figure 5). High radioactivity was found in all roots starting at the 3-h treatment with ^{14}C-RA. Thin-layer chromatography showed no radioactivity for ^{14}C-RA with any of the solvent extracts in roots, stems, or leaves.

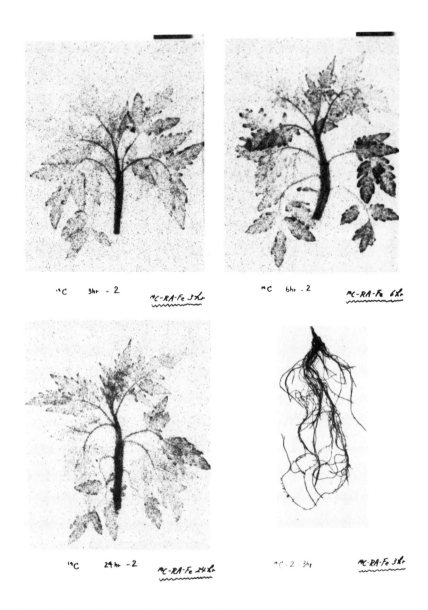

Figure 4.   Autoradiograms of tomato leaves and stems exposed to ^{14}C-RA-Fe after 3, 6, and 24 h and roots after 3 h.

Total Radioactivity in Solution

Table 5 is a summary of the amount of radioactivity found in the tissue and calculated to be taken up from the nutrient solution. This is compared to the measured radioactivity remaining in solution for ^{59}Fe-RA, RA-^{14}C, and

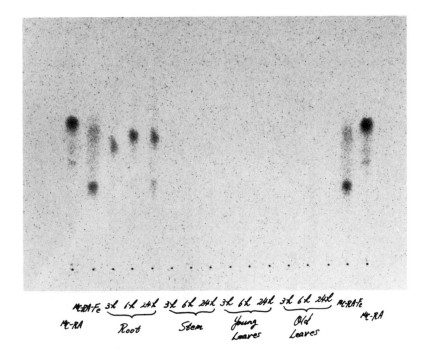

Figure 5. Autoradiogram of the methanol extract from tomato treated with ^{14}C-RA-Fe.

[14]C-RA-Fe. For ^{59}Fe-RA, 86.6, 74.7, and 65.9% of the total radioactivity added remained in the growing medium after 6, 24, and 48 h, respectively. This shows an accumulation of ^{59}Fe in the tissue of 13.4, 25.3, and 34.1% after 6, 24, and 48 h, respectively. With RA-^{14}C and ^{14}C-RA-Fe there was no marked increase in radioactivity with time. The total radioactivity remaining in solution was 78 to 87% over the 6- to 48-h time period. Unlike plants treated with ^{59}Fe, there was no accumulation over time.

The spent nutrient solution after 3-, 6-, and 24-h growth of plants in the ^{59}Fe-RA or ^{14}C-RA-Fe was chromatographed for ^{14}C-RA-Fe, ^{14}C-RA, and ^{59}Fe-RA. As shown in Figure 6 with ^{59}Fe-RA, as expected, ^{59}Fe-RA was the only radioactive compound found over the 24-h time period. When the treatment containing ^{14}C-RA-Fe was analyzed it was obvious that RA-Fe was being converted to RA. Much more RA was found after 24 h than at 3 h. Iron was being released from the RA-Fe molecule giving ^{14}C-RA.

Aliquots of the nutrient solution in which plants had been growing in ^{59}Fe-RA showed a decrease in the ^{59}Fe-RA over time to about 66% (Table 5). Ferric perchlorate was added to this solution to convert any RA to Fe-RA and the concentrations of Fe-RA measured spectrophotometrically at 480 nm.[20] The amount of Fe-RA increased from 66 to over 90% upon the addition of ferric perchlorate. This indicates that much of the Fe-RA after 48 h of plant growth was converted to RA. No decrease was observed in solutions in the absence of growing plants.

Table 4. Radioactivity from ^{14}C-Rhodotorulate in Tomato Tissue after Different Time Exposures

Tissue	Treatment time (h)	Dry wt. (g)	Total tissue	% Total dose	dpm/g dry wt × 10^{-4}			
					Ether extract	Methanol extract	Water extract	Residue
Root	6	0.777	23.00	(15.6)	0.16	0.43	5.17	12.67
	24	0.585	37.07	(18.9)	0.52	1.35	6.10	23.98
	48	0.611	37.96	(19.7)	0.48	1.03	5.16	25.72
Stem	6	0.811	1.32	(0.3)	0.07	3.79	0.39	0.76
	24	0.629	3.24	(1.8)	0.00	2.60	0.58	1.35
	48	0.734	1.21	(0.8)	0.07	0.33	0.36	0.79
Leaves	6	1.063	0.30	(0.3)	0.00	0.00	0.24	0.68
	24	0.934	1.09	(0.9)	0.07	0.30	1.15	0.92
	48	0.841	0.68	(0.5)	0.23	0.37	0.98	0.91

Note: Total radioactivity to nutrient solution (500 ml) 11.49 × 10^5 dpm as ^{14}C-rhodotorulic acid.

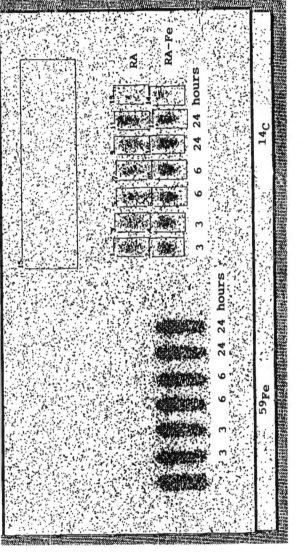

Figure 6. Autoradiogram showing the conversion of ^{14}C-RA-Fe to ^{14}C-RA in the nutrient solution with growing plants.

Table 5. Residual Radioactivity Remaining in Solution and Total Calculated Uptake from Solution

Treatment	Time (h)	Calculated uptake from nutrient solution (dpm $\times 10^{-6}$)	% of Total dose	Residual radioactivity remaining in solution (dpm $\times 10^{-6}$)	% of Total dose
^{59}Fe-RA[a] (dose 9.92 $\times 10^6$ dpm)	6	1.30	(13.4)	8.60	(86.6)
	24	2.50	(25.3)	7.40	(74.7)
	48	3.40	(34.1)	6.54	(65.9)
RA-^{14}C[b] (dose 1.15 $\times 10^6$ dpm)	6	0.19	(16.9)	0.96	(83.1)
	24	0.25	(21.7)	0.90	(78.3)
	48	0.23	(19.9)	0.92	(80.1)
^{14}C-RA-Fe[c] (dose 0.97 $\times 10^6$ dpm)	6	0.14	(14.4)	0.83	(85.6)
	24	0.18	(19.0)	0.79	(81.0)
	48	0.13	(13.3)	0.84	(86.7)

[a] ^{59}Fe-rhodotorulate
[b] ^{14}C-rhodotorulate-Fe
[c] ^{14}C-rhodotorulic acid

Reduction of Fe Chelates by FER

Roots of FER tomato were analyzed for Fe-reductase activity using the procedures outlined by Römheld and Marschner.[14] EDTA-Fe and Fe-RA were used as substrates. Roots had the capacity to reduce Fe^{3+} from Fe-EDTA (40 µmol Fe^{2+}/g FW/h) or Fe-RA (25 µmol Fe^{2+}/g FW/h), thus releasing Fe^{2+} from the chelate.

DISCUSSION

When plants were grown in ^{14}C-RA, ^{14}C-RA-Fe, or ^{59}Fe-RA, only with ^{59}Fe-RA did radioactivity increase over time in the tissue. Radioactivity was found in the tissue when plants were grown in nutrient solution containing either ^{14}C-RA-Fe or ^{14}C-RA, but no accumulation occurred over time in any tissue. Extracts of methanol, ether, and water from plant tissue showed radioactivity of RA or Fe-RA only in root extracts associated with RA or Fe-RA. The spent nutrient solution from ^{59}Fe-RA decreased over 35% in radioactivity over 48 h. Chromatography showed conversion of Fe-RA to RA. Spectrophotometric determination at 480 nm showed with that the addition of Fe to an aliquot of nutrient solution, the Fe-RA increased from 66 to over 90%, showing the presence of free RA. Tomato roots have the capacity to reduce Fe from Fe-EDTA or Fe-RA and thus release Fe^{2+} from the chelate.

It seems clear from experiments described in this research that Fe from Fe-RA is taken up into the plant as the free Fe^{2+}, leaving the ligand, RA, in the nutrient solution. The radioactivity found in the tissue when ^{14}C-RA-Fe or ^{14}C-RA was added to the growing medium probably was taken up by the roots as [14]CO_2 or some degradation product of RA. Once into the plant, it could be metabolized into various compounds (PEP carboxylase catalyzed) and respired as [14]CO_2. This would account for the radioactivity in the tissue but lack of accumulation over time. There was probably little or no uptake by the roots of intact Fe-RA or RA. Iron uptake by tomato with this natural Fe chelate follows the same mechanism found with Fe-EDTA, Fe-EDDHA, or other synthetic chelates, and also that found by Carrano and Raymond[5] for the uptake of Fe from Fe-RA in *Rhodotorula pilimanae*.

REFERENCES

1. Emery, T., The storage and transport of iron, in *Metal Ions in Biologic Systems*, Sigel, H., Ed., Marcel Dekker, New York, 1977.
2. Bezkorovany, A., Microbial iron uptake and antimicrobial properties of transferrins, in *Biochemistry of Nonheme Iron*, Plenum Press, New York, 1980.
3. Neilands, J.B., Microbial iron transport compounds (siderochromes), in *Inorganic Biochemistry*, Eichorn, G.L., Ed., Elsevier, New York, 1973, 167.

4. Miller, G.W., Pushnik, J.C., Brown, J.C., Emery, T.E., Jolley, V.D., and Warnick, K.Y., Uptake and translocation of iron from ferreted rhodotorulic acid in tomato, *J. Plant Nutr.*, 8, 249, 1985.

5. Carrano, C.J. and Raymond, K.N., Coordination chemistry of microbial iron transport compounds. Rhodotorulic acid and iron uptake in *Rhodotorula pilimanae*, *J. Bacteriol.*, 136, 69, 1978.

6. Miller, G.W., Shigematsu, A., Motoji, N., and Shibabe, S., Absorption and translocation of ^{59}Fe and ^{14}C-Rhodotorulate in iron-stressed tomato, *J. Plant Nutr.*, 13, 201, 1990.

7. Shenker, M., Oliver, I., Hehmam, M., Hadar, Y., and Chen, Y., Utilization by tomatoes of iron mediated by siderophore produced by *Rhizopus arrhizus*, *J. Plant Nutr.*, 15, 2173, 1992.

8. Bienfait, H.F., Mechanisms in Fe-efficiency reactions of higher plants, *J. Plant Nutr.*, 11, 605, 1988.

9. Bienfait, H.F., Prevention of stress in iron metabolism of plants, *Acta Bot. Neerl.*, 38, 105, 1989.

10. Tiffin, L.O. and Brown, J.C., Selective absorption of iron from iron chelates by soybean plants, *Plant Physiol.*, 36, 710, 1961.

11. Bienfait, H.F., Duivenvoorden, J., and Verwerke, W., Ferric reduction of chlorotic bean plants. Indications for an enzymatic process, *J. Plant Nutr.*, 5, 451, 1982.

12. Chaney, R.L., Brown, J.C., and Tiffin, L.O., Obligatory reduction of ferric chelates in iron uptake by soybeans, *Plant Physiol.*, 50, 208, 1972.

13. Leong, J. and Neiland, J.B., Mechanisms of siderophore iron transport in enteric bacteria, *J. Bacteriol.*, 126, 823, 1976.

14. Römheld, V. and Marschner, H., Mechanisms of iron uptake by peanut plants. 1. Fe III reduction, chelate splitting and release of phenolics, *Plant Physiol.*, 71, 949, 1983.

15. Hopkins, B.G., Jolley, V.D., and Brown, J.C., Plant utilization of iron solubilized by oat phytosiderophore, *J. Plant Nutr.*, 15, 1599, 1992.

16. Arceneaux, J.E.L., Davis, W.B., and Downer, D.N., Fate of labeled hydroxamates during iron transport from hydroxamate iron chelates, *J. Bacteriol.*, 115, 919, 1973.

17. Emery, T., Role of ferrichrome 25a ferric-ionophore in *Ustilago sphaerogena*, *Biochemistry*, 10, 1483, 1971.

18. Atkin, C.L., Neilands, J.B., and Phaff, H.J., Rhodotorulic acid from species of *Leucosporidium, Rhodosporidium, Rhodotorula, Sporidiobolus* and *Sporobolomyces*, and a new alanine-containing ferrichrome from *Cryptococcus melibiosum, J. Bacteriol.*, 103, 722, 1970.

19. Hoagland, D.R. and Arnon, D.I., The Water Culture Medium For Growing Plants Without Soil, Circ. 347 Calif. Agric. Stn., Berkeley, 1950.

20. Atkin, C.L. and Neilands, J.B., Rhodotorulic acid, diketopiperazine dihydroxamic acid with growth factor activities. I. Isolation and characterization, *J. Biochem.*, 7, 3734, 1968.

CHAPTER 18

Enzymatic Iron Reduction at the Root Plasma Membrane: Partial Purification of the NADH-Fe Chelate Reductase

Marcia J. Holden, Douglas G. Luster, and Rufus L. Chaney

INTRODUCTION

Iron (Fe) is an essential micronutrient required for growth and normal function of plant cells, particularly as a component of enzymes. However, Fe can have toxic effects due to stimulation of free radical reactions.[1] Once taken up into the plant cell, Fe must be incorporated into a functional form or bound in a storage form and sequestered into a compartment. By necessity, Fe uptake must be a highly regulated process.

Iron deficiency in plant tissues is more common than Fe toxicity. While Fe is an abundant mineral in the earth's crust, it is frequently of limited availability in aerated and alkaline soils where the most abundant forms are insoluble ferric hydroxides. Thus, suboptimal concentrations of available Fe leading to deficiency are frequently encountered in agricultural and horticultural production, resulting in loss of yield.[2] To cope with conditions of Fe deficiency, plants have devised strategies to increase the supply of soluble Fe and to facilitate transport into roots.[3]

The primary mechanism (Strategy I)[3] adopted by many plants (dicots and nongraminaceous monocots) to deal with Fe-deficiency stress includes both morphological and biochemical components. Under conditions of Fe deficiency in tomato (*Lycopersicon esculentum* Mill.), which is used in our experiments, priority in vegetative growth is given to new roots including development of short lateral roots covered with root hairs. These new laterals and root hairs are enriched in the enzyme, NADH-dependent Fe-chelate reductase, which performs the function of reducing soil Fe^{3+} (ferric) to Fe^{2+} (ferrous) in tomato.[4,5] The reduction step is essential to the transport of Fe across the plasma membrane and into the root cell.[6] In addition, these same new roots acidify the rhizosphere, which has the effect of solubilizing Fe from insoluble soil complexes, making Fe more available for reduction and uptake. Acidification is likely due to the enhanced activity of the plasma membrane, H^+-ATPase.[7,8] When Fe has been acquired by the roots in quantities sufficient for the plant's metabolic needs, there is a decrease in the rhizosphere acidification rate and Fe reductase activity. Tomato provides a good model system for the response to Fe deficiency, because it is very efficient with regard to this mechanism.

Little is known about the molecular components of this system — how intracellular Fe concentrations are perceived by plant cells, by what mechanism this information is transmitted to root cells, or how the activity or expression of this multicomponent response is regulated in a positive or negative manner. The reduction of apoplastic Fe is known to be due primarily to a plasma membrane enzyme using cytoplasmic reductants.[9] We have studied this activity in plasma membranes isolated from both Fe-deficient and Fe-sufficient tomatoes.[8,10] Several forms of this activity are found in tomato root plasma membrane and these are separable by isoelectric focusing.[10] The activity of these isoforms is dependent on NADH as reductant and all are enhanced under conditions of Fe deficiency. We are purifying a NADH-dependent ferric-chelate reductase (or simply, Fe-chelate reductase) from the plasma membranes of roots from Fe-deficient tomatoes as a first step to understanding how it functions and how its activity is regulated on a molecular level.

MATERIALS AND METHODS

Tomato Root Plasma Membrane Isolation

Plasma membranes were isolated from root tissue of tomato plants grown in solution culture as described in Buckhout et al.[8] and as modified by Holden et al.[10] In brief, plants in hydroponic culture were deprived of Fe for 1 week prior to harvest at 4 weeks of age. The roots were homogenized and microsomes were harvested by differential centrifugation. Plasma membranes were purified from microsomes by aqueous two-phase partitioning and were stored at –80°C. Fe-chelate reductase was solubilized from purified plasma membranes using the detergents octyl-β-D-glucopyranoside (or octylglucoside), as previously described,[10] or Triton X-100 (at a ratio of 6:1 detergent to protein).

Purification and Assays of Fe-Chelate Reductase

A dye-ligand mimetic chromatography medium was synthesized by linking Procion red H-8B (ICI America, Ltd.) to cross-linked Sepharose (Sepharose CL-4B, Pharmacia LKB Biotechnology) using published methodologies.[11] The dye was covalently linked to the Sepharose in the presence of 2% NaCl (w:v) and 1% $NaCO_3$ during incubation at 37°C for 48 h. The resulting matrix was washed extensively with ultrapure water, alternating with 1 M KCl after synthesis and again just before use. Columns (1 cm × 14 cm) were packed and equilibrated with buffer containing 25 mM Hepes, 26 mM octylglucoside, and 50 µM FAD, pH 7.0. Octylglucoside-solubilized plasma membrane proteins were diluted, if necessary, to reduce the octylglucoside concentration to approximately 30 mM, and loaded onto the column. The column was washed with the buffer described above with the addition of 100 mM KCl. Fe-chelate reductase activity was eluted with a simultaneous gradient of NADH (0 to 10 mM) and KCl (100 to 600 mM).

Chromatofocusing chromatography was performed on a 1 cm column packed with PBE 94 (Pharmacia LKB) which was equilibrated with 25 mM imidazole buffer, 0.1% Triton X-100, and 50 µM FAD, pH 7.4. Triton-solubilized plasma membrane proteins in the same buffer were loaded onto the column and the Fe-chelate reductase activity was eluted with Polybuffer (Pharmacia LKB) adjusted to pH 4.8.

NADH-dependent Fe-chelate reductase activity was assayed in plasma membrane and chromatography fractions using the ferrous chelator, 4,7-di(4-phenylsulfonate)-1,10 phenanthroline (BPDS) as previously described.[10] Assays were also conducted using ferricyanide as the electron acceptor.[10]

Electrophoresis

Sodium dodecyl sulfate-polyacrylamide gel electrophoresis in a discontinuous buffer system was done according to Laemmli[12] in a mini-gel format (Bio Rad, Richmond, CA) with 12% total monomers in the running gel. Blotting of proteins to nitrocellulose was performed using a high-glycine buffer system developed for transfer of membrane proteins[13] and using a transfer voltage of 60 V for a period of 3 h. Glycoproteins were detected on blots by enzyme immunoassay of digoxigenin-labeled glycoconjugates (DIG glycan detection, Boehringer Mannheim).

Native isoelectric focusing gels were run as previously described using ampholytes in the range of pH 5 to 7.[10] Duplicate gel halves were either stained for heme using the method of Thomas et al.,[14] or for Fe-chelate reductase activity by running the enzyme assay described above but substituting the ferrous chelator, 3-(2-pyridyl)-5,6-bis(4-phenylsulfonic acid)-1,2,4-triazine (PDTS or ferrozine) for BPDS.[10]

RESULTS

Partial Purification of Fe-Chelate Reductase and Identification of a Relevant Polypeptide

Ferric-chelate reductase can be solubilized from plasma membranes in an active form using nonionic detergents such as Triton X-100 or octylglucoside, with recovery of 90+% of the activity and about 50% of the protein.[10] We had previously tested the ability of triazine dyes to inhibit this enzyme[10] as they have been shown to be inhibitors of other NADH-dependent oxidoreductases. The dyes exhibit competitive inhibition of NADH which probably occurs through structural mimicry.[15] Several of these dyes proved to be inhibitory to the Fe-chelate reductase activity in tomato root plasma membrane, but less inhibitory to the analogous activity using ferricyanide as the electron acceptor.[10] We used this inhibitory property of the triazine dyes as a tool for the purification of the Fe-chelate reductase enzyme in dye-ligand mimetic chromatography. The dye, Procion red H-8B, was covalently linked to a Sepharose matrix. Solubilized Fe-chelate reductase activity was found to bind to the red Sepharose column and it could be eluted using a simultaneous gradient of NADH and KCl. A single peak of Fe-chelate reductase activity eluted near the end of the gradient (Figure 1), whereas two peaks of NADH-dependent ferricyanide reductase activity were found. One of the peaks of ferricyanide reductase activity was coincident with the Fe-chelate reductase activity, while the other peak eluted during a wash phase of the chromatography program prior to the gradient elution. This was evidence for more than one electron transport activity in the root plasma membrane, one of which can use ferric chelates and ferricyanide as electron acceptors, while the other can reduce only ferricyanide. These results were supported by data from preparative isoelectric focusing which demonstrated ferricyanide reductase activity with two different isoelectric points, only one of which was contiguous with the Fe reductase isoforms.[10] Thus red Sepharose chromatography proved useful both for the purification of the Fe-chelate reductase, with recovery of up to 80% of the activity, as well as for the separation from another electron transport system of unknown function. The differential binding of the two enzymes to the chromatography matrix further explains the different patterns of inhibition by various triazine dyes of Fe-chelate reductase and ferricyanide reductase activity in native plasma membranes.[10]

As previously reported,[10] isoelectric focusing permitted separation of solubilized Fe-chelate reductase from other plasma membrane proteins. However, the yield of Fe-chelate reductase activity was low using this procedure, making it less desirable as a purification tool. We substituted chromatofocusing, a form of chromatography that utilizes principles similar to isoelectric focusing, as a purification step. Fe-chelate reductase activity was solubilized from tomato root plasma membrane using Triton X-100 and loaded onto the chromatofocusing column (PBE 94) which was previously equilibrated with an imidazole buffer of pH 7.4. The column was then washed with a pH 4.8 solution (Pharmacia LKB, Polybuffer 74) which results in the establishment of a continuous gradient of pH (from 7.4 to 4.4) in the buffer eluting from the column. In this methodology, bound proteins will dissociate from the column material when the pH of the buffer reaches the isoelectric point of the protein. Fe-chelate reductase eluted from the chromatofocusing column as a single broad peak.

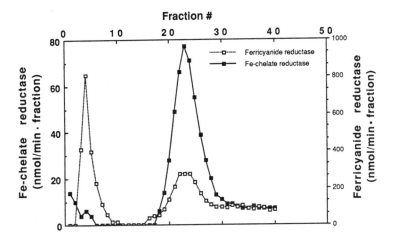

Figure 1.   Chromatography of solubilized plasma membrane on a column of red Sepharose. NADH-dependent Fe-chelate reductase and NADH-ferricyanide reductase activity were measured in all fractions. After loading the protein, the column was washed with a buffer containing 100 mM KCl followed by a simultaneous gradient of KCl and NADH starting at fraction 11 which ran to fraction 26.

This procedure was unable to cleanly separate the several isoforms (data not shown) that were separated with the native isoelectric focusing gels.[10] When the intensity of the silver staining on the gel was compared with the enzyme activity in the fractions (Figure 2B), a polypeptide of 34 to 36 kDa correlated well with the activity. A polypeptide of this molecular weight was also prominent in gel slices from regions of native isoelectric gels showing Fe-chelate reductase activity.[16] In addition, it was a major component in other protein purification protocols that have been used, such as red Sepharose chromatography and gel filtration.[24] Other polypeptides were also associated with the Fe-chelate reductase activity, such as the 27 and 29 kDa bands seen in Figure 2B, but the chromatofocusing experiments demonstrated that these did not correlate well with activity. It is probable that the tomato root Fe-chelate reductase is an enzyme that consists of more than one polypeptide, as do many electron transport systems. To date, evidence suggests that the 35-kDa polypeptide is likely to be a catalytic component.

Is the Fe-Chelate Reductase of Tomato Root Plasma Membrane Glycosylated or a Heme Protein?

We investigated whether the Fe-chelate reductase of tomato root plasma membrane was a glycoprotein. Glycosylation is a common modification of plasma membrane proteins. Glycosyl modification was detected by covalent attachment of oxidized sugar residues to digoxigenin-succinyl-ε-amido-caproic acid hydrazide and then immunodetection with anti-digoxigenin linked to alkaline phosphatase. This sensitive method resulted in the labeling of numerous

A

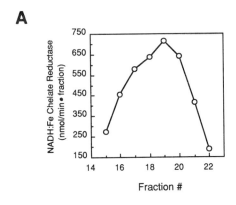

Fraction #

B

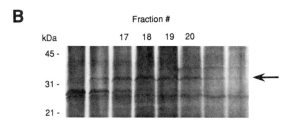

Figure 2. A. Fe-chelate reductase activity profile from chromatofocusing of Triton X-100-solubilized plasma membrane protein. B. Silver-stained SDS-PAGE analysis of MeOH:acetone (1:1) precipitated chromatofocused fractions containing Fe-chelate reductase activity. Relative mobility of molecular weight markers in kDa are indicated to the left of gel. Large arrow indicates the position of the polypeptide which correlated strongly in staining intensity with Fe-chelate reductase activity.

polypeptides in the tomato root plasma membrane separated by SDS-PAGE, but labeling of only a 64 to 66 kDa polypeptide in the partially purified reductase preparation. This result suggests that the Fe-chelate reductase does not have glycosylated polypeptides.

Until the Fe-chelate reductase is purified to homogeneity it will not be possible to determine with any certainty the cofactors that might be involved in the electron transport activity from NADH to Fe^{3+}. It is possible that Fe-S centers, heme groups, or cofactors such as FAD or FMN might be critical components of the reductase. With the ability to separate and detect several Fe-chelate reductase isoforms in native isoelectric focusing gels using activity staining,[10] it is also possible to ask whether heme groups are integral to the protein. Native isoelectric focusing gels were run, and identical halves of gels were either stained for the Fe-chelate reductase activity or the presence of heme using tetramethylbenzidine-H_2O_2, a sensitive method for the detection of cytochromes. This made it possible to ascertain whether heme groups were associated with the Fe-chelate reductase activity (under nondenaturing conditions, where heme groups would not be dissociated from the polypeptide). Results of the tetramethylbenzidine-H_2O_2 staining indicated that numerous heme-containing polypeptides could be solubilized from tomato root plasma membrane. Fe-chelate reductase isoforms with isoelectric point values of 5.5, 5.8, and 6.0 to 6.2 had been previously reported.[10] Four to five distinct heme-stained bands were detected with isoelectric points ranging from 6.5 to 7.3 and at least three others between pH 4 to 5. There was also one band or sometimes a diffuse staining region that was seen in the pH range of 6.1 to 6.4. In any given experiment, this band did not overlap with the regions of the gel that showed Fe-chelate reductase activity. Thus, heme staining did not correlate with Fe-chelate reductase activity. This evidence does not support a cytochrome nature for the plasma membrane Fe-chelate reductase of tomato roots.

DISCUSSION

To date, there have been no reports in the literature of the purification of plasma membrane Fe-chelate reductases from any plant source. Our work has resulted in a partial purification of the reductase in tomato root plasma membrane and has implicated a 35-kDa polypeptide as a putative component of the reductase. While the molecular nature of the electron transport function of this enzyme is as yet unknown, evidence presented here indicates that the purified reductase has neither heme-containing nor glycosylated components.

While reductases with roles in Fe reduction have been detected in many organisms from microbes to mammals, little is known of the enzymes. In addition to plant studies, significant information is available on the Fe-uptake mechanisms in the yeast, *Saccharomyces cerevisiae*. Both reductive and nonreductive (uptake of some Fe-siderophore complexes) strategies are utilized by *S. cerevisiae*. Similar to the tomato system, this yeast appears to rely primarily on a plasma membrane Fe-chelate reductase to supply ferrous ions for uptake.[17-19] Fe-chelate reductase activity in yeast increases in response to Fe deficiency.[17,19]

Comparison of information available on plant and yeast Fe-chelate reductases reveals both some similarities and also some significant differences between the systems. Lesuisse et al.[20] have reported four Fe reductases from yeast which they describe as being localized in the cytosol, mitochondria, and plasma membrane. One of the cytosolic enzymes, a probable flavoprotein of 40 kDa, reduces Fe-EDTA in the presence of NADPH, but not ferrioxamine B. A membrane-bound enzyme that they attribute to the plasma membrane also preferentially uses NADPH. Only the NADPH-dependent Fe-chelate reductase activity was enhanced when the cells were grown under conditions of Fe deficiency.[20] This contrasts with the tomato[8,21] or *Plantago lanceolata*[22] reductases, which preferentially utilized NADH over NADPH, and only the NADH-dependent activity was enhanced when tomatoes were grown under Fe-deficient culture conditions.

The partially purified yeast plasma membrane reductase utilized Fe-EDTA and ferrioxamine B equally well and Fe-citrate almost not at all,[20] although whole yeast cells reduced Fe-citrate.[17] The yeast Fe-chelate reductase had a pH optimum of 7.5 and low substrate specificity. The tomato Fe-chelate reductase has a pH optimum of 6.0 and also a low substrate specificity, but prefers organic acid-chelated Fe, such as citrate, over synthetics like EDTA.[10] Because FMN enhanced the activity of the partially purified yeast enzyme about 11-fold, Lesuisse et al.[20] postulated that the yeast reductase has a weakly or noncovalently bound flavin as a prosthetic group. Heme-deficient yeast mutants did not respond to Fe deficiency by increasing reductase activity, leading Lesuisse et al. to suggest that heme may also be involved in the yeast reductase.[18] There is no evidence yet for involvement of heme groups in the tomato reductase. The yeast reductase has not been purified to homogeneity and may consist of more than one polypeptide.[20]

A gene for a yeast Fe-chelate reductase has been isolated using complementation cloning of a mutant with a normal ferrous uptake mechanism, but deficient in plasma membrane reductase activity.[19] This cloned gene encodes for a 78-kDa polypeptide with some sequence similarity to cytochrome b558, a component of a phagocyte membrane oxidoreductase.[23] Analysis of the amino acid sequence predicts membrane spanning regions suggesting that this gene encodes for a plasma membrane reductase rather than a regulator protein. The molecular weight of this putative yeast reductase is twice that of the 35-kDa polypeptide implicated in the tomato root Fe-chelate reductase activity. Amino acid sequence data of the 35-kDa polypeptide will provide important information as to the possible relationships between the reductase components of these heterologous systems.

The mechanism of the biochemical response to Fe deficiency in plants that involves an enhancement of Fe-chelate reductase activity is similar to that described for the yeast, *S. cerevisiae*. At the present level of knowledge, plant and yeast systems share little in common on a biochemical and molecular level, except the use of cytoplasmic reductants to reduce external Fe. Our efforts continue in the purification of the tomato NADH-dependent Fe-chelate reductase. A purified reductase will be a powerful tool for investigating the biochemical mechanism of action and the molecular characterization of the genes involved in Fe uptake by plants.

REFERENCES

1. Halliwell, B. and Gurreridge, J.M.C., Iron as a biological pro-oxidant, in *ISI Atlas of Science: Biochemistry,* 1988, 48.
2. Lindsay, W.L. and Schwab, A.P., The chemistry of iron in soils and its availability to plants, *J. Plant Nutr.,* 5, 821, 1982.
3. Romheld, V. and Marschner, H., Mobilization of iron in the rhizosphere of different plant species, *Adv. Plant Nutr.,* 2, 155, 1986.

4. Brown, J.C. and Ambler, J.E., Iron-stress response in tomato *(Lycopersicon esculentum)*. I. Sites of Fe reduction, absorption and transport, *Physiol. Plant.,* 31, 221, 1974.

5. Bell, P.F., Chaney, R.L., and Angle, J.S., Staining localization of ferric reduction on roots, *J. Plant Nutr.,* 11, 1237, 1988.

6. Chaney, R.l., Brown, J.C., and Tiffin, L.O., Obligatory reduction of ferric chelates in iron uptake by soybeans, *Plant Physiol.,* 50, 208, 1972.

7. Romheld, V., Muller, C., and Marschner, H., Localization and capacity of proton pumps in roots of intact sunflower plants, *Plant Physiol.,* 76, 603, 1984.

8. Buckhout, T.J., Bell, P.F., Luster, D.G., and Chaney, R.L., Iron stress-induced redox activity in tomato (*Lycopersicon esculentum* Mill.) is localized on the plasma membrane, *Plant Physiol.,* 90, 151, 1989.

9. Bienfait, J.F., Duivenvoorden, J., and Werkerke, W., Ferric reduction by roots of chlorotic bean plants: indications for an enzymatic process, *J. Plant Nutr.,* 5, 451, 1982.

10. Holden, M.J., Luster, D.G., Chaney, R.L., Buckhout, T.J., and Robinson, C., Fe^{3+}-chelate reductase activity of plasma membranes isolated from tomato (*Lycopersicon esculentum* Mill.) roots. Comparison of enzymes from Fe-deficient and Fe-sufficient roots, *Plant Physiol.,* 97, 537, 1991.

11. Lowe, C.R. and Pearson, J.C., Affinity chromatography on immobilized dyes, *Methods Enzymol.,* 104, 97, 1984.

12. Laemmli, U.K., Cleavage of structural protein during the assembly of the head of bacteriophage T_4, *Nature,* 227, 680, 1970.

13. Small, G.M., Imanaka, T., and Lazarow, P.B., Immunoblotting of hydrophobic integral membrane proteins, *Anal. Biochem.,* 169, 405, 1988.

14. Thomas, P.E., Ryan, D., and Levin, W., An improved staining procedure for the detection of the peroxidase activity of cytochrome P-450 on sodium dodecyl polyacrylamide gels, *Anal. Biochem.,* 75, 168, 1976.

15. Thompson, S.T. and Stellwagen, R., Binding of cibacron blue F3GA to proteins containing the dinucleotide fold, *Proc. Natl. Acad. Sci. U.S.A.,* 73, 361, 1976.

16. Holden, M., Luster, D.G., Chaney, R.L., and Buckhout, T.J., Enzymology of ferric chelate reduction at the root plasma membrane, *J. Plant Nutr.,* 15, 1667, 1992.

17. Lesuisse, E., Raguzzi, F., and Crichton, R.R., Iron uptake by the yeast *Saccharomyces cerevisiae:* Involvement of a reduction step, *J. Gen. Microbiol.,* 133, 3229, 1987.

18. Lesuisse, E. and Labbe, P., Reductive and non-reductive mechanisms of iron assimilation by the yeast *Saccharomyces cerevisiae, J. Gen. Microbiol.,* 135, 257, 1989.

19. Dancis, A., Klausner, R.D., Hinnebusch, A.G., and Barriocanal, J.G., Genetic evidence that ferric reductase is required for iron uptake in *Saccharomyces cerevisiae, Mol. Cell. Biol.,* 10, 2294, 1990.

20. Lesuisse, E., Crichton, R.R., and Labbe, P., Iron-reductases in the yeast *Saccharomyces cerevisiae, Biochim. Biophys. Acta,* 1038, 253, 1990.

21. Brüggemann, W., Moog, P.R., Hakagawa, H., Janiesch, P., and Kuiper, P.J.C., Plasma membrane-bound NADH: Fe^{3+}-EDTA reductase and iron deficiency in tomato *(Lycopersicon esculentum)*. Is there a turbo reductase?, *Physiol. Plant.,* 79, 339, 1990.

22. Schmidt, W., Janiesch, P., and Brüggemann, W., Fe-EDTA reduction in roots of *Plantago lanceolata* by a NADH-dependent plasma membrane-bound redox system, *J. Plant Physiol.,* 136, 51, 1990.

23. Dancis, A., Roman, D.G., Anderson, G.J., Hinnebusch, A.G., and Klausner, R.D., Ferric reductase of *Saccharomyces cerevisiae.* Molecular characterization, role in iron uptake, and transcriptional control by iron, *Proc. Natl. Acad. Sci. U.S.A.,* 89, 3869, 1992.

24. Holden, M.J., Luster, D.G., and Chaney, R.L., unpublished data.

CHAPTER **19**

Iron Uptake in *Arabidopsis thaliana*

Ying Yi, Jennifer A. Saleeba, and Mary Lou Guerinot

INTRODUCTION

Iron is an essential nutrient for all plants, with iron-containing compounds figuring prominently in the electron transport systems of photosynthesis and respiration. For the synthesis of these and other iron-containing components, plants must obtain an adequate supply of iron. Although iron is abundant in the soil, ranking fourth among all elements on the surface of the earth, the availability of iron is problematic due to its low solubility in aerobic environments at biological pH. Under such conditions, iron tends to precipitate, forming oxyhydroxide polymers of the general composition FeOOH. The concentration of ferric iron, Fe(III), in equilibrium solution at pH 7 is approximately $10^{-18}\ M$, a value much lower than that required to support maximum growth rates for plants $(10^{-7}\ M)$.[1] Iron availability is particularly problematic in calcareous soils, which cover approximately one-third of the earth's surface. Iron deficiency can result in severe yield losses for agriculturally important crops including citrus, chickpea, beans, and peanut. Plants grown under a limited iron supply exhibit marked changes in the structure and function of the chloroplast, the most obvious change being a decrease in pigment concentration.[2] Thus, young leaves from a plant deficient in iron have low levels of chlorophyll and appear yellow or chlorotic. Symptoms of iron deficiency are also associated with high soil pH, excess soil moisture, and with irrigation water that is high in bicarbonates.[3] High levels of nitrate, potassium, or phosphate can also contribute to iron deficiency.[3]

Iron also poses a problem when present in excess; iron toxicity has been documented for a number of species.[4] Environmental factors which can lead to iron toxicity include acid soil pH, waterlogging, and mineral imbalances. Tolerance to Fe toxicity appears to coincide with tolerance to waterlogging in soils. There is evidence that roots of plant species tolerant to wet soils oxidize Fe more effectively than plants sensitive to wet soils.[4]

Although a great deal of knowledge has accumulated at the physiological level regarding iron uptake in plants, there is almost no information at the molecular level. *Arabidopsis thaliana* offers an opportunity to extend the analysis of plant iron nutrition to the molecular level. To provide perspective for our ongoing work on characterization of iron uptake in *Arabidopsis,* we will first review what is known about iron uptake in plants, including a brief description of plant varieties/mutants with known altered uptake phenotypes. We will then review the advantages that *Arabidopsis* offers for both genetic and molecular genetic approaches, and we will outline how we are applying these approaches to a study of iron uptake.

IRON UPTAKE IN PLANTS

Iron Deficiency Responses in Plants

To overcome iron deficiency, plants have evolved various adaptive mechanisms to acquire iron. These mechanisms have been divided into two broad strategies.[5] One strategy appears to be confined to graminaceous monocots. This strategy (Strategy II) is characterized by a release of specific Fe(III) chelating compounds, termed phytosiderophores,[6] and by the induction of high affinity uptake systems for the ferric phytosiderophores.[7] The phytosiderophores which have been identified are structurally related to nicotianamine, an amino acid thought to be a ferrous iron, Fe(II), carrier in plants.[8] Recent work suggests that nicotianamine is an intermediate in the formation of the mugineic acid family of phytosiderophores.[9] Dicots and nongraminaceous monocots respond to iron deficiency by a very different strategy (Strategy I). In Strategy I plants, iron deficiency results in morphological and physiological changes in subapical regions of the roots. Such changes include increased ferric reduction capacity,[10,11] increased ability to acidify the rhizosphere,[12,13] and, in certain cases, transformation of epidermal cells into transfer cells,[14] and release of reductant compounds.[15] These changes serve to make iron more available for uptake by the root.

Enhanced Fe(III) reduction under iron deficiency is the most typical feature of Strategy I plants and is probably the primary factor in making iron available for absorption. Plants are better able to take up Fe(II).[16] Indeed, as early as 1972, Chaney et al.[10] showed that ferric reduction was an obligatory step in iron uptake in soybean. The reduction of iron has been shown to take place at the plasma membrane of root cells, presumably via a transmembrane redox system which can be activated when iron is in short supply.[17] For example, iron-deficiency stress results in a greater than 7-fold increase in ferric reductase activity in tomato,[17] a 6-fold increase in bean,[11] and up to a 20-fold increase in peanut.[18] The presence of a ferric reductase system associated with the plasma membrane has been demonstrated in intact roots and in purified plasma membrane preparations.[17,19] Bienfait[20] proposed that roots contain two ferric reductase activities: one that is capable of reducing ferric chelates and ferricyanide and another that can only reduce ferricyanide. The inducible ferric reductase system (termed the turbo reductase) was thought to be expressed in the epidermal cells of young lateral roots grown under iron deficiency, while the ferricyanide reductase (termed the standard reductase) was thought to be constitutively expressed in all root cells. However, recent work by Buckhout et al.[17] does not support this view. Both ferricyanide and ferric reductases were found to increase approximately twofold under iron-deficiency conditions in plasma membranes purified from tomato. Buckhout et al.[17] also have determined that the preferred electron donor for the iron deficiency-induced ferricyanide and ferric reductases is NADH. Previous work had implicated NADPH as the preferred electron donor.[21] For a discussion of recent work on the purification and further characterization of ferric reductase from tomato roots, the reader is referred to Holden et al.[19]

Iron starvation causes plant roots to excrete protons via a plasma membrane ATPase pump.[22] This has the effect of lowering the rhizosphere pH and results in the solubilization of iron. (For each pH unit increase above 4.0, the solubility of iron decreases by a factor of 1000.) The rate of proton excretion can be quite fast; iron-deficient sunflower plants can lower the pH of an unbuffered solution from 8.0 to 3.7 in a period of a few hours.[23] The capacity of a plant to acidify the rhizosphere in response to iron deficiency may depend on the cation/anion uptake balance and N metabolism.[24]

Our lab has now demonstrated that *Arabidopsis thaliana* exhibits typical Strategy I iron-deficiency responses, showing induction of ferric reductase activity as well as increased acidification of the rhizosphere under iron deficiency. As will be described below, we are using both of these phenotypes in screens to detect mutants of *Arabidopsis* which have altered Strategy I responses. We should note that an earlier study by Tingey et al.[25] had suggested that *Arabidopsis* was indeed an "iron-efficient" plant (see below for definition of "iron-efficient"). This was based on the ability of iron-deficient *Arabidopsis* plants to re-green after an initial period of chlorosis as well as on the ability of the iron-deficient plants to accumulate iron in their tissues.

Plant Varieties/Mutants With Altered Iron Uptake Phenotypes

Different varieties of a given plant species can vary dramatically in their ability to utilize iron from soils of low iron availability. Varieties which can grow well under iron deficiency are termed iron efficient; varieties unable to respond are designated iron inefficient. The strategies utilized by iron-efficient plants to enhance the acquisition of iron were discussed above. Table 1 lists some of the varieties as well as a number of plant mutants which have altered iron uptake abilities. Calling a line a variety or a mutant may be a matter of

Table 1. Plant Varieties Classified as Either Iron Efficient or Iron Inefficient and Mutants With Altered Iron Uptake Phenotypes

Plant species	Gene or variety designation	Phenotype	Ref.
Varieties:			
Soybean	T203	Iron inefficient	26, 27, 29
Soybean	Hawkeye	Iron efficient	26, 27, 29
Oat	TAM 0-312	Iron inefficient; does not release phytosiderophores	29
Oat	Coker 227	Iron efficient; releases phytosiderophores	29
Corn	ys_1	Iron inefficient	29
Corn	WF9	Iron efficient	29
Mutants:			
Pea	brz	Takes up too much iron	31, 32, 33, 36, 37
Tomato	chloronerva	Nicotianamine auxotroph; expresses iron deficiency symptoms regardless of iron status of plant	30, 34, 35
Tomato	fer (T3238fer)	Cannot induce iron deficiency stress responses	38, 39

semantics in some instances. For example, Weiss[26] found a single gene difference between the iron-inefficient soybean line T203 and the iron-efficient line Hawkeye. Subsequent research on soybean has indicated that although one primary gene confers susceptibility to iron deficiency, there are other genes which contribute to the trait.[27] It has recently been hypothesized that the resistance of Hawkeye soybean to iron deficiency is correlated with an ability to accumulate a large pool of extracellular root iron which can be mobilized to shoots as the plants become iron deficient.[28] There is much less known about the differences between the efficient and inefficient varieties of the Strategy II plants, oat and corn. The difference between the iron-efficient and iron-inefficient varieties of oat appears to be the ability to release phytosiderophores under iron deficiency; Coker 227 releases a phytosiderophore and TAM 0-312 does not.[29] A preliminary comparison of ys_1 and WF9 corn did not detect phytosiderophore production by either variety, but rather showed that WF9 corn released hydrogen ions and showed some ability to reduce Fe(III).[29]

There are two well-studied mutants, the chloronerva mutant of tomato[30] and the brz mutant of pea,[31-33] which accumulate more iron than their wild-type parents. The chloronerva mutant is a nicotianamine auxotroph which was spontaneously derived from the cultivar Bonne Beste. This mutant exhibits typical iron-deficiency responses even in the presence of iron. That is, it shows increased ferric reduction, rhizosphere acidification, and accumulation of cit-rate.[30,34] Consequently, it takes up more iron than the wild type, which is seen as an accumulation of iron in older leaves. Application of nicotianamine to the roots or leaves of mutant *chloronerva* plants leads to recovery;[35] however, the link between the lack of nicotianamine and increased iron uptake is still unknown. Because nicotianamine is a chelator for Fe(II) at physiological pH, Pich et al.[34] have recently proposed that iron bound to nicotianamine may be the "active" form of iron that binds to an iron sensor or gene regulator. Thus, the mutant which has no nicotianamine suffers from apparent iron deficiency and fails to repress inducible iron uptake processes.

The *brz* mutant of pea was originally identified in a mutant hunt for nodulation-defective plants among EMS treated seeds.[32] The leaves in the mutant *brz* line, E107, developed bronze necrotic spots and were found to contain 50-fold more iron than leaves from the parental 'Sparkle' plants. The basis for the excess Fe accumulation appears to be higher rates of ferric reduction; E107 plants had high rates of ferric reduction regardless of plant iron status.[33] They also released more protons than the wild type. There does not appear to be any increase in the activity of the iron transport system in E107 relative to the wild type.[36] Guinel and LaRue[37] have recently reported that stress-related ethylene production by E107 may be responsible for the nodula-tion defect. They speculate that ethylene production may be enhanced in the mutant due to its increased capacity to lower the rhizosphere pH. Thus the nodulation defect is probably an indirect effect of the *brz* mutation.

Another interesting mutant is the *fer* mutant of tomato, which develops severe chlorosis on normal soils because it is unable to respond to iron deficiency.[38] Bienfait[39] has proposed that the wild-type gene, *FER*, encodes a positive activator protein. He has identified at least two membrane proteins which are produced under iron-deficiency conditions in the wild type which are not produced in the *fer* mutant. He has hypothesized that proteins such as ferric reductase would be under the control of the FER protein. Whether one of the proteins identified by Bienfait is the plasma membrane ferric reductase remains to be determined.

ARABIDOPSIS THALIANA AS A MODEL SYSTEM TO STUDY IRON UPTAKE

Advantages of Using *Arabidopsis thaliana* as a Model System to Study Iron Uptake

The use of *Arabidopsis thaliana* (L.) Heynh. as a model system for plant classical and molecular genetic studies has many advantages which have been discussed in detail.[40,41] *Arabidopsis* is now widely used as an experimental organism, and elucidation of the structure and the DNA sequence of its genome is the focus of a multinational effort. *Arabidopsis* is a weed in the mustard family. The plant is small and hardy, with a rapid generation time (6 to 8 weeks). It is a self-fertilizing diploid, so segregation of recessive mutations to the homozygous state is straightforward, and it produces abundant seeds (10,000 per plant), facilitating genetic studies entailing large numbers. The genome is small (100 Mb) and simple, with small gene families, few introns, and a small proportion of repetitive DNA, offering advantages over other plant species for molecular genetic studies. The genetic map of *Arabidopsis* includes more than 120 loci, defined by visible and biochemical mutations on five linkage groups;[42,43] physical maps of approximately 360 restriction fragment length polymorphisms (RFLPs) are being integrated with the genetic map,[43-45] and a complete contiguous physical map of the genome is being assembled in several laboratories.[43] In addition, a new map based on randomly amplified polymorphic DNA (RAPD) markers and recombinant inbred lines has been constructed.[46] These maps will facilitate the cloning of specific genes which have been identified by mutation via chromosome walking. In addition, a population of mutants generated by T-DNA insertion[47] are available for screening, and they have been used by several researchers to clone their favorite genes, including the homeotic gene *agamous*.[48] Genomic subtraction is another method available to clone genes identified by deletion mutations in *Arabidopsis*.[49] Tissue culture and regeneration of fertile shoots[50] and transformation by direct DNA transfer[51] and by *Agrobacterium*[52] are now routine.

Molecular Biological Approaches

A gene, designated *FRE1*, which may encode a ferric reductase, has been cloned from the yeast *Saccharomyces cerevisiae*. We will briefly describe what is known about the yeast *FRE1* gene as we are now working with three clones which show homology to *FRE1* (see below for further discussion of the clones). *FRE1* was originally cloned by complementation of a yeast mutant (*fre1-1*) which lacked plasma membrane ferric reductase activity.[53] The *fre1-1* mutant was also found to be deficient in the uptake of Fe(III), but not in the uptake of Fe(II). These phenotypes are consistent with yeast having a ferrous iron transporter. That is, Fe(III) must first be reduced to Fe(II) before it can be transported. A mutant which cannot reduce iron would not produce any ferrous iron for transport. Indeed, genetic analysis demonstrated that both the inability to reduce iron and the inability to transport Fe(III) were due to the mutation of a single gene in the *fre1-1* mutant. Sequence analysis of the clone which could complement the *fre1-1* mutant revealed a single long open reading frame (ORF) of 686 amino acids with a predicted molecular mass of 78.8 kDa.[54] The ORF has limited homology to the large subunit of human cytochrome b_{558}. This cytochrome is a component of the respiratory burst oxidase of neutrophils and is the protein affected in the X-linked form of chronic granulomatous disease. The FRE1 protein is expected to be a membrane-bound protein and the predicted protein has both a leader peptide and potential transmembrane domains. The first 22 amino acids of the predicted FRE1 protein conform to the von Heijne consensus for the leader peptide of a membrane or secreted protein. There are two amino-terminal hydrophobic regions that are strong candidates for trans-membrane domains; there are also five other hydrophobic regions that may cross the membrane. Expression of *FRE1* is negatively regulated by iron at the transcriptional level.[54] A 85-base-pair segment has been identified in the 5' noncoding region of *FRE1* which is involved in iron regulation.[54] There is a direct repeat (TTTTTGCTCAYC) in this 85-bp segment which is proposed to be a candidate binding site for an iron-responsive regulatory protein.

Because a number of *Arabidopsis* genes have been cloned using genes from yeast as hybridization probes,[55-57] we decided to try to clone *Arabidopsis* homologs using the yeast gene as a probe. Using a 2.0-kb *Eco*RI/*Sac*I fragment which includes most of the *FRE-1* coding region as hybridization probe to screen two different lambda libraries of *Arabidopsis* genomic DNA (ecotypes Landsberg and Columbia), we have identified three different clones which show homology to *FRE-1* (Figure 1). λYY1 and λYY2 were only found in the Columbia library, whereas λYY3 was found in both libraries. We have deter-mined the nucleotide sequence of the region in λYY3 which is homologous to *FRE-1,* and have partial sequence information for the regions of λYY1 and λYY2 which are homologous to *FRE-1.* λYY3 carries an open reading frame which encodes 286 amino acids. This deduced amino acid sequence shares 17.5% identity and 49% similarity with the deduced FRE protein. The inserts carried by the three lambda clones show approximately 40% identity to each other. Whether ferric reductase is encoded by a small gene family remains to be determined. We are in the process of identifying cDNA clones correspond-ing to our three genomic clones from root cDNA libraries. This will allow us to determine the exon/intron structure of the gene(s) in question. We are also in the process of determining whether any or all of our three clones are regulated by iron at the level of mRNA abundance.

Genetic Approaches

As discussed previously, several different mutants with altered iron uptake ability have been described to date. We are screening mutagenized populations of *Arabidopsis* for mutants which either have increased or decreased capacity to acquire iron relative to wild type. Many phenotypes can be observed by screening as few as 2000 M2 individuals. (M2 refers to the generation grown from seed collected from a population of mutagenized seed which was allowed to grow and self fertilize. Thus, in the M2 seed, recessive mutations will have segregated to homozygosity in a proportion of the seed.) We are particularly interested in identifying mutants with altered ferric reductase activity because, as discussed earlier, ferric reduction is probably the primary factor in making iron available for absorption. There are three major classes of ferric reductase

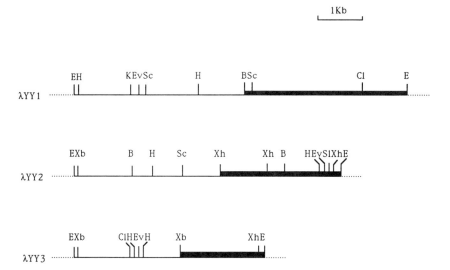

Figure 1.   Restriction endonuclease maps of the relevant *Eco*RI fragments carried by the three
λ clones which were identified using the yeast *FRE-1* gene as a hybridization probe.
The region of cross hybridization is indicated by a black box: B, *Bam*HI; Cl, *Cla*I; E,
*Eco*RI; Ev, *Eco*RV; H, *Hind*III; K, *Kpn*I; Sc, *Sac*I; Sl, *Sal*I; Xb, *Xba*I; Xh, *Xho*I.

mutants we may identify: (1) mutants which have a defect in ferric reductase
itself, (2) mutants which are defective for a gene which controls the reductase,
and (3) mutants which may exhibit some type of root defect which prevents
proper expression of the reductase. We can set up our screens to look for loss
of activity or improper expression of activity relative to the wild type.

We measure ferric reductase activity of *Arabidopsis* via the FerroZine [3-
(2-pyridyl)-5,6-diphenyl-1,2,4-triazine sulphonate] assay. This assay is based
on the affinity of the indicator dye FerroZine for Fe(II).[58] We grow plants under
iron-deficient or iron-sufficient conditions, and then measure ferric reductase
activity by transferring the plants to wells of microtiter dishes which contain
FerroZine and Fe(III)EDTA. FerroZine forms a purplish complex with Fe(II)
but not with Fe(III). Therefore, as the Fe(III)EDTA is reduced, a colored
complex will form. This is a sensitive assay; FerroZine/Fe(II) has a molar
extinction coefficient of 28,600 at 562 n*M*.[58]

To date, we have screened 1500 EMS-mutagenized M2 plants for those
which express ferric reductase in the presence of iron. We have identified two
presumptive mutants (putants) whose progeny we are in the process of re-
screening. We can also screen for ferric reductase mutants which express ferric
reductase regardless of the iron status of the plant by using an "iron overload"
screen. The idea is that plants which express ferric reductase constitutively will
take up too much iron and show symptoms of iron stress, much like the *brz*
mutant of pea. We have characterized the response of wild-type *Arabidopsis* to
increasing levels of iron and have seen leaf "freckling" when iron levels exceed
200 μ*M*. We are now screening for mutants which express freckling or chlorosis

symptoms when wild-type plants do not. We have identified 27 putants from a screen of 3200 M2 plants. We are also screening the T-DNA tagged lines in hopes of identifying a tagged mutant (for a description of creating and tagging mutations via T-DNA see Reference 59). It is important to point out, however, that at present only 10 to 40% of the mutations which have been identified by screening the T-DNA lines have turned out to be tagged.[60]

In addition to the FerroZine and iron overload screens, we are screening for mutants which have lost the ability to acidify the rhizosphere using a simple plate assay. We will also look for mutants which acidify regardless of iron status.

Any confirmed mutants will be crossed with wild type to determine the mode of inheritance as well as the dominance relationship of the mutant alleles. Mutants with recessive mutations will be crossed among themselves in order to determine the number of complementation groups. Selected mutants will also need to be back-crossed to wild type in order to remove any extraneous mutations from the background. Mutations will be mapped relative to known markers and the genes identified by the mutations will be cloned. Even with the short generation time, the genetic analysis is still very time consuming.

SUMMARY

We hope to arrive at a molecular understanding of how ferric reduction in particular is regulated by iron availability, and to determine if and how iron may regulate its own uptake, storage, and utilization. An investigation of the iron nutrition of plants is particularly challenging given the complex chemistry of iron in soil and the numerous factors which are known to influence iron availability. Iron deficiency is a fundamental cause of severe yield losses for many crops and iron overload can be an equally severe problem. By cloning the ferric reductase gene, and by identifying different mutations which affect the plant's ability to respond to iron, we hope to attain a better understanding of plant iron nutrition. A study of iron uptake in *Arabidopsis* should have application to general transmembrane permeation processes and their regulation, and may ultimately lead to the construction of iron-efficient plants.

ACKNOWLEDGMENTS

We thank Carolyn Riley for her help with the development of the iron overload screen and for the countless hours she has spent performing ferric reductase assays. We thank Rob McClung for critically reading the manuscript. The research in our laboratory on iron uptake in *Arabidopsis* is supported by grants from the National Science Foundation (IBN91-10080) and from USDA (91-02987).

REFERENCES

1. Lindsay, W.L. and Schwab, A.P., The chemistry of iron in soils and its availability to plants, *J. Plant Nutr.*, 5, 821, 1982.
2. Morales, F., Abadia, A., and Abadia, J., Characterization of the xanthophyll cycle and other photosynthetic pigment changes induced by iron deficiency in suger beet (*Beta vulgaris* L.) *Plant Physiol.*, 94, 607, 1990.
3. Chaney, R.L., Bell, P.F., and Coulombe, B.A., Screening strategies for improved nutrient uptake and use by plants, *Hortic. Sci.*, 24, 565, 1989.
4. Foy, C.D., Chaney, R.L., and White, M.C., The physiology of metal toxicity in plants, *Annu. Rev. Plant Physiol.*, 29, 511, 1978.
5. Römheld, V., Different strategies for iron acquisition in higher plants, *Physiol. Plant.*, 70, 231, 1987.
6. Takagi, S., Nomato, K., and Takemoto, T., Physiological aspect of mugineic acid, a possible phytosiderophore of gramineceous plants, *J. Plant Nutr.*, 7, 469, 1984.
7. Römheld, V. and Marschner, H., Evidence for a specific uptake system for iron phytosiderophores in roots of grasses, *Plant Physiol.*, 80, 175, 1986.
8. Scholz, G., Becker, R., Stephen, U.W., Rudolph, A., and Pick, A., The regulation of iron uptake and possible functions of nicotianamine in higher plants, *Biochem. Physiol. Pflanz.*, 183, 257, 1988.
9. Stephen, U.W. and Scholz, G., Nicotianamine concentrations in iron sufficient and iron deficient sunflower and barley roots, *J. Plant Physiol.*, 136, 631, 1990.
10. Chaney, R.L., Brown, J.C., and Tiffin, L.O., Obligatory reduction of ferric chelates in iron uptake by soybeans, *Plant Physiol.*, 50, 208, 1972.
11. Bienfait, H.F., Bino, R.J., van der Bliek, A.M., Duivenvoorden, J.F., and Fontaine, J.M., Characterization of ferric reducing activity in roots of Fe deficient *Phaseolus vulgaris*, *Physiol. Plant.*, 59, 196, 1983.
12. Römheld, V., Müller, C., and Marschner, H., Localization and capacity of proton pumps in roots of intact sunflower plants, *Plant Physiol.*, 76, 603, 1984.
13. Zocchi, G. and Cocucci, S., Fe uptake mechanism in Fe-efficient cucumber roots, *Plant Physiol.*, 92, 908, 1990.
14. Landsberg, E.C., Transfer cell formation in the epidermis: a prerequisite for Fe-efficiency, *J. Plant Nutr.*, 5, 415, 1982.
15. Landsberg, E.C., Regulation of iron stress response by whole plant activity, *J. Plant Nutr.*, 7, 609, 1984.
16. Olsen, R.A., Brown, J.C., and Bennett, J.N., Reduction of Fe^{3+} as it relates to Fe chlorosis, *J. Plant Nutr.*, 5, 433, 1982.
17. Buckhout, T.J., Bell, P.F., Luster, D.G., and Chaney, R.L., Iron stress-induced redox activity in tomato (*Lycopersicum esculentum* Mill.) is localized on the plasma membrane, *Plant Physiol.*, 90, 151, 1989.
18. Römheld, V. and Marschner, H., Mechanism of iron uptake by peanut plants. I. Fe(III) reduction, chelate splitting and release of phenolics, *Plant Physiol.*, 71, 949, 1983.
19. Holden, M.J., Luster, D.G., and Chaney, R.L., Enzymatic iron reduction at the root plasma membrane: partial purification of the NADH Fe chelate reductase, in *Biochemistry of Metal Micronutrients in the Rhizosphere*, Lewis Publishers, Chelsea, MI, 1994, chap. 18.

20. Bienfait, H.F., Regulated redox processes at the plasmalemma of plant root cells and their function in iron uptake, *J. Bioenerg. Biomember.*, 17, 73, 1985.

21. Sijmons, P.C., Bienfait, H.F., and van den Briel, W., Cytosolic NADPH is the electron donor for extracellular Fe(III) reduction in iron-deficient bean roots, *Plant Physiol.*, 75, 219, 1984.

22. Alcéntera, E., de la Guardia, M.D., and Romera, F.J., Plasmalemma redox activity and H^+ extrusion in roots of Fe-deficient cucumber plants, *Plant Physiol.*, 96, 1034, 1991.

23. Olsen, R.A., Clark, R.B., and Bennett, J.H., The enhancement of soil fertility by plant roots, *Am. Sci.*, 69, 378, 1981.

24. Marschner, H. and Römheld, V., *In vivo* measurement of root-induced pH changes at the soil-root interface: effect of plant species and nitrogen source, *Z. Pflanzenphysiol.*, 111, 241, 1983.

25. Tingey, D.T., Raba, S., Rodecap, K.D., and Wagner, J.J., Vermiculite, a source of metals for *Arabidopsis thaliana, J. Am. Soc. Hortic. Sci.*, 107, 465, 1982.

26. Weiss, M.G., Inheritance and physiology of efficiency in iron utilization in soybeans, *Genetics,* 28, 253, 1943.

27. Fehr, W.R., Current practices for correcting iron deficiency in plants with emphasis on genetics, *J. Plant Nutr.*, 7, 347, 1984.

28. Longnecker, N. and Welch, R.M., Accumulation of apoplastic iron in plant roots: a factor in the resistance of soybeans to iron deficiency induced chlorosis?, *Plant Physiol.*, 92, 17, 1990.

29. Brown, J.C. and Jolley, V.D., Plant metabolic responses to iron deficiency stress, *BioScience,* 39, 546, 1989.

30. Stephen, U.W. and Grün, M., Physiological disorders of the Nicotianamine-auxotroph tomato mutant *chloronerva* at different levels of iron nutrition. II. Iron deficiency response and heavy metal metabolism, *Biochem. Physiol. Pflanz.*, 185, 189, 1989.

31. Welch, R.M. and LaRue, T.A., Physiological characteristics of Fe accumulation in the 'bronze' mutant of *Pisum sativum* (L.), CV 'sparkle' E107 *(brz brz), Plant Physiol.*, 93, 723, 1990.

32. Kneen, B.E., LaRue, T.A., Welch, R.M., and Weeden, N.F., Pleiotropic effects of *brz,* a mutation in *Pisum sativum* (L.) CV 'sparkle' conditioning decreased nodulation and increased iron uptake and leaf necrosis, *Plant Physiol.*, 93, 717, 1990.

33. Grusak, M.A., Welch, R.M., and Kochian, L.V., Physiological characterization of a single-gene mutant of *Pisum sativum* exhibiting excess iron accumulation. I. Root iron reduction and iron uptake, *Plant Physiol.*, 93, 976, 1990.

34. Pich, A., Scholz, G., and Seifert, K., Effect of nicotianamine on iron uptake and citrate accumulation in two genotypes of tomato, *Lycopersicon esculentum* Mill, *J. Plant Physiol.*, 137, 323, 1991.

35. Scholz, G., Schlesier, G., and Seifert, K., Effects of nicotianamine on iron uptake by the tomato mutant 'chloronerva', *Physiol. Plant.*, 63, 99, 1985.

36. Grusak, M.A., Welch, R.M., and Kochian, L.V., Does iron deficiency in *Pisum sativum* enhance the activity of the root plasmalemma iron transport protein?, *Plant Physiol.*, 94, 1353, 1990.

37. Guinel, F.C. and LaRue, T.A., Ethylene inhibitors partly restore nodulation to pea mutant E107 *(brz), Plant Physiol.*, 99, 515, 1992.

38. Brown, J.C., Chaney, R.L., and Ambler, J.E., A new tomato mutant inefficient in the transport of iron, *Physiol. Plant.*, 25, 48, 1971.

39. Bienfait, H.F., Proteins under the control of the gene for Fe efficiency in tomato, *Plant Physiol.*, 88, 785, 1988.

40. Meyerowitz, E.M., *Arabidopsis thaliana, Annu. Rev. Genet.*, 21, 93, 1987.

41. Meyerowitz, E.M., *Arabidopsis,* a useful weed, *Cell,* 56, 263, 1989.

42. Koornneef, M., *A Compilation of Linkage and Restriction Maps of Genetically Studied Organisms,* O'Brian, S.J., Ed., Cold Spring Harbor Laboratory, Cold Spring Harbor, N.Y., 1987.

43. Anon., The Multinational Coordinated *Arabidopsis thaliana* Genome Research Project, Progress Report, Year Two, National Science Foundation, Washington, D.C., 1992.

44. Chang, C., Bowman, J.L., DeJohn, A.W., Lander, E.S., and Meyerowitz, E.M., Restriction fragment length polymorphism linkage map for *Arabidopsis thaliana, Proc. Natl. Acad. Sci. U.S.A.*, 85, 6856, 1988.

45. Nam, H.G., Giraudat, J., den Boer, B., Moonan, F., Loos, W.D.B., Hauge, B.M., and Goodman, H.M., Restriction fragment length polymorphism linkage map of *Arabidopsis thaliana, Plant Cell,* 1, 699, 1989.

46. Reiter, R.S., Willaims, J.G.K., Feldmann, K.A., Rafalski, J.A., Tingey, S.V., and Scolnik, P.A., Global and local genome mapping in *Arabidopsis thaliana* by using recombinant inbred lines and random amplified polymorphic DNA, *Proc. Natl. Acad. Sci. U.S.A.*, 89, 1477, 1992.

47. Feldmann, K.A. and Marks, M.D., Agrobacterium-mediated transformation of germinating seeds of *Arabidopsis thaliana:* a non-tissue culture approach, *Mol. Gen. Genet.*, 209, 1, 1987.

48. Yanofsky, M.F., Ma, H., Bowman, J.L., Drews, G.N., Feldmann, K.A., and Meyerowitz, E.M., The protein encoded by the *Arabidopsis* homeotic gene *agamus* resembles transcription factors, *Nature,* 346, 35, 1990.

49. Straus, D. and Ausubel, F.M., Genomic subtraction for cloning DNA corresponding to deletion mutations, *Proc. Natl. Acad. Sci. U.S.A.*, 87, 1889, 1990.

50. Estelle, M.A. and Somerville, C.R., The mutants of *Arabidopsis, Trends Genet.*, 2, 89, 1986.

51. Damm, B., Schmidt, R., and Willmitzer, L., Efficient transformation of *Arabidopsis thaliana* using direct gene transfer to protoplasts, *Mol. Gen. Genet.*, 217, 6, 1989.

52. Valvekens, D., Van Montegu, M., and Van Lijsebettens, M., *Agrobacterium tumefaciens* — mediated transformation of *Arabidopsis thaliana* root explants by using kanamycin selection, *Proc. Natl. Acad. Sci. U.S.A.*, 85, 5536, 1988.

53. Dancis, A., Klausner, R.D., Hinnebusch, A.G., and Barnocenal, J.G., Genetic evidence that ferric reductase is required for iron uptake in *Saccharomyces cerevisiae, Mol. Cell Biol.*, 10, 2294, 1990.

54. Dancis, A., Roman, D.G., Anderson, G.J., Hinnebusch, A.G., and Klausner, R.D., Ferric reductase of *Saccharomyces cerevisiae:* molecular characterization, role in iron uptake, and transcriptional control by iron, *Proc. Natl. Acad. Sci. U.S.A.*, 89, 3869, 1992.

55. Mazur, B.J., Chui, C.-F., and Smith, J.K., Isolation and characterization of plant genes coding for acetolactate synthase, the target enzyme for two classes of herbicides, *Plant Physiol.*, 85, 1110, 1987.

56. Learned, R.M. and Fink, G.R., 3-Hydroxy-3-methylglutaryl-coenzyme A reductase from *Arabidopsis thaliana* is structurally distinct from the yeast and animal enzymes, *Proc. Natl. Acad. Sci. U.S.A.,* 86, 2779, 1989.

57. Berlyn, M.B., Last, R.L., and Fink, G.R., A gene encoding the tryptophan synthase beta subunit of *Arabidopsis thaliana, Proc. Natl. Acad. Sci. U.S.A.,* 86, 4604, 1989.

58. Gibbs, C.R., Characterization and application of FerroZine iron reagent as a ferrous iron indicator, *Anal. Chem.,* 48, 1197, 1976.

59. Koncz, C., Schell, J., and Redei, G.P., T-DNA transformation and insertion mutagenesis, in *Methods in Arabidopsis Research,* Koncz, C., Chua, N.-H., and Schell, J., Eds., World Scientific Pub., River Edge, N.J., 1992, 224.

60. Koncz, C., Neweth, G.P., and Schell, J., T-DNA insertional mutagenesis in *Arabidopsis, Plant Mol. Biol.,* 20, 963, 1992.

CHAPTER 20

The Role of Root Apoplast Acidification by the H⁺ Pump in Mineral Nutrition of Terrestrial Plants

Jean-Baptiste Thibaud, Hervé Sentenac, and Claude Grignon

INTRODUCTION: CONCENTRATION GRADIENTS IN THE RHIZOSPHERE

As a result of the net exchange of materials between roots and their external medium, steady-state concentration gradients are set up perpendicular to the root surface. Both the depletion of some ions, and the accumulation of others in the soil around roots are well documented.[1] Depletion results from the net uptake of solutes with a low concentration and/or a low diffusion rate in the soil. Accumulation may be due either to excess transport to the root surface by convection (due to the transpiration flow), or to the restriction of diffusion in the soil of solutes excreted by the roots. These phenomena, especially depletion zones, are known to limit root performance in tapping the soil for mineral resources.[2]

In plant cells, a pH gradient is maintained between the medium and the cytoplasm by active transport systems which use metabolic energy to extrude H⁺ from the cells. This pH gradient is central to plant mineral nutrition, because it is the energy source used by most membrane systems (the so-called H⁺-cotransporters) to drive solute transport. Since the cytoplasmic pH is strictly controlled,[3] and its variations are normally restricted to a rather narrow range,[4] the size of the pH gradient at the plasma membrane is determined by the external pH.

This is especially true for the cells at the root surface which are surrounded by the soil solution. Since these cells apparently rely on the transmembrane pH gradient to drive their transporters, how do they cope with external pH variations?

0-87371-942-5/94/$0.00+$.50

In contrast to animal cells, plant cells do not live in an environment homeostatically controlled by the organism. However, in both the shoots and the roots of terrestrial plants the cells are in contact with a local environment they modify, and may perhaps control. This environment comprises the unstirred layer (UL) of solution inside the cell walls, and in the case of roots, the region of the rhizosphere which is modified by root activity. Measurements of the pH of the rhizosphere (outside the cell walls) with selective electrodes, or pH dyes, indicate that it may differ notably from its value in the bulk soil solution. An abundant literature is available on the pH profiles along the roots,[5-8] and on their changes in response to nitrogen uptake,[9] mineral deficiency,[10] etc. The pH of the rhizosphere is considered to be a key factor for root uptake efficiency for two reasons: it may affect mobilization of soil mineral resources,[11] and it may control the activity of the membrane transporters.[12]

In solid or gelosed media, the pH around the root is determined by the net exchange of H^+ equivalents between the root and the medium. Thus, the acidification or alkalinization of the rhizosphere (typically over a distance of one to several millimeters) depend on whether there is a net production or consumption of H^+ by root metabolism.[13] Exploration of regions much closer to the root surface (typically a few micrometers) has been made possible with new ion-selective microelectrodes.[14] In liquid medium, provided that convection is reduced (e.g., no stirring nor percolation), an unstirred layer of solution is established around the roots, extending up to several tens of micrometers from their surfaces. By placing ion-specific microelectrodes within the UL one can measure local ionic activities. From such data, applying Fick's law to describe the ion diffusion within the UL, several authors have derived the net flux rate of the ion of interest (H^+ and K^+,[15,16] and NO_3^- and NH_4^+[17,18]). Mature zones generally present a net H^+ excretion, while meristematic zones are sinks for H^+. Thus, a continuous cycling of H^+ may allow an acidification of the root surface in mature regions, even in the absence of net H^+ production at the root level.

The internal frontier of the rhizosphere is inside the cell walls of the external cortical root cells, at the outer surface of their plasma membrane. Ion concentration gradients steeper than those in the rhizosphere are expected in the apoplast, due to the restriction of diffusion by the solid phase of cell walls, and to electrostatic interactions with their immobilized charges.[19] However, this region is not accessible to microelectrodes, and indirect approaches have been necessary to probe the pH in the vicinity of the membrane transporters.

A first indication of the importance of surface acidification was obtained by studying Pi transport. Corn roots absorb Pi probably via a $H^+/H_2PO_4^-$ cotransporter.[20] Measurements of Pi influx under conditions where H^+, but not $H_2PO_4^-$ were limiting in the medium showed that the $H_2PO_4^-$ influx was strongly inhibited if the proton pump was blocked.[21] Since this treatment did not modify $H_2PO_4^-$ uptake at low pH values, it was concluded that the H^+ used by $H_2PO_4^-$transport system could originate either from the medium (when abundant in this phase), or from the H^+ pump. This implies that the pH at the proton pump maintains a low pH at the membrane surface, even if the medium pH is high.

PROBING THE pH WITHIN THE APOPLAST

In order to probe the pH at the cell surface (inside the apoplast), we measured the influx rate of acetic acid (Jaa). The basis of this method has been detailed elsewhere.[22] Briefly, we first showed that this weak acid is absorbed in the neutral protonated form, termed AH (thus no effect on the membrane potential). The acetic acid influx is not saturable (probably because it is purely diffusive), and is probably a linear function of the concentration of AH at the cell surface. Therefore given a fixed bulk acetic acid concentration, any steady-state value of Jaa is related to a single set of values of surface concentration of AH and surface pH.

Figure 1 summarizes the three steps for evaluation of the surface pH shift due to H^+ excretion. First (top panel, "control"), root cell H^+-ATPases excrete H^+ (black dots); the bulk pH is 6, and the buffering power of the medium is negligible. H^+ accumulate within the cell walls, so the surface concentration of AH is increased and Jaa is high (thick inward arrow). Then (central panel, "vanadate"), vanadate is added to the medium to inhibit the H^+-ATPases *without changing anything else in the medium* (pH, acetic acid concentration, buffering power), H^+ are no longer accumulated within the cell walls, the surface concentration of AH is low, and Jaa is reduced (thin inward arrow). Finally (bottom panel, "vanadate and low pH in the medium"), in the presence of vanadate, the bulk medium pH is progressively lowered until Jaa is restored to its initial value (that in the absence of vanadate). It is inferred that, at this point, surface concentration of AH and surface pH have been artificially restored to the value that was maintained by active H^+ extrusion. In this example, the bulk pH value in the bottom panel is 4.8. So we assume that in the top panel the surface pH shift due to H^+ excretion by H^+-ATPases is equal to $6 - 4.8$, i.e., to 1.2 pH units.

Figure 2A shows values of Jaa obtained at different bulk pH values in the absence or presence of vanadate. Curves fit the data by polynomial adjustment. The so-called δpH (cell surface pH shift due to H^+ excretion by H^+-ATPase) at any bulk pH value is given by the difference, along the x-axis direction, between the two curves. Figure 2B shows δpH as a function of bulk pH. Numbered circles indicate experimental conditions shown in Figure 1, and dashed lines show the graphic estimation of δpH at bulk pH 6. It is clear that if bulk pH is sufficiently low then, as expected, H^+ excretion by H^+-ATPases does not result in significant acidification of the apoplast. However, there is a positive pseudolinear relationship between the bulk pH value and δpH — the line in Figure 2B having a mean

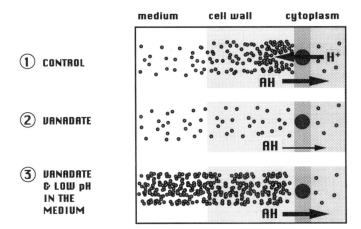

Figure 1. Schematic representation of the procedure for estimating the surface pH shift at cell surface due to H⁺ excretion by the H⁺-pump. There are three panels from top to bottom, depicting by a change in shading the external medium, the cell wall, the plasma membrane, and the cytoplasm. H⁺ are represented by black dots. The plasma membrane has an H⁺-pump shown with a shaded circle. In the top panel ("control"), pH is 6.0 in the bulk medium. The H⁺-pump excretes H⁺ which accumulate in the cell wall at steady state. This so-called δpH results in a high concentration of the protonated form of acetic acid (AH) at the membrane surface. The acetic acid influx rate (Jaa) is high (thick inward arrow). In the middle panel, the bulk pH is again 6.0, the H⁺-pump is inhibited by vanadate, and H⁺ are no longer accumulated in the cell wall. Surface AH concentration is lowered and Jaa is reduced (thin inward arrow). In the bottom panel, the H⁺-pump is inhibited with vanadate, but as bulk pH was lowered to 4.8, a new steady state is reached for Jaa and AH concentration at the cell surface. The former, being the same as in the top panel, it is assumed that δpH in the top panel was of 1.2 pH units (difference between bulk pH of bottom and top panels).

slope close to 1 (in our experimental conditions, δpH exceeded 2 pH units for bulk pH > 7). Thus, provided that there is no buffer in the medium and that H⁺-pumps operate freely, the cell surface pH would never exceed a limit value of around 5. It is likely that such an "homeostasis" in cell surface pH has important consequences for plant mineral nutrition.

STUDYING THE EFFECT OF APOPLASTIC pH ON ABSORPTION OF MINERAL NUTRIENTS BY ROOTS

In order to study the effect of δpH on ion transport, it is necessary to vary this parameter without changing the rate of H⁺ extrusion by cells. Proton buffers are expected to dissipate the pH gradient between the membrane surface and the bulk phase (inward diffusion of the unprotonated buffer form, and outward diffusion of the protonated form across the cell walls would act as a shuttle for H⁺).[23] Adding buffers to the medium while maintaining the bulk pH constant within 0.01 of a pH unit with an automatic pH-stat system allowed us to verify this prediction.

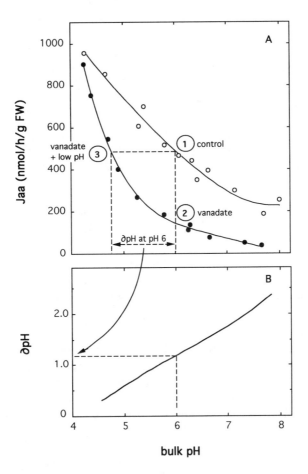

Figure 2.   Determination of the surface pH shift due to H⁺ excretion in excised corn roots. **A** (top
panel): effects of vanadate on acetic acid influx (Jaa) into corn roots. All media
contained 0.2 mM CaSO₄, 50 µM K₂SO₄, and 50 µM acetic acid. [2-¹⁴C]acetic acid
was added to the medium 5 min after the excised roots had been introduced into the
medium. Incorporation was carried out for 10 min. (○), control treatment: neither
vanadate nor buffers. (•), vanadate (NaVO₃) was added at 0.2 mM, and media
contained 5 mM Mes-Tris (pH<6.8), or 5 mM Hepes-Tris (pH>6.8). The pH was kept
constant with an automatic pH-stat system. The data are fitted to polynomial curves.
Numbered circles point out experimental situations sketched in Figure 1 (1 to 3 for top
to bottom). **B** (bottom panel): surface pH shift due to H⁺ excretion (δpH) expressed
as a function of the bulk external pH. δpH was estimated from the shift of the Jaa curve
along the x-axis due to addition of vanadate + buffers. (Data from Thibaud, J.-B. et
al., *Plant Physiol.,* 88, 1469, 1988 and Sentenac, H. and Grignon, C., *Plant Physiol.,*
84, 1367, 1987, with permission.)

It is important to note that upon addition of buffers in the medium, the net
H⁺ efflux was slightly increased and the membrane potential was slightly more
negative;[21] thus, the activity of the electrogenic H⁺-pump was probably not
reduced.

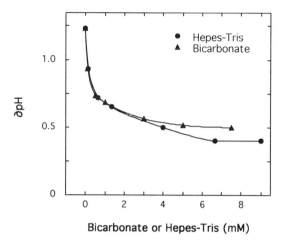

Figure 3. Effect of bicarbonate and of Hepes-Tris on surface pH shift due to H^+ excretion (δpH) in excised corn roots. δpH was estimated at pH 7.0, by comparing Jaa values at pH 7.0 in the presence of HCO_3^- (▲) or Hepes-Tris (•) to those measured at different pH values in the presence of vanadate, following the procedure described in Figures 1 and 2 and Reference 22. (Data from Toulon, V., Sentenac, H., Thibaud, J.-B., Soler, A., Clarkson, D. T., and Grignon, C., *Planta*, 179, 235, 1989, with permission.)

Figure 3 shows that increasing bulk concentrations of either Hepes (pKa $\approx$ 7.55) or HCO_3^- (pKa $\approx$ 6.35), while keeping a constant bulk pH at 7.00, did decrease δpH. A 70% inhibition of δpH was obtained at a 10 mM concentration of either of these buffers.[21,24]

Using vanadate (no H^+ excretion) or buffers (high H^+ excretion but accelerated H^+ diffusion to the medium) to inhibit δpH, we demonstrated the physiological importance of δpH on transport systems in root cells.[21] For instance, when $(^{32}P)H_2PO_4^-$ influx was measured at various bulk pH values and at a constant concentration of $H_2PO_4^-$, roots could absorb Pi at a relatively high rate at pH values greater than 7 when the H^+-pump was fully activated, but not when it was inhibited by vanadate (Figure 4). The same result could be obtained by using buffers instead of vanadate, indicating that Pi transport depended on the availability of H^+ at the cell surface, rather than on the activity of the H^+-pump. These results, and data from studies of other transport systems,[21] suggest that plasma membrane-bound H^+-cotransport systems experience an almost constant cell surface pH rather than the bulk pH which we varied. Thus, the shift in pH at the root cell surface, due to the operation of the H^+-pump, ensures the energetic coupling of H^+-cotransport systems. This may be beneficial to plant mineral nutrition, particularly in alkaline conditions.

STUDYING FE REDUCTASE ACTIVITY

The findings described above prompted the study of Fe reductase activity for three main reasons.

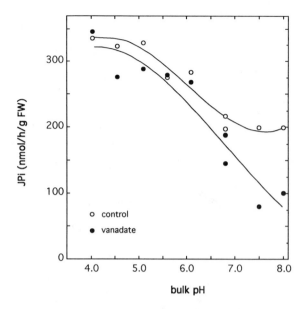

Figure 4. Effect of pH on Pi influx in the presence of vanadate and buffers. The solutions contained 50 μM K_2SO_4 and 0.2 mM $CaSO_4$. Pi was added as KH_2PO_4 to obtain 50 μM $H_2PO_4^-$ at all pH values (see Reference 20). ^{32}Pi was added to the medium 5 min after excised roots had been introduced. Uptake was carried out for 10 min. (O), control treatment: neither vanadate nor buffer. (•), vanadate ($NaVO_3$) was at 0.2 mM, and the solutions were buffered with 5 mM Mes-Tris (pH<6.8), or 5 mM Hepes-Tris (pH>6.8). The pH was kept constant with an automatic pH-stat system. Data are fitted to polynomial curves. (Data from Thibaud, J. C., Davidian, J.-C., Sentenac, H., Soler, A., and Grignon, C., *Plant Physiol.*, 88, 1469, 1988, with permission.)

Firstly, not only transport processes but also enzymatic activities may be controlled by local pH resulting from H^+-pump activity; extracellular Fe(III) reductase activity seemed a good candidate for studying this hypothesis.

Secondly, in the case of Fe depletion, the roots of Fe-efficient plants are able to enhance *both* (1) their capability to reduce Fe(III) and (2) the acidification of the rhizosphere.[11] The latter factor has been shown to improve the mobilization of ferric pools in the soil.[25] However, one may assume some physiological relationship between the two processes at the membrane/apoplast level, and postulate that an apoplastic pH shift is involved.

Finally, it is well known that in calcareous soils (characterized by both high bulk pH and high bicarbonate concentration), many plants encounter some difficulties in their Fe nutrition. It may be postulated that the apoplastic pH shift and buffer effects are involved.

Most of our results on this topic have recently been published,[26] and are summarized below.

In the following discussion, δpH is still defined as the shift in apoplastic pH due to H^+-pumping activity of H^+-ATPases. Preliminary evidence of an effect of δpH on Fe reductase activity was obtained from the measurement of the

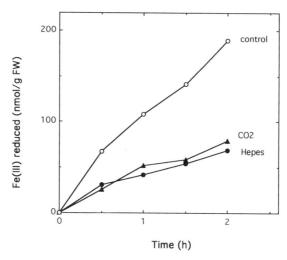

Figure 5.   Comparison between the inhibitory effect of Hepes and of CO_2/HCO_3^- on Fe(III)EDTA
reduction by Fe-deficient rape roots. Seedlings were grown for 3 d in an iron-depleted
nutrient solution. They were then pretreated in 0.2 mM $CaSO_4$ for 24h. Roots were
then excised and the rate of Fe(III)EDTA reduction was measured in a solution (40
ml/g root FW) containing 0.1 mM Fe(III)NaEDTA, 0.4 mM BPDS, 0.2 mM $CaSO_4$,
5mM K_2SO_4, 10mM glucose, and 50g.m^{-3} chloramphenicol. In each treatment, the pH
was maintained at 8.0 with an automatic pH-stat system. In the control (○) experiment,
Fe(III) reduction was measured in an unbuffered solution that was bubbled with CO_2-
free air. In the Hepes (•) treatment, 10 mM Hepes was added, and the pH was initially
adjusted to 8.0 by adding KOH. In the CO_2 (▲) treatment, 10 mM $KHCO_3$ was added,
and the partial pressure of CO_2 in the air entering the solution was held at 0.6%. Mean
values ± SD (n=3). (Data from Toulon, V., Sentenac, H., Thibaud, J.-C., Davidian,
J.-C., Moulineau, C., and Grignon, C., *Planta*, 186, 212, 1992, with permission.)

reduction of Fe(III)EDTA at pH8, by Fe-stressed rape roots, with or without
buffers added to the medium (Figure 5). It is clear that, with no time lag, the
presence of either Hepes or CO_2 (pCO_2 was 0.6%, so the calculated steady-state
HCO_3^- concentration was about 10mM) significantly decreased the reduction of
Fe(III)EDTA. It should be remembered that in each of the three cases (control,
Hepes, CO_2) bulk pH was held at 8.00 with the help of an automatic pH-stat.
The action of the buffers on Fe reduction arose from a direct effect on reduction
itself, or indirectly from their effect described above on δpH. In order to gain
evidence for the latter, we studied the buffer effect on Fe reductase activity in
a wide pH range, from 4 to 9 (Figure 6). In the absence of buffer, the rate of
Fe(III)EDTA reduction was not strongly dependent on pH in the 4 to 7 range.
However, it was increasingly inhibited when pH was increased above 7. The
inhibition of the rate of Fe(III)EDTA reduction in the presence of buffers was
calculated. Inhibition was observed at pH >6, it was near 50% at pH 7, and
exceeded 80% at pH 8. Therefore, buffers inhibited the reduction of Fe(III)EDTA
only in the pH range within which δpH was expected (see Figure 2B); furthermore, the higher the expected value of δpH, the higher the inhibition. It was

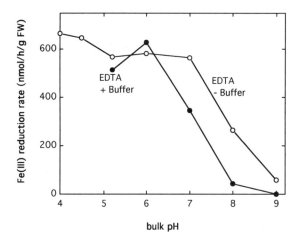

Figure 6.   Effect of pH and buffers on Fe(III)EDTA reduction by excised roots from Fe-deficient
rape. The reduction rate was measured with roots excised on seedlings pretreated for
24 h in 0.2 mM CaSO$_4$. Protocols for culture and pretreatment: see legend to Figure
5. Fe(III)EDTA reduction rate was measured in the solution described in legend to
Figure 5. "–Buffer" (○): the experimental solutions were unbuffered and their pH was
maintained at the desired value with an automatic pH-stat system. "+Buffer" (•): the
experimental solutions were buffered with 10 mM Mes at pH 5.0 or pH 6.0, with 10
mM Hepes at pH 7.0 or pH 8.0, and with 10 mM BTP at pH 9.0. Data are means of
two values. (Data from Toulon, V., Sentenac, H., Thibaud, J.-B., Davidian, J.-C.,
Moulineau, C., and Grignon, C., *Planta,* 186, 212, 1992, with permission.)

concluded that buffers inhibited Fe(III)EDTA reduction indirectly because of
their effect on δpH, rather than directly. Indeed, the inhibition was observed
with all buffers tested (HCO$_3^-$, Hepes, Tris, BTP, see Figures 5 and 6): a direct
effect of all these different molecules on Fe(III)EDTA reduction was unlikely.

In the presence of buffers, and hence with no δpH, the effect of pH on the
rate of Fe(III) reduction may be studied alone: the data relative to Fe(III)EDTA
(Figure 5) suggested that pH had an effect on the reductase itself and/or on the
concentration of the ferric substrate at the reducing site. In order to further
elucidate this question, we compared the reduction rate of Fe(III)citrate to that
of Fe(III)EDTA as a function of pH and in presence of buffers (Figure 7).
Fe(III) reduction rate depended on the chelate present in the reducing system,
whatever the pH effect on the system itself. The titration curves of the two
ferric chelates shown in Figure 8 differ in form, reflecting the different pKa
values. For example, Fe(III)EDTA has an apparent pKa near 7.2. So, when the
pH was increased from 6.2 to 8.2, this ferric chelate became more negative. The
acquisition of this charge probably led to an increased electrostatic repulsion
between the chelate and the fixed negative charges of cell walls. This may
explain, at least in part, the decreased Fe(III) reduction rates measured at pH >6
(Figure 7). An analysis of the curves in both Figures 7 and 8 led to similar
conclusions for Fe(III)citrate.

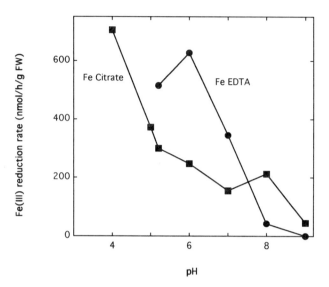

Figure 7. Effect of pH on reduction rate of Fe(III)EDTA or Fe(III)citrate by excised roots from Fe-deficient rape, in presence of buffers. Protocols for culture and pretreatment: see legend to Figure 5. Protocol for reduction rate measurement: see legend to Figure 6. Buffers were present in all solutions. Both Fe(III)citrate (■) and Fe(III)EDTA (•) were used at 0.1 mM concentration. Data are means of two values. (Data from Toulon, V., Sentenac, H., Thibaud, J.-B., Davidian, J.-C., Moulineau, C., and Grignon, C., *Planta*, 186, 212, 1992, with permission.)

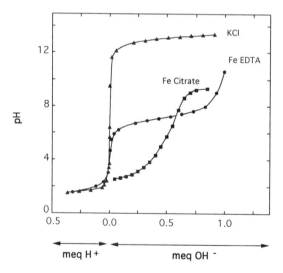

Figure 8. Titration curves of Fe(III)EDTA and of Fe(III)citrate. Titrating solutions were 1N HCl or 1N KOH. The volume of titrated solution was 50ml. "KCl" (▲), titration curve of 1mM KCl. "Fe EDTA" (•), titration curve of 1mM KCl + 10mM Fe(III)EDTA. "Fe citrate" (■), titration curve of 1mM KCl + 10 mM Fe(III)citrate. (Data from Toulon, V., Sentenac, H., Thibaud, J.-B., Davidian, J.-C., Moulineau, C., and Grignon, C., *Planta*, 186, 212, 1992, with permission.)

Fe(OH)$_3$ precipitation is a sink for iron in nutrient solutions. Chelators (EDTA, and to a lesser extent citrate) are able to keep iron available for plant nutrition provided that pH value is <6. At higher pH values and at equilibrium, significant amounts of iron may precipitate, depending on the chelator used.[27] However, Fe(OH)$_3$ formation from iron chelates is very slow, and its precipitation is a particular cause for concern for those who have to grow plants hydroponically for several days. Our experimental solutions were prepared extemporaneously, and no immediate precipitation was observed unless above pH 9 for iron citrate or above pH 11 for iron EDTA (this is the reason why titration curves are truncated in Figure 8). It is likely, however, that progressive precipitation of iron occurred during our experiments, especially for the higher pH values and more rapidly from iron citrate than from iron EDTA. As Fe(OH)$_3$ precipitation liberates H$^+$, if 10% of iron precipitated during our experiments this would have produced quantities of H$^+$ of the same order of magnitude as H$^+$ exchanged by roots with the solution. This would have increased significantly the amount of H$^+$ neutralized by the pH-stat system. Actually, net H$^+$ flux was not changed in the presence of iron chelates with respect to its control value in absence of iron chelates (not shown). Thus, if it occurred, iron precipitation was negligible within the duration of our experiments (30 min).

To summarize, (1) the pH dependency of Fe(III)EDTA or Fe(III)citrate reduction rates originated *at least in part* from variations of the electrostatic repulsion between those Fe(III) chelates and the cell wall or plasma membrane fixed charges, and (2) the pH shift within cell walls due to H$^+$ excretion (δpH) stimulated Fe(III) reduction by diminishing this electrostatic repulsion. To eliminate these electrostatic effects, an *in vitro* study was performed to address the intrinsic pH sensitivity of the reducing system (plasma membrane fractions purified from rape or corn roots).[36] When the rate of Fe(III)EDTA reduction was plotted against pH, a classical bell-shaped curve was obtained, with maximal reductase activity near pH 6 (rape roots) or 6.5 (corn roots).[28]

The conclusion of this study was that buffers inhibited Fe reductase activity via their effect on δpH, because both the enzyme activity and the electrostatic interactions between ferric chelates and apoplastic fixed charges are pH sensitive. This mechanism may be involved in ferric chlorosis in calcareous soils containing the H$_2$CO$_3$/HCO$_3^-$ buffer system.

DISCUSSION AND CONCLUSION

The role of buffers is central to the investigation of δpH, its consequence on membrane activities at the root cell surface, and the proposed effect of CO$_2$ on ferric chlorosis. For this reason, this point merits further attention. The requirement of the maintenance of electrical neutrality in both the medium and the cells, including at the membrane level, means that any net transport of H^+ is balanced by charge-compensating transport of other ions. For instance, in the case of a root exchanging H^+ and K^+, local pH within the apoplast results as well from local steady-state K^+ depletion as from H^+ excretion. Therefore, the restricted diffusion rate of K^+ (and other ions) in the apoplast is involved in δpH as well as the restricted diffusion rate of H^+ themselves. By the same token, inhibition of δpH by buffers may be easily understood: buffers compensate imbalance of strong ions thus avoiding local variation of pH. Excreted H^+ have to reach the bulk solution, however, and buffers actually participate in their diffusion.[23]

It should be noted that when δpH occurs, the electrochemical potential of H^+ at the cell surface is actually increased (so is the proton motive force, which drives H^+-cotransports). This contrasts with the relative acidity of apoplast due to electrostatic interactions (Donnan equilibrium), which does not result in an increase in proton motive force because H^+ remains at thermodynamic equilibrium.[19]

Plant roots resort to an "H^+-fueled strategy" to absorb mineral nutrients. On one hand, our results suggest that a local pH value close to the cortical cell membrane is a key factor in this strategy. On the other hand, they predict that the presence in soils of high bicarbonate concentrations may inhibit rhizosphere/apoplast acidification, thus questioning the reliability of the H^+-fueled strategy. However, it must be pointed out that our estimations of δpH are mean values, averaging possible radial heterogeneities[29] and well-demonstrated longitudinal heterogeneities[6-8] at the surface of roots. So, even in alkaline conditions, with high bicarbonate concentration/buffering capacity, there may be a pH shift at specific locations of the root system. Indeed, specialized structures originating from anatomical differentiations may allow high local acidification. They combine a high-rate excretion of H^+ with unstirred layers highly resistant to H^+ diffusion toward the bulk. The charasomes described in acid bands of *Chara* cells[30] are examples of such structures. They consist of multiple invaginations of the plasma membrane allowing local accumulation of excreted H^+. Analogous structures have been described in plants: transfer cells in *Elodea* and *Potamogeton* leaves,[31] swollen root tips and rhizodermal transfer cells of Fe-stressed plants,[32] and proteoid roots.[33] Similarly, the "sealed" apoplast at the symbiont interface in ectomycorrhiza could play an equivalent function.[12,34,35] One may ask whether such structures are induced by medium conditions, as are charasomes, which are formed under alkaline conditions.

ACKNOWLEDGMENTS

We thank Dr. Siobhan Staunton for her kind help.

REFERENCES

1. Nye, P. H. and Tinker, P. B., Solute movement in the soil-root system, *Studies in Ecology*, Vol. 4, Blackwell Scientific, Oxford, 1977, 342.
2. Nye, P. H., The rate-limiting step in plant nutrient absorption from soil, *Soil Sci.*, 123, 292, 1977.
3. Kurkdjian, A. and Guern, J., Intracellular pH: measurement and importance in cell activity, *Annu. Rev. Plant Physiol. Plant Mol. Biol.*, 40, 271, 1989.
4. Felle, H., Short-term pH regulation in plants, *Physiol. Plant.*, 74, 583, 1988.
5. Römheld, V., Müller, C., and Marschner, H., Localization and capacity of proton pumps in roots of intact sunflower plants, *Plant Physiol.*, 76, 603, 1984.
6. Marschner, H., Römheld, V., and Ossenberg-Neuhaus, H., Rapid method for measuring changes in pH and reducing processes along roots of intact plants, *Z. Pflanzenphysiol.*, 105, 407, 1982.
7. Mulkey, T. J. and Evans, M. L., Geotropism in corn roots: evidence for its mediation by differential acid efflux, *Science*, 212, 70, 1981.
8. Pilet, P. E., Versel, J. M., and Mayor, G., Growth distribution and surface pH patterns along maize roots, *Planta*, 158, 398, 1983.
9. Marschner, H. and Römheld, V., *In vivo* measurements of root-induced pH changes at the soil-root interface: effect of plant species and nitrogen source, *Z. Pflanzenphysiol.*, 111S, 241, 1983.
10. Hedley, M. J., Nye, P. H., and White, R. E., Plant-induced changes in the rhizosphere of rape (*Brassica napus* var. Emerald) seedlings. II. Origin of the pH change, *New Phytol.*, 91, 31, 1982.
11. Marschner, H., Römheld, V., Horst, W. J., and Martin, P., Root-induced changes in the rhizosphere: importance for the mineral nutrition of plants, *Z. Pflanzenernaehr. Bodenkd.*, 149, 441, 1986.
12. Clarkson, D. T., Factors affecting mineral nutrient acquisition by plants, *Annu. Rev. Plant Physiol.*, 36, 77, 1985.
13. Raven, J. A., pH regulation in plants, *Sci. Prog. (Oxford)*, 69, 495, 1985.
14. Lucas, W. J. and Kochian, L. V., Ion transport processes in corn roots: an approach utilizing microelectrode techniques, in *Advanced Agricultural Instrumentation: Design and Use*, NATO Ser. E, No. III, Gensler, W. G., Ed., Martinius Nijhoff, Boston, 1986, 402.
15. Newman, I. A., Kochian, L. V., Grusak, M. A., and Lucas, W. J., Fluxes of H^+ and K^+ in corn roots. Characterisation and stoichiometries using ion selective microelectrodes, *Plant Physiol.*, 84, 1177, 1987.
16. Kochian, L. V., Shaff, J. E., and Lucas, W. J., High affinity K^+ uptake in maize roots. A lack of coupling with H^+ efflux, *Plant Physiol.*, 91, 1202, 1989.
17. Henriksen, G. H., Bloom, A. J., and Spanswick, R. M., Measurement of net fluxes of ammonium and nitrate at the surface of barley roots using ion-selective microelectrodes, *Plant Physiol.*, 99, 734, 1992.
18. Henriksen, G. H., Raman, D. R., Walker, L. P., and Spanswick, R. M., Measurement of net fluxes of ammonium and nitrate at the surface of barley roots using ion-selective microelectrodes. II. Patterns of uptake along the root axis and evaluation of the microelectrode flux estimation technique, *Plant Physiol.*, 99, 734, 1992.

19. Grignon, C. and Sentenac, H., pH and ionic conditions in the apoplast, *Annu. Rev. Plant Physiol. Plant Mol. Biol.*, 42, 103, 1991.

20. Sentenac, H. and Grignon, C., Effect of pH on orthophosphate uptake by corn roots, *Plant Physiol.*, 77, 136, 1985.

21. Thibaud, J.-B., Davidian, J.-C., Sentenac, H., Soler, A., and Grignon, C., H^+ cotransports in roots as related to the surface pH shift induced by H^+ excretion, *Plant Physiol.*, 88, 1469, 1988.

22. Sentenac, H. and Grignon, C., Effect of H^+ excretion on the surface pH of corn root cells evaluated by using weak acid influx as a pH probe, *Plant Physiol.*, 84, 1367, 1987.

23. Price, G. D. and Badger, M. R., Inhibition by proton buffer of photosynthetic utilization of bicarbonate in *Chara corallina*, *Aust. J. Plant Physiol.*, 12, 257, 1985.

24. Toulon, V., Sentenac, H., Thibaud, J.-B., Soler, A., Clarkson, D. T., and Grignon, C., Effect of HCO_3^- concentration in the absorption solution on the energetic coupling of H^+-cotransports in roots of *Zea mays* L., *Planta*, 179, 235, 1989.

25. Bienfait, H. F., Mechanisms in Fe-efficiency reactions of higher plants, *J. Plant Nutr.*, 11, 605, 1988.

26. Toulon, V., Sentenac, H., Thibaud, J.-B., Davidian, J.-C., Moulineau, C., and Grignon, C., Role of apoplast acidification by the H^+ pump. Effect on the sensitivity to pH and CO_2 of iron reduction by roots of *Brassica napus* L., *Planta*, 186, 212, 1992.

27. Chaney, R. L., Plants can utilize iron from Fe-N,N'-di-(2-hydroxybenzoyl)-ethylenediamine-N,N'-diacetic acid, a ferric chelate with 10^6 greater formation constant than Fe-EDDHA, *J. Plant Nutr.*, 11, 1033, 1988.

28. Toulon, V., Contrôle du pH Apoplasmique de la Racine et Conséquences Sur les Activités de la Membrane Plasmique, PhD Thesis, Université Montpellier II, France, 1992.

29. Sentenac, H., Thibaud, J.-B., and Grignon, C., Surface pH, bicarbonate and H^+-cotransport coupling in corn roots, in *Plant Membrane Transport: The Current Position*, Dainty, J., de Michelis, M. I., Marrè, E., and Rasi-Caldogno, F., Eds., Elsevier, Amsterdam, 1989, 611.

30. Price, G. D., Badger, M. R., Basset, M. E., and Whitecross, M. I., Involvement of plasmalemmasomes and carbonic anhydrase in photosynthetic utilization of bicarbonate in *Chara corallina*, *Aust. J. Plant Physiol.*, 12, 241, 1985.

31. Prins, H. B. A., Snel, J. F. H., Zanstra, P. E., and Helder, R. J., The mechanism of bicarbonate assimilation by the polar leaves of *Potamogeton* and *Elodea*. CO_2 concentration at the leaf surface, *Plant Cell Environ.*, 1, 231, 1982.

32. Röhmeld, V. and Marschner, H., Iron deficiency stress induced morphological and physiological changes in root tips of sunflower, *Physiol. Plant.*, 53, 354, 1981.

33. Gardner, W. K., Parbery, D. G., and Barber, D. A., The acquisition of phosphorus by *Lupinus albus* L. I. Some characteristics of the soil/root interface, *Plant Soil*, 68, 19, 1982.

34. Ashford, A. E., Allaway, W. G., Peterson, C. A., and Cairney, J. W. G., Nutrient transfer and the fungus-root interface, *Aust. J. Plant Physiol.*, 16, 85, 1989.
35. Smith, S. E. and Smith, F. A., Structure and function of the interface in biotrophic symbioses as they relate to nutrient transport, *New Phytol.*, 114, 1, 1990.
36. Toulon, V., Sentenac, H., and Thibaud, J.-B., unpublished data.

CHAPTER 21

Salinity Alters Root Cell Wall Properties and Trace Metal Uptake in Barley

Charles G. Suhayda, Robert E. Redmann, and Xiaoyan Wang

INTRODUCTION

Salinity impacts the trace element composition of plants by (1) changing the available concentration of these elements in the soil, (2) altering their uptake by plant roots, or (3) inhibiting cell expansion and plant growth. Irrigation can enhance dissolution of trace elements into the soil solution in arid and semiarid soils. This leads to the enrichment of subsurface drainage water with certain elements,[1] compounding existing salinity problems and placing additional constraints on crop production and management of subsurface drainage water.[2] For example, in the San Joaquin Valley of California the accumulation of selenium and molybdenum by crop plants and the bioavailability of these elements in the food chain are significant environmental concerns.[2]

Salinity stress can cause deficiency or toxicity imbalances in the trace element composition of plants with a potential for reduction of yield.[3] Manganese is required for plant growth, but can be toxic at higher concentrations in the plant. Solubility of manganese in the soil is influenced by salinity. Khattak and Jarrell[4] reported a salt-dependent increase in the Mn concentration of the soil saturation extracts in eight California soils treated with NaCl-CaCl$_2$ salinities. A follow-up study by these authors established a positive correlation between Mn accumulation by sugar beet plants and salt-dependent Mn leaching from these soils.[5] Similarly, Hassan et al.[6] demonstrated an increase in the exchangeable Mn of a Nebraska silt loam soil salinized with the sulfate salts of Na and Mg, and CaCl$_2$. Barley plants grown to maturity in this system showed increasing levels of Mn and Zn and decreasing levels of Fe and Cu in vegetative tissues, with increasing salinity.

A number of hydroponic studies have examined the effect of salinity on trace element uptake by plants and sought to evaluate the effects of altered trace

0-87371-942-5/94/$0.00+$.50

metal content of plant tissues on growth.[7-9] In general, these studies indicate that the relationship between salinity and trace element nutrition is complex, with the interaction dependent upon the level and type of salts used, crop species tested, and the micronutrient concentration of the nutrient solution.

Calcium/total cation ratios are known to significantly influence dry matter production and salt tolerance of several grass species. Foxtail barley *(Hordeum jubatum)*, a native halophyte, has greater salt tolerance under adverse Ca/total cation ratios than a commercial barley cultivar.[10] Foxtail barley can withstand high levels of Na and Mg sulfate salinity under conditions where plant-available Ca is extremely limited.[11] Differences in tissue concentrations of Na, K, Ca, Mg, Cl, and S between salt-stressed foxtail barley and commercial barley indicate genotype-specific differences in ion accumulation.[10] These observations raise the possibility of genotypic differences in salinity-trace metal interactions in these barley taxa.

The objectives of this study were (1) to examine the effects of salinity on the accumulation and distribution of Fe, Mn, Zn, and Cu provided at concentrations sufficient for growth, in barley cultivars and several foxtail barley populations, (2) to determine the effects of different salt treatments and Ca supply on the trace metal content of these barley species, and (3) to examine the effects of salinity on the chemical composition of the root cell wall. This information will enhance our present understanding of salinity-micronutrient interactions and lay the groundwork for further studies on the ecological, physiological, and cellular mechanisms controlling trace metal uptake in the Gramineae.

MATERIALS AND METHODS

Comparison of Cultivated and Wild Barley: Experiment 1

Seeds of the malting barley *(Hordeum vulgare* L.) cultivars 'Harrington' and 'Bonanza' were obtained from field plots at the University of Saskatchewan. Foxtail barley *(Hordeum jubatum* L.) seeds from a population designated 'University' were collected from a nonsaline site on the University of Saskatchewan campus. Germination procedures and the hydroponic growth system employed are described in Suhayda et al.[10]

The control nutrient solution used for hydroponics had the following composition (μmol l^{-1}): $NH_4H_2PO_4$, 200; $NaNO_3$, 2000; $CaCl_2$, 2500; K_2SO_4, 750; $MgSO_4$, 1800; Fe (as Fe-EDDHA), 54; H_3BO_4, 6.25; $MnSO_4$, 0.5; $ZnSO_4$, 0.5; $CuSO_4$, 0.125; Na_2MoO_4, 0.125; and Na_2SiO_3, 33. The pH of fresh nutrient solution was 6.8.

Two salt treatments consisted of a mixture of Na_2SO_4, $MgSO_4$, and $CaCl_2$ and one of two Ca concentrations. High Ca salt treatments (HCa) had final concentrations of 80 mmol l^{-1} Na, 80 mmol l^{-1} Mg, and 16 mmol l^{-1} Ca in the control nutrient solution. Low Ca salt treatments (LCa) had 80 mmol l⁻¹ Na, 80 mmol l⁻¹ Mg, and 4 mmol l⁻¹ Ca (final concentration) in the control solution. Final salt concentrations were obtained by incremental additions of salt over a period of five consecutive days to prevent salt shock. All plants received the initial salt treatment followed by three periodic solution changes during the course of the experiment. Each treatment was replicated three times. Plants were harvested 21 days after the start of salt treatments. At harvest, Harrington and Bonanza plants were 23 days old. Foxtail barley plants were given a 14-day-longer growth period than commercial barley prior to initiation of salt stress to compensate for slower growth. Foxtail barley plants were 36 days old at harvest.

All plants were maintained in a growth chamber at 21/18°C day/night temperatures under a 16/8 h light/dark photoperiod. The average photon irradiance at the tops of the plants was 530 μmol m⁻² s⁻¹.

At harvest, the roots were separated from the plants and given two 30-s rinses in 1.2 l of ice-cold deionized water to remove surface salts. Shoots and roots were oven dried at 80°C, dry weights were determined, and the samples ground for ion analysis. Ion analyses of plant tissues were performed by the Saskatchewan Soil Testing Laboratory. Samples were digested using nitric acid:perchloric acid (2:1 v/v); inductively coupled plasma emission spectroscopy was used to determine Fe, Mn, Zn, and Cu concentrations.

Comparison of Foxtail Barley Populations: Experiment 2

Seeds of foxtail barley populations were collected from three sites in central Saskatchewan. The 'University' population was taken from a nonsaline site, and the 'Floral' and 'Elstow' populations were taken from saline sites that differed in salt composition.[11] The hydroponic system, control nutrient solution, germination, and growth of plants were as described for Experiment 1. Two different salt treatments, which varied only in Na:Mg ratio, with Ca held constant, were applied to seedlings. The high Na salt stress (HNa), Na:Mg = 10:1, contained 202 mmol l⁻¹ Na, 20 mmol l⁻¹ Mg, and 2.5 mmol l⁻¹ Ca (final concentrations) in the control nutrient solution. The high Mg salt stress (HMg), Na:Mg = 1:1, contained 102 mmol l⁻¹ Na, 100 mmol l⁻¹ Mg, and 2.5 mmol l⁻¹ Ca (final concentrations) in the control nutrient solution. Plant harvest and chemical analyses were the same as in Experiment 1.

Analysis of variance (ANOVA) for all variables which incorporated the effects of treatments, species or populations, and their interactions were performed using the general linear model procedure in MINITAB. Results are presented for differences significant at $p < 0.05$. A protected least significant difference (LSD) test was used for comparison of means ($p < 0.05$).

Isolation of Cell Wall Material

The harvested roots obtained as described for Experiment 1 were frozen in liquid nitrogen. Using a mortar and pestle, 10 g of frozen tissue was ground to a powder in liquid N_2. The powdered root material was added to 30 ml of deionized water and homogenized, using two 1-min bursts with a polytron. The homogenate was centrifuged at 1000 G for 10 min. The resulting pellet was resuspended in deionized water and centrifuged as before. Cytoplasmic and membrane-associated materials were removed using the procedure of Codignola et al.[12] The crude cell wall pellet was washed three times with ten volumes of $CHCl_3/CH_3OH$ (1:1), acetone, and diethyl ether. Following each addition of solvent, the wall material was first resuspended then collected by filtration on a fritted glass filter. The resulting cell wall fraction was dried under vacuum, then stored desiccated in the dark.

Diffuse Reflectance Infrared Fourier Transform Spectroscopy (DRIFT)

Dispersions of cell wall material were prepared by mixing powdered cell wall fractions with spectroscopic grade KBr at a 1:10 (w/w) ratio. Spectra were recorded on the BioRad mid-infrared FTIR, model FTS-40. Interference from water vapor and atmospheric CO_2 was minimized by purging the sample chamber with dry N_2 gas. Percent reflectance was obtained from the ratio of the cell wall-KBr mixture to a KBr reference. Percent reflectance spectra were transformed to Kubelka-Munk units to correct for the wavelength dependence of infrared absorption. Linear baseline correction of spectra was obtained by assigning null values at 4000, 2000, and 860 cm^{-1}. To facilitate the quantitative comparison of spectra between cell walls isolated from different treatments, all spectra were normalized to a Kubelka-Munk value of 1.0 at the 2900 cm^{-1} absorption band (aliphatic C–H stretch).

RESULTS

Plant Growth and Trace Metal Composition Under Salinity Stress

Salinity stress caused a greater reduction of dry matter growth in Harrington and Bonanza barley than in the University population of foxtail barley (Figure 1). For both species, root growth was more sensitive to salinity than shoot growth. The foxtail barley population had significantly higher root biomass than commercial cultivars when stress was applied as salt treatments varying in Ca level (Experiment 1). For salt treatments varying in Na:Mg ratio (Experiment 2), root growth showed greater sensitivity to salt stress than shoot growth for the Elstow and University populations, while the opposite was true for the

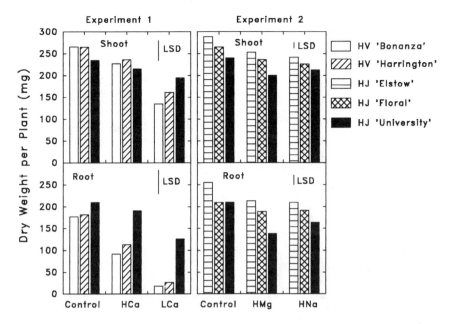

Figure 1.   Shoot and root dry weights of cultivated barley (HV) and foxtail barley populations
(HJ). High Ca salt stress (HCa), low Ca salt stress (LCa), high Mg salt stress (HMg),
high Na salt stress (HNa).

Floral population (Figure 1). The University population of foxtail barley, a
population obtained from a nonsaline site, showed greater growth reduction
under salt stress, especially high Mg salt stress, than Elstow and Floral populations derived from saline habitats.

Major differences in Fe uptake under saline conditions were confined to roots
(Figure 2). A significant species × treatment interaction was evident in Experiment 1; the Fe levels of Bonanza and Harrington roots were substantially greater
than foxtail barley under low Ca salt stress. Roots of Bonanza and Harrington
showed a reddish-brown discoloration, suggesting accumulation or precipitation
of Fe or the Fe-EDDHA complex. Iron concentrations of roots from the three
foxtail barley populations were reduced by both salt treatments relative to
controls in Experiment 2, but there was no treatment × population interaction.

Manganese levels in roots of Bonanza and Harrington were inversely related to the Ca concentration of the salt treatments in Experiment 1 (Figure 3).
The high Ca salt treatment slightly increased shoot manganese levels relative
to the other treatments. In Experiment 2, root Mn concentrations were significantly increased only by the high Mg salt treatment (Na:Mg = 1:1) for all
foxtail barley populations. ANOVA indicated a significant population × treatment interaction for shoot Mn. Manganese concentrations in shoots increased
more under high Mg than under high Na salt treatments. University plants
accumulated less shoot Mn than Elstow or Floral plants under salt stress.

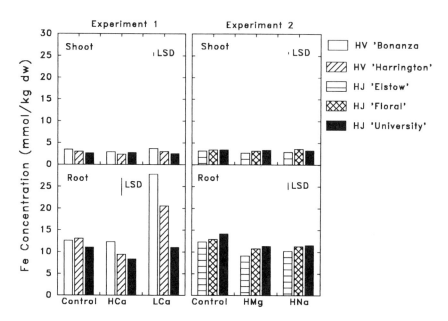

Figure 2.   Iron content of shoot and root tissue of cultivated barley (HV) and foxtail barley populations (HJ). High Ca salt stress (HCa), low Ca salt stress (LCa), high Mg salt stress (HMg), high Na salt stress (HNa).

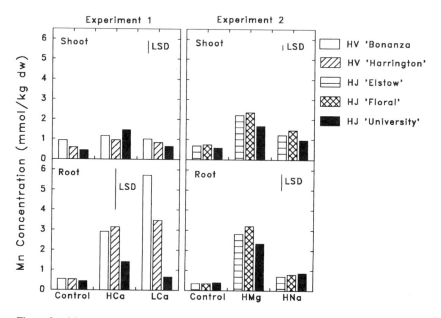

Figure 3.   Manganese content of shoot and root tissue of cultivated barley (HV) and foxtail barley populations (HJ). High Ca salt stress (HCa), low Ca salt stress (LCa), high Mg salt stress (HMg), high Na salt stress (HNa).

Zinc levels of plant roots in Experiment 1 showed a significant species ×
treatment interaction; there were no significant differences in shoot Zn levels
(Figure 4). Zinc concentrations of Bonanza roots increased as the Ca level of
the salt treatments decreased; increased Zn in Harrington roots was only
significant under low Ca salt treatments. In Experiment 2, there was a statistically significant treatment effect on shoot Zn concentration, with zinc levels
slightly increasing in the order: control < high Mg < high Na.

Copper levels in shoot tissue of plants in Experiment 1 showed a significant
difference among taxa; foxtail barley had the highest shoot Cu concentrations and
Bonanza had the lowest under salt stress (Figure 5). Salt treatments reduced the Cu
content of roots from all three foxtail barley populations in Experiment 2.

DRIFT Analysis of Salt-Induced Changes in Root Cell Wall Composition

The midrange infrared spectra of control and salt-stressed Harrington root cell
walls (RCW) are shown in Figure 6. Chemical identities for the major absorption
bands were assigned by comparison with known standards,[13] experimentally
derived spectra of humic acids,[14] or composted organic matter.[15] Root cell walls
from control and salt treatments displayed the same qualitative DRIFT spectra,
however major changes in the intensity of the various absorption bands were
evident. The group frequencies of interest include: 3400 to 3300 cm^{-1} (O–H
stretch, H–bonded OH groups); 2980 to 2850 cm^{-1} (aliphatic C–H stretch); 1725
to 1720 cm^{-1} (C=O stretch of COOH); 1660 to 1630 cm^{-1} (C=O stretch of amides
[amide I band], aromatic C=C, COO^-, H–bonded C=O); 1620 to 1600 cm^{-1}
(aromatic C=C); 1565 to 1475 cm^{-1} (NH stretch of secondary amides [amide II
band]); 1520 cm^{-1} (C=C of aromatic rings); 1470 to 1450 cm^{-1} (CH_2 deformation); 1440 to 1400 cm^{-1} (COO^-, C–O stretch, O–H deformation); 1400 to 1360
cm^{-1} (CH_3 deformation); 1400 to 1310 cm^{-1} (salts of carboxylic acids); 1280 to
1200 cm^{-1} (C–O stretching and OH deformation of COOH); 1170 to 950 cm^{-1}
(C–O stretch of polysaccharides, C–O–C stretch).

Minor differences in RCW were found between control and high Ca salt-
stressed Harrington barley plants. The most apparent difference was the 11%
increase in the absorption band centering near 1650 cm^{-1} (Figure 6). The
intensity of most infrared absorption bands was reduced in low Ca salt-stressed
RCW relative to other treatments. Increased intensities were found in the 1700
to 1520 cm^{-1} region of low Ca salt-stressed RCW; the band centered around
1650 cm^{-1} increased by 37% over control (Figure 6). The difference spectrum
between low Ca and control RCW is given in Figure 7. Six group frequency
intensities decreased in low Ca salt-stressed RCW relative to control. These
include: 3429 cm^{-1} (H–bonded OH); 1427 cm^{-1} (COOH); 1321 cm^{-1} (salts of
COOH); 1248 cm^{-1} (C–O stretch and OH deformation of COOH); 1160 cm^{-1}
(C–O stretch and C–O–C stretch); and 1056 cm^{-1} (C–O stretch of polysaccharides). The peak centered at 1056 cm^{-1} was only 55% of the control value. The

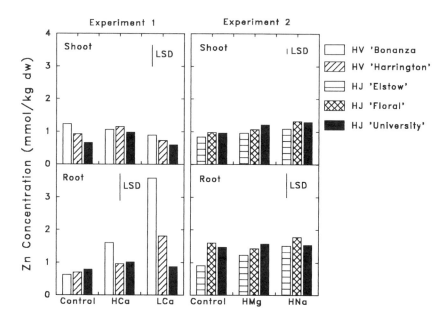

Figure 4.   Zinc content of shoot and root tissue of cultivated barley (HV) and foxtail barley populations (HJ). High Ca salt stress (HCa), low Ca salt stress (LCa), high Mg salt stress (HMg), high Na salt stress (HNa).

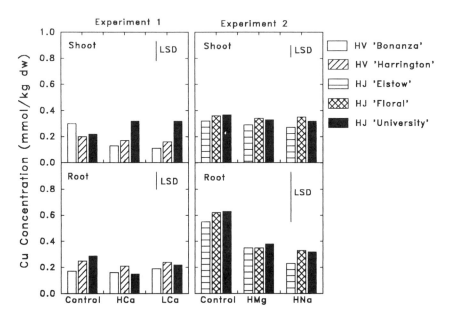

Figure 5.   Copper content of shoot and root tissue of cultivated barley (HV) and foxtail barley populations (HJ). High Ca salt stress (HCa), low Ca salt stress (LCa), high Mg salt stress (HMg), high Na salt stress (HNa).

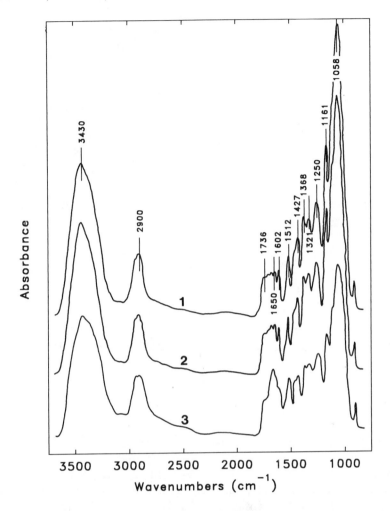

Figure 6. Diffuse reflectance FTIR spectra of Harrington root cell walls (RCW) isolated from control (1), high Ca salt-stressed (2), and low Ca salt-stressed roots (3).

intensities of the 1659 cm^{-1} (C=O stretch of amides [amide I band], aromatic C=C) and 1538 cm^{-1} (NH deformation of secondary amides [amide II band]) group frequencies increased by 36 and 29%, respectively, in low Ca salt-stressed RCW as compared to control RCW.

DRIFT spectra of foxtail barley RCW displayed the same qualitative absorption bands as Harrington, however some quantitative differences were evident (Figure 8). No change of intensity was observed in the 1660 to 1650 cm^{-1} region between control and low Ca salt-stressed RCW. The ratio of absorption intensities at 1650 cm^{-1} for control RCW of foxtail barley to Harrington RCW from different treatments gave the following results: control, 1.19; high Ca salt stress, 1.08; low Ca salt stress, 0.87. In foxtail barley RCW,

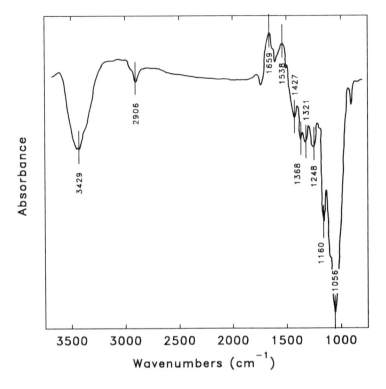

Figure 7. Difference spectrum between low Ca salt-stressed and control Harrington root cell walls (RCW).

as in Harrington, the 1060 cm^{-1} band (C–O stretch of polysaccharides) decreased with low Ca salt stress. There were significant differences between species: (1) the absorption intensity of control RCW was greater in the 1060 cm^{-1} region for foxtail barley than Harrington, and (2) the low Ca salt stress treatment decreased the intensity of this band to 75% of control RCWin foxtail barley, whereas in this same region Harrington was reduced to 55% of control. Decreases in absorption intensities in low Ca salt-stressed foxtail barley RCW were also found at 3435, 1427, 1368, 1321, and 1160 cm^{-1}; these changes coincided with those observed for low Ca salt-stressed RCW of Harrington. The difference spectrum between low Ca salt-stressed RCW of foxtail barley and Harrington is given in Figure 9.

DISCUSSION

Impact of Salinity on Plant Growth and Trace Metal Distribution

Interactions between salinity and trace element accumulation under saline conditions have received far less attention than interactions between salinity

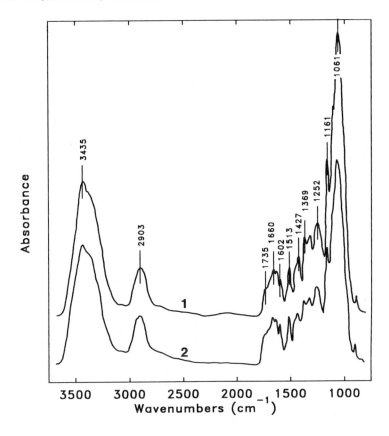

Figure 8.   Diffuse reflectance FTIR spectra of foxtail barley root cell walls (RCW) isolated from control (1), and low Ca salt-stressed roots (2).

and calcium. Data from our experiments indicate that the Fe content of salinized roots is influenced both by Ca supply and genotype (Figure 2). The high concentrations of Fe, and the reddish-brown discoloration in Bonanza and Harrington, but not foxtail barley, roots may indicate altered cell wall properties that bind or limit Fe movement through the apoplast and affect uptake by cells. Salinity alters cell wall properties of cotton roots by delaying wall thickening[16] and inhibiting biosynthesis of cellulosic and noncellulosic polysaccharides.[17] Halophytes, in contrast to glycophytes, can produce large amounts of cell wall material relative to protoplasm and have greater wall plasticity under saline conditions, thus increasing the Ca-binding capacity of the wall and maintaining cell growth.[18] The root apoplasm of wheat plants functions as a storage compartment for Fe from which Fe can be mobilized into the symplast via phytosiderophores.[19] We hypothesize that Fe or the Fe-EDDHA chelate are adsorbed to exchange sites on the cell walls of Bonanza and Harrington roots salinized under low Ca conditions, and that phytosiderophore-mediated transport of Fe into root cells is impaired in these cultivars.

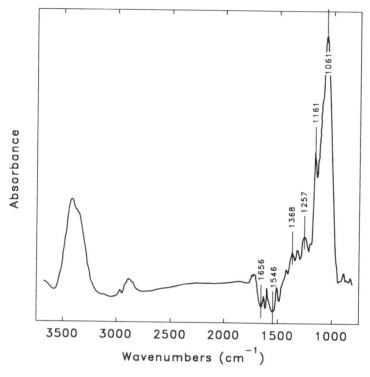

Figure 9.   Difference spectrum between low Ca salt-stressed foxtail barley root cell walls and low Ca salt-stressed Harrington root cell walls (RCW).

Alternatively, increased numbers of mitochondria in root tissue may partially account for the observed increase in root Fe levels of Bonanza and Harrington plants salt-stressed under low Ca conditions. Mitochondria are rich in iron-sulfur proteins. Ultrastructural studies of salinized corn root tissue have demonstrated the presence of greater numbers of mitochondria in xylem parenchyma cells of roots treated with Na_2SO_4 relative to control or NaCl-treated roots.[20]

Literature reports differ with respect to the influence of salinity on Mn accumulation by crop plants. Maas et al.[7] reported salinity-induced reductions in root Mn concentration for tomato, squash, and soybean at high salt concentrations (100 mM NaCl, 2 mM Ca). Salinity increased the concentration of Mn in shoot tissues of tomato and soybean, whereas a reduction of shoot Mn was found in squash. Cramer et al.[9] found a salinity-dependent reduction in shoot Mn of the barley cultivar 'CM 72' salinized with 125 mM NaCl or 125 mM KCl at Ca concentrations of 0.4 or 10 mM. Our results, which show that both root and shoot Mn concentrations of cultivated and wild barley species can increase with salinity stress, agree with Hassan et al.[6] who observed increased shoot Mn concentrations in salt-stressed barley, and Khattak and Jarrell[4] who reported increases of 27 to 97% in shoot Mn concentrations for sugar beets irrigated with saline water. Differences in the type of salt treatments applied in these studies appear to have a significant effect upon tissue Mn concentrations. Our salt treatments were high in Mg, and in some cases, Ca. Hassan et al.[6] salinized barley plants with equal parts Na_2SO_4, $MgSO_4$, and $CaCl_2$, and Khattak and Jarrell[5] used salt treatments containing $NaCl-CaCl_2$ on a 2:1 molar basis. The high divalent-to-monovalent cation ratios of the salt treatments used in these studies contrasts with the low divalent-to-monovalent cation ratios used by Maas et al.[7] and Cramer et al.[9] Taken together, these studies suggest that a saline environment rich in divalent cations enhances the accumulation of Mn in plant tissues, whereas a saline environment rich in monovalent cations inhibits Mn accumulation in plant tissues.

Phytotoxicity and stunting of plant growth can result from accumulation of high concentrations of Mn in plant tissues. A threshold level of leaf Mn considered phytotoxic to most crops is 1.82 mmol/kg, while deficiency occurs below 0.27 mmol/kg.[3] A 10% reduction in wheat dry matter production was established for a Mn tissue concentration of 6.9 mmol/kg.[21] For barley shoots, Mn concentrations below 0.24 mmol/kg are indicative of deficiency.[9] Our data give no indication of Mn deficiency, but do provide evidence for Mn toxicity caused by salinity. The tissue concentrations of Mn in roots of Bonanza and Harrington under salt stress, and in roots and shoots of foxtail barley populations given the high Mg salt stress, exceed levels presumed to be toxic. Manganese phytotoxicity may contribute to growth reduction in these cases.

Sodium chloride salinity increases Zn concentrations in both the roots and shoots of tomato, squash, and soybean.[7] For barley, shoot Zn levels increase under $NaCl$[8] and a mixture of Na_2SO_4 and $MgSO_4$[6] salinity. In our study, only Bonanza root tissues subjected to low Ca salt treatments showed increased Zn levels.

Sulfate salts of Na and Mg reduce the Cu content of mature barley shoots.[6] Leaf copper levels of 0.4 mmol/kg are considered toxic to most crop plants, whereas deficiency occurs below Cu concentrations of 0.08 mmol/kg.[3] In our experiments, shoot Cu levels were within the range considered adequate for growth, except in salt-stressed Bonanza barley where shoot Cu concentrations approached deficiency levels.

Significant reductions in the root and shoot growth of Bonanza and Harrington barley, relative to foxtail barley, occurred at lower total salt levels that contained higher levels of Ca. In contrast, foxtail barley populations maintained high root and shoot biomass under high salt, low Ca conditions (Figure 5). The interaction of plant growth and nutrient uptake under environmental stress is complex. However, when changes in yield, total element accumulation, and element concentration are taken into account then stress-dependent effects (antagonism, synergism, concentration, or dilution effects) on nutrient uptake are discernable.[22]

Analysis of our data indicates an antagonistic effect between salinity stress and Cu accumulation in Bonanza shoots in Experiment 1, and Fe accumulation in roots of the three foxtail barley populations in Experiment 2. In these cases, salinity interferes with the translocation or uptake of these elements.

The concentration effect, where growth declines more rapidly than the rate of element uptake, is a common occurrence in salt-stressed plants.[22] Bonanza and Harrington plants concentrated Fe, Mn, and Zn in root tissue; Mn was concentrated in the roots of Elstow and University foxtail barley populations subjected to high Mg salt treatments, and shoot Mn was concentrated in all three populations under both high Mg and high Na salt stress.

Synergism, where salinity enhances trace metal uptake, was observed for Mn accumulation in shoots of both commercial cultivars and foxtail barley subjected to high Ca salt stress in Experiment 1, and in shoots of the Floral population of foxtail barley stressed with the high Mg salt treatment in Experiment 2.

Root Cell Wall Composition

Infrared spectra of barley root cell walls (RCW) showed qualitative similarities to Type III humic acids[14] and composted organic matter derived from cattle manure.[15] Infrared bands for RCW were identified using data from this type of plant organic matter and from IR spectra of organic functional groups.[13]

Absorption intensities of seven bands decreased (3429, 1427, 1368, 1321, 1248, 1160, and 1056 cm^{-1}) and two increased (1659 and 1538 cm^{-1}) in RCW of low Ca salt-stressed Harrington barley relative to control plants, indicating significant differences in chemical composition (Figures 6 and 7). Only a small increase in the IR band centered near 1650 cm^{-1} was noted for high Ca salt-stressed RCW of Harrington (Figure 6). Changes in RCW chemical composition resulting from low Ca salt stress include:

1. A decrease in polysaccharide content (decreasing intensities at 1160 and 1056 cm^{-1}, C–O–C stretch and C–O stretch of polysaccharide)
2. A decrease in carboxyl groups (decreasing intensities at 1427, 1321, and 1248 cm^{-1}, various deformations of COOH)
3. An increase in cell wall protein (increased intensity at 1659 and 1538 cm^{-1}, amide I and II bands)
4. An increase in aromatic compounds (increased intensity at 1659 cm^{-1}, aromatic C=C and mono- or disubstituted C=C)

The decrease at 3429 cm^{-1} (Figure 6) may reflect a decrease in water molecules tightly bound to the cell wall matrix; this is probably a secondary effect linked to the decreased polysaccharide content of the wall.

Differences between Harrington barley and foxtail barley are evident in comparisons between low Ca salt-stressed RCW (Figure 9). Although this treatment reduced the polysaccharide content of foxtail barley RCW, the wild species still retained a higher relative content of cell wall polysaccharide than Harrington. No apparent changes in protein or aromatic content of RCW of foxtail barley were induced by low Ca salt stress (Figure 8).

The infrared data on low Ca salt-stressed RCW of Harrington corroborate our data on trace metal content of root tissue. The increased levels of Fe in Harrington roots are due to adsorption and absorption phenomena (Figure 2). Iron adsorption to cell walls would be enhanced by complexing with wall-bound phenolic acids. The increased relative abundance of aromatic compounds in RCW, as indicated by IR, suggests a higher phenolic acid content. Ferulic acid, diferulic acid, and *p*-coumaric acid comprise the group of hydroxycinnamic acids prevalent in the cell walls of several monocot families.[23] The most abundant of these, ferulic acid, is a component of arabinoxylans in hemicellulose and is associated with the unlignified cell walls of grasses, including barley.[23-25] The ability of Fe to complex with phenolic acids of the root cell wall is exemplified by the use of 10% aqueous $FeSO_4$ solution as a standard technique for localization of cell wall-bound phenolics in plant tissue by transmission electron microscopy.[12] We postulate that in low Ca salt-stressed Harrington and Bonanza RCW, adsorption of Fe onto the wall matrix occurs via complex formation with phenolic acids, and that adsorption is the primary mechanism responsible for the increased Fe content of these roots. This same adsorption mechanism most likely contributes to the higher levels of Mn and Zn associated with low Ca salt-stressed Harrington and Bonanza roots.

Covalently linked ferulic acid in pectin and in the arabinoxylan fraction of hemicellulose is thought to play a role in wall extensibility.[26] Peroxidases catalyze the oxidative reaction that link ferulic acid monomers into diferulate cross-links between polysaccharide chains. The cell wall glycoprotein, extensin, can also be cross-linked by the oxidative coupling of tyrosine residues, forming isodityrosine or trityrosine bridges between protein strands.[26] Covalent bonding of wall residues is one proposed mechanism by which cell wall extensibility is controlled.[27] The relative enrichment of protein and aromatic compounds, as indicated by IR in cell walls isolated from Harrington roots subjected to low Ca salt stress, increase the potential for the oxidative coupling of protein and phenolic entities. A portion of the growth inhibition observed in Harrington roots under low Ca salt stress may result from the increased cross-linkage of wall polymers.

The large decrease in relative polysaccharide content of Harrington cell walls isolated from low Ca salt-stressed roots is of considerable interest (Figures 6 and 7). Salinity and Ca supply significantly influence the biosynthesis of cell wall polymers, wall metabolism, and wall composition. Sodium chloride salinity inhibits the incorporation of [^{14}C]glucose into cellulosic and noncellulosic fractions of cotton seedling roots.[17] In cotton roots, a high Ca/Na ratio reverses salt inhibition of the biosynthesis of cellulosic materials. Biosynthesis of noncellulosic polysaccharides (pectin, hemicellulose) is inhibited by NaCl salinity and does not respond to Ca/Na ratio. Cucumber roots grown in solution culture under conditions of Ca deficiency show quantitative and qualitative changes in the pectic polysaccharide fraction of the cell wall.[28] Calcium deprivation reduces the quantity of extractable pectic polysaccharides from cucumber cell walls (33% of control), alters the branching pattern, chemical composition, and molecular size of pectins, and increases polygalacturonase 4.1-fold in the cell wall fraction.

These studies indicate that the biosynthesis and chemical structure of cell wall pectins is sensitive to salinity stress and Ca supply. Sodium sulfate and $MgSO_4$ salinity stress reduces Ca ion activity in nutrient or soil solutions by complex formation and ion pair interactions, thus limiting available Ca for plant growth.[10] The combined effects of salinity stress and salt-induced Ca deficiency probably account for the large decrease in polysaccharide content of low Ca salt-stressed Harrington RCW (55% of control). Although it is not possible to distinguish between cellulosic and noncellulosic polysaccharide in the RCW by infrared techniques, based upon the available literature we postulate that the low Ca salt stress causes a large decrease in the pectic polysaccharide content of Harrington RCW. In contrast, the low Ca salt-stressed RCW of foxtail barley had a polysaccharide content that was 39% greater than the comparable treatment in Harrington. The Ca content of salt-stressed foxtail barley roots has been shown to be two times greater than that found in Harrington.[10] These results suggest that foxtail barley may have a higher pectic polysaccharide content in the RCW. The pectic fraction is a Ca-rich region of the cell wall where polygalacturonan regions are cross-linked by Ca ions.[26] The differences in cell wall composition between foxtail barley and Harrington barley are likely correlated with the fact that wild barley has evolved in, and adapted to, a highly saline Ca-depleted environment.

SUMMARY

The influence of salinity on the content and distribution of trace metals in roots and shoots was examined in two barley *(Hordeum vulgare)* cultivars and three foxtail barley *(Hordeum jubatum)* populations. Salt stress caused greater imbalances in the trace metal content of roots than shoots, and the commercial barleys accumulated greater levels of trace metals than foxtail barley. Trace metal uptake was influenced by the divalent cation content of the salt treatments. Salt treatments at low Ca levels induced Fe accumulation in the roots of commercial lines, whereas high Mg salt treatments enhanced the uptake of Mn into roots and shoots of foxtail barley. Manganese concentrations were near toxic levels in roots of the commercial cultivars under low Ca salt stress conditions. In addition to salinity, Mn toxicity may contribute to growth reduction in cultivated barley.

Major changes in root cell wall composition in Harrington barley were induced by low Ca salt stress conditions. Relative decreases in cell wall polysaccharide and carboxyl group content, and increases in protein and aromatic groups, were the major salt-induced changes identified by diffuse reflectance infrared Fourier transform (DRIFT) spectroscopy in Harrington RCW.

Foxtail barley RCW contained 39% greater polysaccharide content than Harrington under low Ca salt stress.

Growth data, tissue trace metal content, and root cell wall composition indicate that foxtail barley is better adapted to salinity stress, especially to Ca-limiting growth conditions, than commercial barley.

ACKNOWLEDGMENTS

This work was supported by Saskatchewan Agriculture through the Agriculture Development Fund (Contract R-88-05-0386). We thank Mr. Ken Thoms from the Department of Chemistry, University of Saskatchewan, for assistance with diffuse reflectance infrared Fourier transform spectroscopy.

REFERENCES

1. Lee, E.W., Drainage water treatment and disposal options, *Agricultural Salinity Assessment and Management,* Tanji, K.K., Ed., American Society of Civil Engineers, New York, 1990, 450.

2. Grattan, S.R. and Rhoades, J.D., Irrigation with saline ground water and drainage water, *Agricultural Salinity Assessment and Management,* Tanji, K.K., Ed., American Society of Civil Engineers, New York, 1990, 432.

3. Page, A.L., Chang, A.C., and Adriano, D.C., Deficiencies and toxicities of trace elements, *Agricultural Salinity Assessment and Management,* Tanji, K.K., Ed., American Society of Civil Engineers, New York, 1990, 138.

4. Khattak, R.A. and Jarrell, M.W., Salt-induced manganese solubilization in California soils, *Soil Sci. Soc. Am. J.,* 52, 1606, 1988.

5. Khattak, R.A. and Jarrell, W.M., Effect of saline irrigation waters on soil manganese leaching and bioavailability in sugar beet, *Soil Sci. Soc. Am. J.,* 53, 142, 1989.

6. Hassan, N.A.K., Drew, J.V., Knudsen, D., and Olson, R.A., Influence of soil salinity on production of dry matter and uptake and distribution of nutrients in barley and corn. I. Barley (*Hordeum vulgare* L.), *Agron. J.,* 62, 43, 1970.

7. Maas, E.V., Ogata, G., and Garber, M.J., Influence of salinity on Fe, Mn, and Zn uptake by plants, *Agron. J.,* 64, 793, 1972.

8. Mozafar, A. and Oertli, J.J., Multiple stress and growth of barley. Effect of salinity and temperature shock, *Plant Soil,* 128, 153, 1990.

9. Cramer, G.R., Epstein, E., and Läuchli, A., Effects of sodium, potassium and calcium on salt-stressed barley. II. Elemental analysis, *Physiol. Plant.,* 81, 197, 1991.

10. Suhayda, C.G., Redmann, R.E., Harvey, B.L., and Cipwynk, A.L., Comparative response of cultivated and wild barley species to salinity stress and calcium supply, *Crop Sci.,* 32, 154, 1992.

11. Wang, X.Y., Suhayda, C.G., and Redmann, R.E., Identification of physiological ecotypes in *Hordeum jubatum* based on responses to salinity stress, *Can. J. Bot.,* 70, 1123, 1992.

12. Codignola, A., Verotta, L., Spanu, P., Maffei, M., Scannerinni, S., and Bonfante-Fasolo, P., Cell wall bound-phenols in roots of vesicular-arbuscular mycorrhizal plants, *New Phytol.,* 112, 221, 1989.

13. Lambert, J.B., Shurvell, H.F., Verbit, L., Cooks, R.G., and Stout, G.H., *Organic Structural Analysis,* Macmillan, New York, 1976.

14. Stevenson, F.J., *Humus Chemistry,* John Wiley & Sons, New York, 1982.

15. Inbar, Y., Chen, Y., and Hadar, Y., Solid-state carbon-13 nuclear magnetic resonance and infrared spectroscopy of composted organic matter, *Soil Sci. Soc. Am. J.,* 53, 1695, 1989.

16. Gerard, C.J. and Hinjosa, E., Cell wall properties of cotton roots as influenced by calcium and salinity, *Agron. J.,* 65, 556, 1973.

17. Zhong, H. and Läuchli, A., Incorporation of [^{14}C] glucose into cell wall polysaccharides of cotton roots: effects of NaCl and $CaCl_2$, *Plant Physiol.,* 88, 511, 1988.

18. Binet, P., Salt resistance and the environment of the cell wall of some halophytes, *Vegetatio,* 61, 241, 1985.

19. Zhang, F., Roemheld, V., and Marschner, H., Role of root apoplasm for iron acquisition by wheat plants, *Plant Physiol.,* 97, 1302, 1991.

20. Yeo, A.R., Kramer, D., Lauchli, A., and Gullasch, J., Ion distribution in salt-stressed mature *Zea mays* roots in relation to ultrastructure and retention of sodium, *J. Exp. Bot.,* 28, 17, 1977.

21. Keisling, T.C., Thompson, L.F., and Slabaugh, W.R., Visual symptoms and tissue manganese concentrations associated with manganese toxicity in wheat, *Comm. Soil Sci. Plant Anal.,* 15, 537, 1984.

22. Jarrell, W.M. and Beverly, R.B., The dilution effect in plant nutrition studies, *Adv. Agron.,* 34, 197, 1981.

23. Harris, P.J. and Hartley, R.D., Detection of bound ferulic acid in cell walls of *Gramineae* by ultraviolet fluorescence microscopy, *Nature,* 259, 508, 1976.

24. Smith, M.M. and Hartley, R.D., Occurrence and nature of ferulic acid substitution of cell wall polysaccharides in graminaceous plants, *Carbohydr. Res.,* 118, 65, 1983.

25. Yamamoto, E. and Towers, G.H.N., Cell wall bound ferulic acid in barley seedlings during development and its photoisomerization, *J. Plant Physiol.,* 117, 441, 1985.

26. Fry, S.C., Cross-linking of matrix polymers in the growing cell walls of angiosperms, *Annu. Rev. Plant Physiol.,* 37, 165, 1986.

27. Taiz, L., Plant cell expansion. Regulation of cell wall mechanical properties, *Annu. Rev. Plant Physiol.,* 35, 585, 1984.

28. Konno, H., Yamaya, T., Yamasaki, Y., and Matsumoto, H., Pectic polysaccharide breakdown of cell walls in cucumber roots grown with calcium starvation, *Plant Physiol.,* 76, 633, 1984.

CHAPTER **22**

Soil Factors Influencing the Mobilization of Iron in Calcareous Soils

Richard H. Loeppert, Liang-Chou Wei, and William R. Ocumpaugh

INTRODUCTION

Iron-deficiency chlorosis is usually associated with calcareous soils, which occupy approximately 30% of the Earth's total land area;[1] however, neither all calcareous soils nor all crops on a given soil are susceptible to Fe-deficiency chlorosis. The Fe-deficiency stress environment of the plant is characterized by a number of factors, including indigenous soil composition (e.g., carbonate and Fe oxide contents and reactivities, and clay and organic matter contents); environmental variables (e.g., soil water relations, gas-phase composition and temperature); soil solution composition (e.g., bicarbonate concentration, salinity, and ionic strength); and nutritional status (e.g., phosphate availability, N source and concentration, K status, and trace metal composition). These factors, along with the effectiveness of the plant Fe-deficiency stress-response mechanism, determine the susceptibility of a given genotype on a given soil to Fe-deficiency chlorosis. The objective of this paper is to discuss the soil and environmental factors which influence: (1) the occurrence of Fe-deficiency stress on calcareous soils, and (2) the mobilization and utilization of Fe by plants under Fe-deficiency stress conditions. This discussion is preceded by a brief review of the Fe-deficiency stress-response mechanism of plants, which is an essential factor in the mobilization and utilization of soil Fe.

IRON-DEFICIENCY STRESS-RESPONSE MECHANISMS

Any plant in a calcareous soil is in a potential Fe-deficiency stress environment; however, plant species and genotypes differ in their Fe-chlorosis susceptibilities, which occur as a continuum: from chlorosis resistant to chlorosis susceptible. Chlorosis-resistant cultivars have effective Fe-deficiency stress-response mechanisms. It has been established that there are two predominant mechanisms (or strategies) of Fe-deficiency stress response.[2]

Strategy I occurs in nongraminaceous plants and involves exudation of H^+ and enhancement of Fe^{3+} reduction at the plasma membrane,[3-5] and possibly the exudation of organic complexing and reducing agents,[6-10] the formation of rhizodermal transfer cells,[11,12] or changes in root morphology.[7,12] Iron-deficiency stress-response processes are related to plant metabolism, in which certain enzymes play key roles. For example, H^+ release is controlled by H^+-ATPase on the plasma membrane.[13] Substantial indirect[5,14] and direct[15,16] evidence indicates that the Fe^{3+} reduction by roots is an enzymatic process.

Strategy II, which occurs in grasses, is characterized by the release of phytosiderophore, a plant produced Fe^{3+} chelator, and a regulated membrane transport system.[17] The predominant phytosiderophores are nonproteinaceous, N-containing compounds of the mugineic acid family.[18-20] The phytosiderophores have a high affinity for Fe^{3+}, and can facilitate the dissolution of soil Fe and increase the availability of Fe to plant roots.[21,22] The phytosiderophore-releasing ability is variable among genotypes, and decreases as follows among plant species: barley > wheat > oat > corn >> sorghum > rice.[20,23] Although there are significant differences in the abilities of plant species to release phytosiderophores, the capacity to take up ferrated phytosiderophores is not significantly different.[23]

INDIGENOUS SOIL COMPOSITION AND IRON CHLOROSIS

Soil Carbonate

Properties and Reactions of Soil Carbonate

From chemical and nutritional perspectives, a calcareous soil may be defined as a soil in which pH is controlled by solid-phase carbonates. Most calcareous soils have pH values which fall within the range of 7.5 to 8.5, though pH values as high as 9 may be attained in soils which contain appreciable quantities of dissolved $NaHCO_3$. The dominant soil carbonate is calcite ($CaCO_3$). Appreciable substitution of Mg for Ca can occur in the crystal lattice, and Mg/(Ca + Mg) molar ratios ranging from 0 to 0.10 are common.[24,25] Dolomite [$CaMg(CO_3)_2$] occurs in some soils. Concentrations of solid-phase carbonate in soil range from traces to >80%. Soil carbonates may have very high effective reactivities,[25,26] and a considerable quantity of the soil carbonate may exist in the clay particle-size fraction.[24] Soil carbonates differ considerably in relative reactivity due to differences in particle-size distribution, mineralogy, and surface morphology; therefore, a reactivity parameter, e.g., particle-size distribution, active carbonate,[27] or relative reactivity to acid[26] may be a more reasonable index with which to compare carbonates from different soils than is the total carbonate content.[28,29]

Soil solution pH and HCO_3^- concentration are controlled by the equilibrium relations of the solid phase carbonate (Figure 1). Bicarbonate concentration in equilibrium with calcite and atmospheric CO_2 ($P_{CO_2} = 0.00035$ atm) is approximately 1 mM. With increasing P_{CO_2}, pH decreases and HCO_3^- concentration increases. High bicarbonate concentrations are especially prevalent in the rhizosphere of wet soils;[30,31] Inskeep and Bloom[31] observed values as high as 10 mM. This phenomenon has been attributed to the decreased rate of diffusion of CO_2 from the roots in wet soils, and resulting increases in CO_2 partial pressure and bicarbonate activity.[1,32,33] The total HCO_3^- concentration of the soil solution is also influenced by supplemental carbonates which have entered the system, e.g., from the application of HCO_3^--rich irrigation water. The soil solution is frequently observed to be supersaturated with respect to calcite, due to kinetic inhibitions to $CaCO_3$ precipitation which may occur in soils.[24] For example, in a field study of Calciaquolls of Minnesota, Inskeep and Bloom[24] obtained saturation parameters, IAP/K_{sp}, which ranged from 3 to 40, where IAP is the experimental $(Ca^{2+})(CO_3^{2-})$ ion activity product and K_{sp} is the theoretical ion activity product. An IAP/K_{sp} of 1 indicates equilibrium with respect to $CaCO_3$. Therefore, in the field, actual bicarbonate levels may be considerably higher than those predicted from equilibrium calculations.

Carbonate Reactivity and Fe Chlorosis

Yaalon[34] observed a relationship between incidence of Fe chlorosis and carbonate reactivity, and reported that 10% active carbonate by the Drouineau[27] procedure was the critical level for sensitive crops. Similarly, Carter[35] concluded that active carbonate rather than total carbonate provided a more effective site index for several tree species in Canada. The importance of carbonate reactivity has been verified, although a single critical level of active carbonate may not be reasonable due to the different Fe-deficiency stress-response mechanisms and the wide range of tolerances of crops to carbonate reactivity. In a study of 23 diverse calcareous soils of Texas, Morris et al.[28] grew soybean in sand/soil mixes under soil water conditions that would ensure good aeration; the severity of Fe chlorosis was strongly correlated with clay-size carbonate, total surface area of the carbonate phase, and carbonate reactivity. The carbonate reactivity parameters were more highly correlated with incidence of chlorosis than was total carbonate content. Using these same 23 soils, Loeppert and Hallmark[36] determined that carbonate content or reactivity were not major factors influencing the severity of Fe chlorosis of sorghum. The different behaviors of sorghum (monocot, strategy II) and soybean (dicot, strategy I) reflect the different Fe-deficiency stress-response mechanisms of these two crops. For crops such as soybean, in which acidification of the rhizosphere is an important component of the Fe-deficiency stress response, a

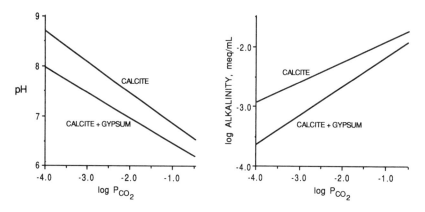

Figure 1. Influence of partial pressure of CO_2 on pH and total dissolved inorganic alkalinity of soil solutions in equilibrium with calcite and calcite + gypsum. Alkalinity increases and pH decreases with increasing partial pressure of CO_2. Also, soil solution pH and alkalinity are reduced with increasing Ca^{2+} activity of the soil solution, due to the presence of gypsum.

strong influence of carbonate reactivity on severity of Fe chlorosis is observed. The actual rhizosphere pH is determined by the two competing processes of H^+ exudation by the plant and alkalinity production by the dissolving $CaCO_3$ (Figure 2).[28,37] On the other hand, the utilization of Fe by sorghum is not strongly influenced by carbonate reactivity, since acidification of the rhizosphere is not an important component of the Fe-deficiency stress-response mechanism. For sorghum, it is likely that the equilibrium pH resulting from the presence of carbonate is a more important factor than carbonate reactivity in the overall Fe-deficiency stress condition.

Bicarbonate Concentration and Fe Chlorosis

Solution bicarbonate concentration has been correlated to the incidence of Fe chlorosis. For example, increased bicarbonate concentration resulted in increased chlorosis of kidney beans in sand culture,[32] and of bean,[38] mustard,[39] and soybean[40,41] in nutrient solution. The relative role of pH and bicarbonate was corroborated by Porter and Thorne[42] in experiments in which pH and bicarbonate concentrations of nutrient solutions were varied by controlling CO_2 partial pressure; the intensity of chlorosis of bean and tomato increased when pH of the nutrient solution was held constant and concentrations of $NaHCO_3$ and CO_2 were increased. With HCO_3^- concentration held constant at approximately 10 meq l^{-1}, and pH varied by controlling the CO_2 stream, pH did not influence chlorosis significantly within the range of 7.3 to 8.3. Brown et al.[43] provided strong evidence that the primary locus of action of HCO_3^- was either in the external solution or at the plasma membrane, by use of a split medium technique in which the soybean roots were allowed to grow from a calcareous soil (which provided the sole source of Fe) through an air gap into

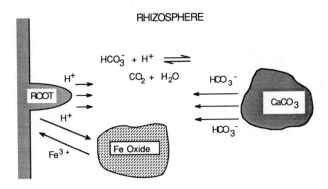

Figure 2.   Influence of $CaCO_3$ reactivity on the Fe-deficiency stress-response mechanism of nongraminaceous plants. Rhizosphere pH and the effectiveness of the Fe-deficiency stress-response mechanism are influenced by two competing processes: (1) the rate of exudation of H^+ by the plant roots, and (2) the rate of release of alkalinity by soil carbonate.[28,37]

nutrient solutions with various bicarbonate concentrations. High HCO_3^- concentrations in the solution did not induce chlorosis or reduce the uptake of Fe from the soil; therefore, the Fe did not appear to be inactivated internally. The primary inhibitory effects of HCO_3^- at the root surface are (1) the neutralization of protons which are excreted as a part of the Fe-deficiency stress-response mechanism of dicots, and (2) a reduction in activity of the plasma-membrane bound reductase attributable to a less favorable pH from (1) above. There is considerable evidence that HCO_3^- may also influence Fe transport in the plant. For example, Baxter and Belcher[44] observed that HCO_3^- was taken up by the root and concluded that it may influence the internal pH of the root and the subsequent stability of Fe^{2+} complexes. Bicarbonate was observed to inhibit the movement of ^{59}Fe to the shoots of orange trees[45] and chrysanthemum.[46] Fleming et al.[41] observed that transport of Fe to soybean shoots was significantly lower in chlorosis-susceptible than in chlorosis-resistant cultivars.

Chlorosis of monocots may also be enhanced by bicarbonate. For example, dallis grass became severely chlorotic in sand culture with a solution containing 12 meq $NaHCO_3$ l^{-1}, but a nutrient solution adjusted to pH 8.0 with NaOH did not result in an enhanced chlorosis.[47] Woolhouse[48] observed that bicarbonate inhibited the uptake and movement of ^{59}Fe to shoots in calcifuge grasses. The mechanism by which HCO_3^- detrimentally affects the Fe-deficiency stress response of grasses has yet to be determined.

Fe Oxide

The predominant Fe oxides in the soil are goethite, hematite, and ferrihydrite. The concentration of solution-phase Fe is controlled by equilibrium relations of the soil Fe-oxide phases. The log K_{sp} (log Fe^{3+} + 3 log OH^-) of soil Fe as determined by Norvell and Lindsay[49] is -39.3. At the pH of a calcareous soil (approximately 8.0), the total concentration of inorganic Fe is approximately 10^{-10} M (Figure 3), which is considerably less than the $10^{-7.7}$ M concentration required to meet the immediate nutritional needs of the plant.[50] Therefore, any plant in a calcareous soil is in a potential Fe-deficiency stress environment. The utilization of Fe by plants and microorganisms is highly dependent on the mobilization of Fe from solid phase Fe sources, predominantly the Fe oxides. Iron oxides in the rhizosphere are dissolved by either of two processes: (1) complex (chelate)-mediated dissolution, or (2) reductive dissolution.[51] For strategy II plants, it is likely that the release of Fe occurs predominantly via a complexation reaction between the phytosiderophore and soil Fe oxide (Figure 4). Reductive dissolution likely plays a relatively more important role for the strategy I plants. Within the rhizosphere, or in microenvironments of abundant microbial activity, it is likely that the concentration of total dissolved Fe is considerably higher than in the bulk soil due to the action of plants and microorganisms to meet their nutritional needs and the presence of Fe^{3+} and Fe^{2+} organic complexes.

The incidence of chlorosis of sorghum[36] and soybean[28] grown on well-drained sand/soil mixes of calcareous soils was negatively correlated and the chlorophyll contents of plants were positively correlated with the amorphous Fe oxide content of the soil as determined by the ammonium oxalate extraction procedure of McKeague and Day.[52] The amorphous Fe oxide component has a high surface area and surface reactivity and is therefore the most readily mobilized of the inorganic Fe sources within the soil.[53] Even though the amorphous Fe oxides may be present in small quantities in some soils, they represent the predominant form of labile Fe in calcareous soils due to this relatively high reactivity (Figure 5). Vempati and Loeppert[54] successfully utilized synthetic amorphous Fe oxides as an Fe amendment to sorghum in a low Fe calcareous soil, which corroborates the probable role of naturally occurring amorphous Fe oxides in Fe mobilization by plants and microorganisms.

Organic Matter

Even though soil humic and fulvic acids can retain Fe in exchangeable or complexed forms at low pH, there is little or no unambiguous evidence for the existence of Fe^{3+}-humic chelates at pH >7.0.[55,56] As pH increases to values of 4.0 and above, the large majority of exchangeable Fe is hydrolyzed and precipitated. Negative correlation coefficients have been obtained between soil organic matter content and the incidence of Fe-deficiency chlorosis;[28,36] however, these relationships may be largely indirect due to the influence of organic matter in stabilizing amorphous Fe oxides. For example, Schwertmann[57] observed that soil organic matter may prevent the recrystallization of ferrihydrite to more crystalline oxides under alkaline conditions.

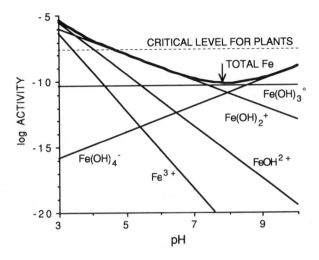

Figure 3. Activities of Fe(III) hydrolysis species and total dissolved Fe(III) in equilibrium with soil Fe oxide, as influenced by pH. At pH 8 (the approximate pH of a calcareous soil), the concentration of total dissolved Fe is considerably less than that required to meet the nutritional needs of the plant.[50]

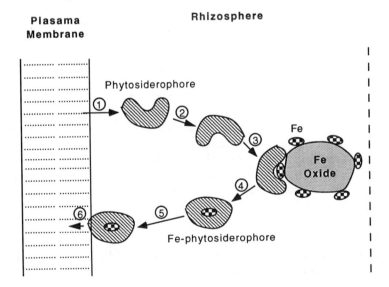

Figure 4. The mobilization and uptake of Fe by strategy II plants may be considered to involve six steps: (1) release of phytosiderophore by the plant, (2) transport of phytosiderophore from the root to the Fe-oxide surface, (3) surface adsorption of phytosiderophore by the Fe oxide, (4) desorption of Fe^{3+}-phytosiderophore from the Fe-oxide surface, (5) transport of Fe^{3+}-phytosiderophore from the Fe-oxide surface to the root, and (6) uptake of Fe^{3+}-phytosiderophore at the plasma membrane. The disruption of any of these steps could influence the effectiveness of the Fe-deficiency stress response.

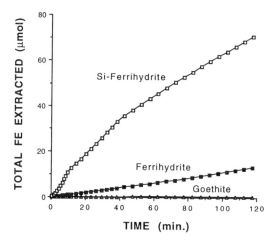

Figure 5. The relative rates of reaction of ferrihydrite (a noncrystalline or amorphous Fe oxide) and goethite (a crystalline Fe oxide) with DTPA. Even though ferrihydrite is present in most soils in relatively low concentrations, it usually controls the availability of iron in soils due to its high reactivity.

Iron-complexing organic compounds do exist in the soil, even at pH values greater than 7.0, but these are usually direct products of plant and microbial exudation, e.g., siderophores and simple organic acids such as citric acid. Concentrations of these compounds may be several orders of magnitude higher in the rhizosphere than in the bulk soil.

The reactions of organic matter which has been applied to the soil are complex (e.g., see Chen and Barak[1]), and the influence on Fe mobilization and uptake is usually indirect. For example, the addition of organic matter would help to support an active microbial population which may enhance the mobilization of Fe within the rhizosphere due to the production of siderophores. If the organic matter is banded into the soil, then microsites of lower pH or redox potential may be created in which the total solubility of Fe is greater than in the bulk soil. Organic matter added to a wet or flooded soil may actually enhance chlorosis, especially of dicots, due to the increased consumption of oxygen and accumulation of CO_2 and the resulting increased equilibrium bicarbonate levels within the rhizosphere. Organic matter and clay also provide sites for the adsorption of plant and microbially produced chelates, which influences the transport of Fe in the soil.[58]

Clay Minerals

Negative correlation coefficients between the incidence of chlorosis and the layer silicate content of the soil have been obtained.[28,36] At the pH of a calcareous soil, clay minerals are not likely to retain Fe^{3+} as an exchangeable ion, since the Fe is readily hydrolyzed and precipitated as Fe oxyhydroxides. However, layer silicates can play an indirect role in the availability of Fe due to the adsorption of poorly crystalline Fe oxide, either as hydroxy-interlayer complexes[59] or externally adsorbed phases, and its influence on prevention of reprecipitation of these surface-adsorbed phases as more highly crystalline Fe oxides.[53]

ENVIRONMENTAL FACTORS

Soil Water Content and Gas-Phase Composition

Numerous workers in field and greenhouse studies have observed a relationship between high soil moisture and incidence of Fe chlorosis.[30-33,60] Inskeep and Bloom[31] concluded that air-filled porosity and matric potential were important parameters influencing gas-phase composition and soil solution HCO_3^- concentration, and that they therefore had a significant impact on the incidence of Fe-deficiency chlorosis. The influence of high soil water contents has been attributed to the reduced diffusion of CO_2 originating from microbial and plant-root respiration, increased concentration of CO_2, and resultant increased equilibrium HCO_3^- concentrations within the rhizosphere.[32,61,62]

There has been considerable discussion concerning the relative roles of CO_2 accumulation and O_2 depletion on Fe-deficiency chlorosis. Lindsay and Thorne[38] observed that chlorosis of bean grown in HCO_3^- enriched nutrient culture was not increased by reductions in O_2 partial pressure, which suggests that the increased chlorosis in wet soils can be attributed primarily to increases in HCO_3^- concentration and not to reductions in O_2 concentration at the root. Under extreme conditions of waterlogging or flooding, O_2 contents and redox potential may be reduced and Fe and Mn solubility may be increased to such an extent that normal root metabolic processes are restricted and Mn and Fe toxicities occur.

Another aspect of soil-water relations and its influence on Fe-deficiency stress deserves mention. Strategy I plants respond to Fe-deficiency stress by H^+ exudation and strategy II plants by phytosiderophore exudation. If the water content of the soil is increased during a rainfall or irrigation event, then the aforementioned stress-response factors may be diluted to an extent that the plant's Fe-stress response will be less effective. This possibility is corroborated by the frequent field observation of a short-term yellowing of new plant tissue immediately following a heavy rain. The short-term chlorosis is observed with sorghum and soybean on well-drained and poorly drained calcareous soils in the coastal bend region of Texas, which indicates that the occurrence is neither a strategy-specific nor a soil-specific phenomenon.

Soil Temperature

As with soil-water relations, soil temperature has been noted from field observations to be a factor which influences Fe-deficiency chlorosis.[63]

Enhancement of chlorosis at low soil temperature has been observed in soybean,[17,64] bushbean,[17] flax,[17] and clover.[65] Wei et al.[65] noted that chlorosis of clover increased at both low and high temperatures and that maximum Fe uptake occurred at approximately 22°C (Figure 6). Clark and Reinhard[66] found that as temperature of a low Fe calcareous soil was increased from 12 to 27°C, leaf chlorosis of sorghum genotypes became more severe. There is evidence that Fe chlorosis of monocots and dicots may be enhanced at either low or high temperatures, but especially at low temperatures.[65]

For strategy I plants, reduced activities of enzymes, e.g., ATPase and those involved in Fe^{3+} reduction, at low (or high) soil temperature may influence the Fe-deficiency stress-response mechanism.[17] Ferric reduction is much lower at low temperatures than at higher temperatures.[67] Reduced root growth has also been suggested as a cause of low-temperature induced chlorosis, based on the observation that Fe-stress response is located in the apical root zone;[17] however, Wei et al.[65] noted that temperatures as low as 8°C had no significant effect on root biomass of clover (a cool season crop).

Soil temperature may exert a number of indirect influences on Fe mobilization and utilization as summarized below. Microorganisms excrete siderophores, which may result in an increased concentration of dissolved Fe in the rhizosphere.[58,68] They also compete with plants for available and labile sources of Fe, and may decompose phytosiderophores. Soil temperature influences microbial activity, which in turn may influence each of the above processes. For example, at higher soil temperatures, phytosiderophores are decomposed more rapidly.[69] Rhizobia activity of legumes may be reduced at either low or high soil temperatures, with resulting decreases in root nodulation. It has been suggested that the H^+ produced during N_2 fixation may enhance Fe mobilization.[70] The observation by Wallace[71] that degree of nodulation was negatively correlated with Fe chlorosis of soybean on calcareous soil supports this hypothesis. Soil temperature also influences soil physical and chemical processes, which may affect Fe chlorosis. Reactions of siderophores at colloidal Fe oxide surfaces are kinetic processes which are influenced by temperature. Also, since iron transport in the soil is diffusion-controlled,[62] soil temperature may affect Fe nutrition by influencing temperature-dependent diffusion processes.

NUTRITIONAL FACTORS

Nitrogen

Ammonium/nitrate ratio of the nutrient solution or soil solution has a strong influence on incidence of Fe chlorosis.[72] For example, the Fe chlorosis of chickpea[73] was increased when NO_3^- was used as the predominant N source. The NH_4^+/NO_3^- ratio may also influence an acidification of the rhizosphere which is independent of any specific Fe-deficiency stress-response mechanism. For example, when sorghum was grown in Steinberg's nutrient solution,

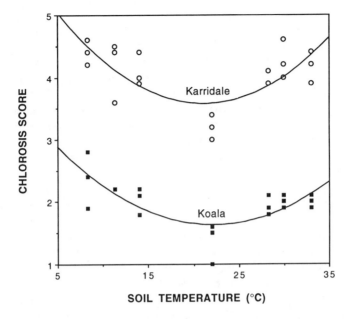

Figure 6. Influence of soil temperature on chlorosis of Koala (Fe-chlorosis resistant) and Karridale (Fe-chlorosis susceptible) clover cultivars.[65] Chlorosis increased both below and above the optimum temperature of approximately 22°C.

which contains NH_4^+ as the sole N source, the pH of the medium decreased from 6.5 to 4.0; in contrast, in Hoagland's #1 solution, which contains NO_3^-, the pH rose from 4.5 to 7.0.[74] Nitrate nutrition results in the release of OH^- by roots, whereas, NH_4^+ has the reverse effect, i.e., the release of H^+.[75] The exudation of protons and the lowering of rhizosphere pH could influence the dissolution of Fe oxides and the utilization of Fe by the plant. Nitrate nutrition also results in the accumulation of organic anions and an increase in alkalinity within the plant.[75,76] A change from NO_3^- to NH_4^+ nutrition resulted in a decrease in alkalinity of all plant parts. These observations suggest that NO_3^- may result in the precipitation of Fe in the root and leaf apoplast and a reduction in the availability of Fe for chlorophyll synthesis.

Phosphorus

Several studies have indicated that there may be a relationship between Fe chlorosis and high soil P.[77-80] Awad et al.[69] found that high P concentration substantially decreased the amount of labeled Fe mobilized by phytosiderophore *in vitro*. They attributed this phenomenon to the decreased accessibility of Fe^{3+} to chelation, due to the adsorption of phosphate at the Fe oxide surface. Chaney and Coulombe[81] found that an increased P concentration significantly inhibited Fe-deficiency stress response of Fe-inefficient species or cultivars, probably because phosphate inhibits reduction of Fe^{3+} to Fe^{2+}.[67,82] Brown et al.[83] concluded that Fe may precipitate within the plant tissue with a resulting decrease in the concentration of active Fe.[78] Several investigators have suggested that the effect of P on Fe absorption and utilization may be due, in part, to competition between P and the organic ligands which function to maintain the Fe in a soluble and mobile form within the plant.[84] Elliot and Laüchli[85] observed that both Fe absorption by roots and partitioning of Fe to shoots in maize decreased with increasing concentration of P in the nutrient solution. They suggested that P might inhibit transport of Fe from roots to shoots by competing with citrate, thus interfering either with the release of Fe from root symplasm to xylem vessels or with translocation of Fe in the xylem. It has also been suggested that phosphate might interfere with photoreduction-degradation of Fe citrate enroute to actively growing regions of the shoot.[86] Mengel et al.[61] suggested that the exudation of H^+ and the lowering of rhizosphere pH of dicots under Fe-deficiency stress conditions could also result in the increased dissolution and uptake of P. They also concluded that the frequently observed high P contents of chlorotic leaves are the consequence and not the cause of Fe chlorosis. Wallace and Cha[87] observed that bicarbonate alone caused severe damage to Fe-efficient Hawkeye soybean, while P alone had little effect; in contrast, P was very damaging with less effect from HCO_3^- with an Fe-inefficient cultivar. In summary, there is ample evidence that high P concentrations may be detrimental to Fe utilization. It is also likely that bicarbonate may strongly influence the antagonistic effect of P on Fe utilization. More research is needed to clarify the exact mechanisms and the HCO_3^-/phosphate interactions involved in Fe utilization.

Potassium

Hughes et al.[88] concluded that an adequate supply of K is critical to the proper function of the Fe-deficiency stress-response mechanisms of monocotyledonous and dicotyledonous plants. McCallister et al.[89] observed some reduction in chlorosis upon the addition of KCl, KNO_3, and K_2SO_4, and attributed this phenomenon to the improvement of cation/anion balance and the exudation of H.[+] On the other hand, high levels of K may be detrimental in a manner similar to Na, due to its influence on the dispersion of montmorillonitic clays and the deterioration of soil structure, soil water relations, and soil aeration under wet conditions. There is strong evidence that in some cases the enhanced chlorosis in chlorotic hot spots in Vertisols in the coastal bend region of Texas has been influenced by the accumulation of K from high fertilization levels.

Trace Metals (Manganese, Copper, Zinc)

Phytosiderophores can mobilize Cu, Mn, and Zn, as well as Fe, from a calcareous soil.[90] The relative affinity of mugineic acid for heavy metal cations decreases in the following order: $Cu^{2+} > Fe^{3+} >> Zn^{2+} >> Mn^{2+}$.[91,92] Zhang et al.[90] observed that mobilization of Fe from Fe hydroxide in aqueous suspensions was not affected by the addition of $MnSO_4$, slightly inhibited by $ZnSO_4$, and strongly inhibited by $CuSO_4$;[90] in a calcareous soil, mobilization of Fe by phytosiderophores remained unaffected by Zn and Mn amendments, but was progressively impaired by increasing levels of Cu.

In wet soils, the uptake of Mn may be increased dramatically, due to the reduction and solubilization of Mn oxides at reduced redox potentials. The overall significance of this reaction, which is dependent on soil manganese mineralogy and redox potential, to Fe chlorosis has not been adequately evaluated.

Salinity

Several researchers have observed a relationship between the incidence of Fe-chlorosis and the concentration of salts in the soil solution.[28,36,79,93] In nutrient culture, high NaCl was also found to induce Fe chlorosis.[11] The mechanism for salinity-induced Fe chlorosis has not been determined; however, some evidence indicates that high salinity may influence the stability of Fe chelates.[94] Also, Fe mobilization by caffeic acid[17] and by phytosiderophores[69] was inhibited at high ionic strength. Römheld and Marschner[17] suggested that high ionic strength impairs strategy I primarily at the mobilization step, and not via an influence on the inducible-reductase or subsequent membrane-transport process; however, the possibility that high ionic strength may affect enzyme activity at the plasma membrane can still not be excluded. High levels of Na or K in the soil may result in dispersion of soil clays, a resultant deterioration of soil structure and soil-water and gas-phase relations, and an increased susceptibility to Fe chlorosis.

There are some reports that salinity either increased or had no effect on Fe concentration in the plant tissue.[95] A possible beneficial effect of salt could be an improvement in cation/anion balance of the soil solution.

CONCLUSIONS

The Fe-deficiency stress response of the plant is influenced by a wide range of very complex soil properties and processes. We have come a long way during the past decade in our understanding of the soil stress environment and the complex soil/plant interactions which influence the mobilization and utilization of Fe. Many questions remain unanswered. But one thing that has been learned is that if we are to understand the plant's stress response, it is essential that we also understand the nature of the stress environment. Hence in looking at the plant, the soil can never be ignored, or vice versa.

REFERENCES

1. Chen, Y. and Barak, P., Iron nutrition of plants in calcareous soils, *Adv. Agron.*, 35, 217, 1982.
2. Marschner, H., Römheld, V., and Kissel, M., Different strategies in higher plants in mobilization and uptake of iron, *J. Plant Nutr.*, 9, 695, 1986.
3. Römheld, V. and Marschner, H., Plant-induced pH changes in the rhizosphere of Fe-efficient and Fe-inefficient soybean and corn cultivars, *J. Plant Nutr.*, 7, 623, 1984.
4. Chaney, R. L., Brown, J. C., and Tiffin, L. O., Obligatory reduction of ferric chelates in iron uptake by soybeans, *Plant Physiol.*, 50, 208, 1972.
5. Bienfait, H. F., Bino, R. J., van der Bliek, A. M., Duivenvoorden, J. F., and Fontaine, J. M., Characterization of ferric reducing activity in roots of Fe-deficient *Phaseolus vulgaris, Physiol. Plant.*, 59, 196, 1983.
6. Brown, J. C. and Ambler, J. E., "Reductants" released by roots of Fe-deficient soybeans, *Agron. J.*, 65, 311, 1973.
7. Brown, J. C. and Ambler, J. E., Iron-stress response in tomato *(Lycopersion esculentum)*. I. Site of Fe reduction, absorption and transport., *Physiol. Plant.*, 31, 221, 1974.
8. Landsberg, E.-Ch., Organic acid synthesis and release of hydrogen ions in response to Fe deficiency stress of mono- and di-cotyledonous plant species, *J. Plant Nutr.*, 3, 579, 1981.
9. Römheld, V. and Marschner, H., Mechanism of iron uptake by peanut plants. I. Fe (III) reduction, chelate splitting, and release of phenolics, *Plant Physiol.*, 71, 949, 1983.
10. Hether, N. H., Olsen, R. A., and Jackson, L. L., Chemical identification of iron reductants exuded by plant roots, *J. Plant. Nutr.*, 7, 667, 1984.
11. Kramer, D., Genetically determined adaptation in root to nutritional stress: correlation of structure and function, *Plant Soil*, 72, 167, 1983.
12. Landsberg, E.-Ch., Transfer cell formation in the root epidermis. A prerequisite for Fe-efficiency?, *J. Plant Nutr.*, 5, 415, 1982.
13. Römheld, V., Müller, C., and Marschner, H., Localization and capacity of proton pumps in roots in intact sunflower plants, *Plant Physiol.*, 76, 603, 1984.
14. Barrett-Lennard, E. G., Marschner, H., and Römheld, V., Mechanism of short term Fe(III) reduction by roots. Evidence against the role of secreted reductants, *Plant Physiol.*, 73, 893, 1983.
15. Brüggemann, W., Moog, P. R., Nakagawa, H., Janiesch, P., and Kuiper, P. J. C., Plasma membrane-bound NADH:Fe^{3+}-EDTA reductase and iron deficiency in tomato *(Lycopersicon esculentum)*. Is there a turbo reductase?, *Physiol. Plant.*, 79, 339, 1990.
16. Holden, M. J., Luster, D. G., Chaney, R. L., Buckhout, T. J., and Robinson, C., Fe^{3+}-chelate reductase activity of plasma membrane isolated from tomato *(Lycopersicon esculentum* Mill.) roots. Comparison of enzymes from Fe-deficient and Fe-sufficient roots, *Plant Physiol.*, 97, 537, 1991.
17. Römheld, V. and Marschner, H., Mobilization of iron in the rhizosphere of different plant species, *Adv. Plant Nutr.*, 2, 155, 1986.
18. Takagi, S., Naturally occurring iron chelating compounds in oat and rice root washings, *Soil Sci. Plant Nutr.*, 22, 423, 1976.

19. Takagi, S., Nomoto, K., and Takemoto, T., Physiological aspect of mugineic acid, possible phytosiderophore of graminaceous plants, *J. Plant Nutr.,* 7, 469, 1984.

20. Kawai, S., Takagi, S., and Sato, Y., Mugineic acid-family phytosiderophores in root-secretions of barley, corn, and sorghum varieties, *J. Plant Nutr.,* 11, 633, 1988.

21. Takagi, S., Kamei, S., and Yu, M.-H., Efficiency of iron extraction from soil by mugineic acid family phytosiderophores, *J. Plant Nutr.,* 11, 643, 1988.

22. Mihashi, S. and Mori, S., Characterization of mugineic-acid-Fe transporter in Fe-deficient barley roots using the multi-compartment transport box method, *Biol. Metals,* 2, 146, 1989.

23. Römheld, V. and Marschner, H., Genotypical differences among graminaceous species in release of phytosiderophores and uptake of iron phytosiderophores, *Plant Soil,* 123, 147, 1990.

24. Inskeep, W. P. and Bloom, P. R., Calcium carbonate supersaturation in soil solutions of Calciaquolls, *Soil Sci. Soc. Am. J.,* 50, 1431, 1986.

25. Bui, E. N., Loeppert, R. H., and Wilding, L. P., Carbonate phases in calcareous soils of the western United States, *Soil Sci. Soc. Am. J.,* 54, 39, 1990.

26. Moore, T. J., Loeppert, R. H., and Hartwig, R. C., Steady state procedure for studying the effective particle size distribution of soil carbonates, *Soil Sci. Soc. Am. J.,* 54, 55, 1990.

27. Drouineau, G., Dosage rapide du calcaire actif de sols, *Ann. Agron.,* 12, 441, 1942.

28. Morris, D. R., Loeppert, R. H., and Moore, T. J., Indigenous soil factors influencing Fe chlorosis of soybean in calcareous soils, *Soil Sci. Soc. Am. J.,* 54, 1329, 1990.

29. del Campillo, M. C., Torrent, J., and Loeppert, R. H., The reactivity of carbonates in selected soils of southern Spain, *Geoderma,* 52, 149, 1992.

30. Boxma, R., Bicarbonate as the most important soil factor in lime induced chlorosis in the Netherlands, *Plant Sci.,* 27, 233, 1972.

31. Inskeep, W. P. and Bloom, P. R., Effects of soil moisture on soil pCO_2, soil solution bicarbonate, and iron chlorosis in soybeans, *Soil Sci. Soc. Am. J.,* 50, 946, 1986.

32. Wadleigh, C. H. and Brown, J. C., The chemical status of bean plants afflicted with bicarbonate-induced chlorosis, *Bot. Gaz.,* 113, 373, 1952.

33. Mengel, K., Breininger, M. T., and Bübl, W., Bicarbonate, the most important factor inducing iron chlorosis in vine grapes on calcareous soil, *Plant Soil,* 81, 333, 1984.

34. Yaalon, D. H., Problems of soil testing on calcareous soils, *Plant Soil,* 8, 275, 1957.

35. Carter, M. R., Association of total $CaCO_3$ and active $CaCO_3$ with growth of five tree species on chernozemic soils, *Can. J. Soil Sci.,* 61, 173, 1981.

36. Loeppert, R. H. and Hallmark, C. T., Indigenous soil properties influencing the availability of iron in calcareous soils, *Soil Sci. Soc. Am. J.,* 49, 597, 1985.

37. Loeppert, R. H., Geiger, S. C., Hartwig, R. C., and Morris, D. R., A comparison of indigenous soil factors influencing the Fe-deficiency chlorosis of sorghum and soybean in calcareous soils, *J. Plant Nutr.,* 11, 1481, 1988.

38. Lindsay, W. L. and Thorne, D. W., Bicarbonate ion and oxygen level as related to chlorosis, *Soil Sci.,* 77, 271, 1954.

39. DeKock, P. C., Iron nutrition of plants at high pH, *Soil Sci.*, 79, 167, 1955.
40. Coulombe, B. A., Chaney, R. L., and Wiebold, W. J., Bicarbonate directly induces iron chlorosis in susceptible soybean cultivars, *Soil Sci. Soc. Am. J.*, 48, 1297, 1984.
41. Fleming, A. L., Chaney, R. L., and Coulombe, B. A., Bicarbonate inhibits Fe-stress response and Fe uptake-translocation of chlorosis-susceptible soybean cultivars, *J. Plant Nutr.*, 7, 699, 1984.
42. Porter, L. K. and Thorne, D. W., Interrelation of carbon dioxide and bicarbonate ions in causing plant chlorosis, *Soil Sci.*, 79, 373, 1955.
43. Brown, J. C., Lunt, O. R., Holmes, R. S., and Tiffin, L. O., The bicarbonate ion as an indirect cause of Fe chlorosis, *Soil Sci.*, 88, 260, 1959.
44. Baxter, P. and Belcher, R., The role of the bicarbonate ion in lime induced chlorosis, *J. Aust. Inst. Agric. Sci.*, 21, 32, 1955.
45. Wallihan, E. F., Effect of sodium bicarbonate on iron absorption by orange seedlings., *Plant Physiol.*, 36, 52, 1961.
46. Rutland, R. B. and Bukovic, M. J., Effect of calcium bicarbonate on iron absorption and distribution by Chrysanthemum morifolium, *Plant Soil*, 35, 225, 1971.
47. Gaugh, H. G. and Wadleigh, C. H., Salt tolerance and chemical composition of Rhodes and Dallis grasses grown in sand culture, *Bot. Gaz.*, 112, 259, 1951.
48. Woolhouse, H. W., The effect of HCO_3^- on uptake of Fe in four related grasses, *New Phytol.*, 65, 372, 1966.
49. Norvell, W. A. and Lindsay, W. L., Estimation of the concentration of Fe^{3+} and the $Fe(OH)_3$ ion product from equilibria of EDTA in soil, *Soil Sci. Soc. Am. J.*, 46, 710, 1982.
50. Lindsay, W. L. and Schwab, A. P., The chemistry of iron in soils and its availability to plants, *J. Plant Nutr.*, 5, 821, 1982.
51. Stumm, W. and Furrer, G., The dissolution of oxides and aluminum silicates. Examples of surface coordination controlled kinetics, in *Aquatic Surface Chemistry*, Stumm, W., Ed., John Wiley & Sons, New York, 1987, 197.
52. McKeague, J. A. and Day, J. H., Dithionite and oxalate extractable Fe and Al as aids in differentiating various classes of soils, *Can. J. Soil Sci.*, 46, 13, 1966.
53. Vempati, R. K. and Loeppert, R. H., Chemistry and mineralogy of Fe-containing oxides and layer silicates in relation to plant available iron, *J. Plant Nutr.*, 11, 1557, 1988.
54. Vempati, R. K. and Loeppert, R. H., Synthetic ferrihydrite as a potential Fe amendment in calcareous soils, *J. Plant Nutr.*, 9, 1039, 1986.
55. Goodman, B. A., The characterization of Fe complexes with soil organic matter, in *Iron in Soils and Clay Minerals*, Stucki, J. W., Goodman, B. A., and Schwertmann, U., Eds., Reidel, Dordrecht, 1987.
56. Goodman, B. A. and Cheshire, M. V., A Mössbauer spectroscopic study of the effect of pH on the reaction between iron and humic acid in aqueous media, *J. Soil Sci.*, 30, 85, 1979.
57. Schwertmann, U., Inhibitory effect of soil organic matter on the crystallization of amorphous ferric hydroxide, *Nature*, 212, 645, 1966.
58. Powell, P. E., Szaniszlo, P. J., Cline, G. R., and Reid, C. P. P., Hydroxamate siderophores in the iron nutrition of plants, *J. Plant Nutr.*, 5, 653, 1982.
59. Carstea, D. D., Harward, M. E., and Knox, E. G., Comparison of iron and aluminum hydroxy interlayers in montmorillonite and vermiculite. I. Formation, *Soil Sci. Soc. Am. Proc.*, 34, 517, 1970.

60. Wallace, A., Romney, E. M., and Alexander, G. V., Lime induced chlorosis caused by excess irrigation water, *Commun. Soil Sci. Plant Anal.*, 7, 47, 1976.

61. Mengel, K., Bübl, W., and Scherer, H. W., Iron distribution in vine leaves with bicarbonate induced chlorosis, *J. Plant Nutr.*, 7, 715, 1984.

62. Chaney, R. L., Diagnostic practices to identify iron deficiency in higher plants, *J. Plant Nutr.*, 7, 47, 1984.

63. Wallace, A. and Lunt, D. R., Iron chlorosis in horticultural plants. A review, *Proc. Am. Soc. Hortic. Sci.*, 75, 819, 1960.

64. Inskeep, W. P. and Bloom, P. R., Effects of soil moisture on soil pCO_2, soil solution bicarbonate, and iron chlorosis in soybeans, *Soil Sci. Soc. Am. J.*, 50, 946, 1986.

65. Wei, L. C., Ocumpaugh, W. R., and Loeppert, R. H., Differential effect of soil temperature on Fe-deficiency chlorosis in susceptible and resistant subclovers, *Crop Sci.*, 35, in press, 1994.

66. Clark, R. B. and Reinhard, N., Effects of soil temperature on root and shoot growth traits and iron deficiency chlorosis in sorghum genotypes grown on a low iron calcareous soil, in *Iron Nutrition and Interactions in Plants*, Chen, Y. and Hadar, Y., Eds., Kluwer Academic, Dordrecht, 1991, 127.

67. Olsen, R. A. and Brown, J. C., Factors related to iron uptake by dicotyledonous and monocotyledonous plants, *J. Plant Nutr.*, 2, 629, 1980.

68. Crowley, D. E., Reid, C. P. P., and Szaniszlo, P. J., Microbial siderophores as iron sources for plants, in *Iron Transport in Microbes, Plants, and Animals*, Winkelmann, G., Van der Helm, D., and Neilands, J. B., Eds., VCH Publishers, Weinheim, Germany, 1987, 375.

69. Awad, F., Römheld, V., and Marschner, H., Mobilization of ferric iron from a calcareous soil by plant-borne chelators, *J. Plant Nutr.*, 11, 701, 1988.

70. Raven, J. A., Franco, A. A., de Jesus, E. L., and Jacob-Neto, J., H^+ extrusion and organic-acid synthesis in N_2-fixing symbioses involving vascular plants, *New Phytol.*, 114, 369, 1990.

71. Wallace, A., Effect of nitrogen fertilizer and nodulation on lime-induced chlorosis in soybeans, *J. Plant Nutr.*, 5, 363, 1982.

72. Wallace, A., Wood, R. A., and Soufi, S. M., Cation-anion balance in lime-induced chlorosis, *Commun. Soil Sci. Plant Anal.*, 7, 15, 1976.

73. Alloush, G. A., Le-Bot, J., Sanders, F. F., and Kirkhy, F. A., Mineral nutrition of chickpea plants supplied with NO_3- or NH_4-N. Ionic balance in relation to iron stress, *J. Plant Nutr.*, 13, 1575, 1990.

74. Esty, J. C., Onken, A. B., Hossner, L. R., and Matheson, R., Iron use efficiency in grain sorghum hybrids and parental lines, *Agron. J.*, 72, 589, 1980.

75. Kirkby, E. A. and Mengel, K., Ionic balance in different tissues of the plant in relation to nitrate, urea, or ammonium nutrition, *Plant Physiol.*, 42, 6, 1967.

76. Mengel, K. and Geurtzen, G., Relationship between iron chlorosis and alkalinity in *Zea mays*, *Physiol. Plant.*, 72, 460, 1988.

77. Watanabe, F. S., Lindsay, W. L., and Olsen, S. R., Nutrient balance involving phosphorus, iron, and zinc, *Soil Sci. Soc. Am. Proc.*, 29, 562, 1965.

78. Naik, G. R., Inactive iron in sugarcane leaves and its influence on enzymatic reactions and chlorophyll metabolism, *J. Plant Nutr.*, 7, 785, 1984.

79. Inskeep, W. P. and Bloom, P. R., A comparative study of soil solution chemistry associated with chlorotic and monchlorotic soybeans in western Minnesota, *J. Plant Nutr.*, 7, 513, 1984.

80. Yen, P. Y., Inskeep, W. P., and Westerman, R. L., Effects of soil moisture and phosphorus fertilization on iron chlorosis of sorghum, *J. Plant Nutr.*, 11, 1517, 1988.

81. Chaney, R. L. and Coulombe, B. A., Effect of phosphate on regulation of Fe-stress response in soybean and peanut, *J. Plant Nutr.*, 5, 469, 1982.

82. Olsen, R. A., Brown, J. C., Bennett, J. H., and Blume, D., Reduction of Fe^{3+} as it relates to Fe chlorosis, *J. Plant Nutr.*, 5, 433, 1982.

83. Brown, J. C., Tiffin, L. D., Holmes, R. S., Specht, A. W., and Resmicky, J. W., Internal inactivation of iron in soybeans as affected by root growth medium, *Soil Sci.*, 87, 89, 1959.

84. Murphy, L. S., Ellis, R., and Adriano, D. C., Phosphorus-micronutrient interaction effects on crop production, *J. Plant Nutr.*, 3, 593, 1981.

85. Elliot, G. C. and Laüchli, A., Phosphorus efficiency and phosphate-iron interaction in maize, *Agron. J.*, 77, 399, 1985.

86. Bennett, J. H., Lee, E. H., Krizek, D. T., Olsen, R. A., and Brown, J. C., Photochemical reduction of iron. II. Plant related factors, *J. Plant Nutr.*, 5, 335, 1982.

87. Wallace, A. and Cha, J. W., Effects of bicarbonate, phosphorus, iron EDDHA and nitrogen sources on soybeans grown in calcareous soil, *J. Plant Nutr.*, 9, 251, 1986.

88. Hughes, D. F., Jolley, V. D., and Brown, J. C., Roles for potassium in the iron-stress response mechanisms of strategy-I and strategy-II plants, *J. Plant Nutr.*, 15 1821, 1992.

89. McCallister, D. J., Wiese, R. A., and Soleman, N. J., Effect of potassium salts on alleviation of lime-induced chlorosis in soybean, *J. Plant. Nutr.*, 12, 1153, 1989.

90. Zhang, F. S., Treeby, M., Römheld, V., and Marschner, H., Mobilization of iron by phytosiderophores as affected by other micronutrients, In *Iron Nutrition and Interactions in Plants*, Chen, Y. and Hadar, Y., Eds., Kluwer Academic, Dordrecht, 1991, 205.

91. Sugiura, Y., Tanaka, H., Mino, Y., Ishida, T., Ota, N., Inoue, M., Nomoto, K., Yoshioka, H., and Takemoto, T., Structure, properties, and transport mechanism of iron (III) complex of mugineic acid, a possible phytosiderophore, *J. Am. Chem. Soc.*, 103, 6979, 1981.

92. Nomoto, K., Sugiura, Y., and Takagi, S., Mugineic acids: studies on phytosiderophores, In *Iron Transport in Microbes, Plants and Animals*, Winkelmann, G., Van der Helm, D., and Neilands, J. B., Eds., VCH Publishers, Weinheim, Germany, 1987, 401.

93. Carter, M. R., Association of cation and organic anion accumulation with iron chlorosis of Scots pine on prairie soils, *Plant Soil*, 56, 293, 1980.

94. Nabhan, H. M., Vanderdeelen, J., and Cottenie, A., Chelate behavior in saline-alkaline soil conditions, *Plant Soil*, 46, 603, 1977.

95. Sepaskhah, A. R., Maftoun, M., and Karimian, N., Growth and chemical composition of pistachio as affected by salinity and applied iron, *J. Hortic. Sci.*, 60, 115, 1985.

Index